OXFORD SURVEYS IN
EVOLUTIONARY BIOLOGY

Volume 7
1990

OXFORD SURVEYS IN EVOLUTIONARY BIOLOGY

EDITED BY

DOUGLAS FUTUYMA AND JANIS ANTONOVICS

Volume 7 1990

OXFORD UNIVERSITY PRESS
1990

Oxford University Press

Oxford New York Toronto
Delhi Bombay Calcutta Madras Karachi
Petaling Jaya Singapore Hong Kong Tokyo
Nairobi Dar es Salaam Cape Town
Melbourne Auckland

and associated companies in
Berlin Ibadan

Oxford is a trade mark of Oxford University Press

Published by Oxford University Press, Inc.
200 Madison Avenue, New York, New York 10016

Library of Congress Cataloging in Publication Data

(LC Card number 86-641417)

ISBN 0-19-506289-2

9 8 7 6 5 4 3 2 1

Printed in the United States of America
on acid-free paper

Preface

Previous volumes of the *Oxford Surveys in Evolutionary Biology* have not focussed on any one theme, but have addressed the great diversity of topics embraced by evolutionary biology. In seeking contributors to this volume, we also chose not to focus on one particular topic, but instead to indulge in the luxury of having a free hand to ask anyone who we felt had "something interesting to say." We urged the authors to take a personal, even provocative, approach, rather than write a comprehensive review article that, although very useful, is also so often stodgily unreadable. We hope that the reader, whose attention will no doubt first be attracted by an article related to his or her own specialty, will take the time to look at other articles. One message of this volume is that evolutionary studies not only embrace all of biology, but that to understand a particular evolutionary phenomenon attention to information, levels of analysis, and approaches that are parts of other specialties are required. These specialties may be either in different areas within evolutionary biology itself, or in fields such as physiology, neurobiology, and biochemistry that traditionally have occupied a somewhat peripheral position in our science.

Richard Hudson summarizes the present status of the rapidly moving field of gene genealogies. The DNA sequence data that are rapidly becoming abundant can provide not only a wealth of data for the systematist interested in phylogeny reconstruction, but they can also provide detailed information about the history of genes themselves and the populations from which they are sampled. Developing models to address these questions is currently an active subject in population genetics.

Among the controversies that seem never to die is the definition and nature of species. The most recent argument is whether the biological species concept should be replaced by a phylogenetic concept. In debates on definitions there is no question of truth or falsehood, right or wrong, but only of utility and consonance. Although John Avise and R. Martin Ball no doubt hope their chapter will provide an ultimate resolution that will serve both population biologists and evolutionary systematists, we suspect it is likely to be one more stage in a perhaps never-ending controversy! The nature of species and speciation is explored further by Richard Harrison in a review of hybrid zones. Hybrid zones have long attracted the attention of systematists, but they also provide opportunity for studying geographic, genetic, behavioral, and ecological aspects of speciation. Harrison's emphasis on the importance of ecology and the dynamic nature of hybrid zones shows that these natural experiments are rich in opportunities for insight.

The two faces of evolutionary biology, the study of history and the study of processes, have usually looked, Janus-like, in opposite directions. Hampton Carson has trained both pairs of eyes, so to speak, in the same direction. Reflecting on some of his many studies of *Drosophila*, Carson points out how a phylogenetic framework can pose questions and provide insights to the population biologist. Among these are his conclusions about the evolutionary importance of sexual selection, a theme that Michael Ryan takes up from a rather different point of view. Bringing neurobiology to bear on the subject, Ryan proposes that sensory systems play an active role in determining the course of the evolution of communication systems. Biases in animals' perceptual capacities may set the framework for sexual selection, with subsequent repercussions for mating systems and implications for speciation.

Over the past 15 years, considerable excitement has been generated by the concept of "selfish DNA," in relation both to the spread through the genome of specific DNA sequences in the form of transposable elements and to the question of levels of selection: are all forms of selection ultimately reducible to differential survival of pieces of DNA? Discussion of these topics has generally been rather abstract, largely because it has been centered heavily either on genetic mechanism (as in the case of transposable elements) or on developing plausible scenarios for the evolution of behavior, with little attention to specifics, genetic or otherwise (as in the case of group or family selection). The chapters from M. W. Shaw and G. M. Hewitt and from D. Couvet and co-workers review two phenomena — B-chromosomes and male-sterility — that have been familiar to biologists for a long time, but that have been somewhat enigmatic and elusive of a ready adaptive explanation. The realization that both these phenomena may be driven by "selfish DNA" has led to new insights and to new empirical approaches for their study. It has also, as the chapter by Couvet *et al.* emphasizes, led to the realization that processes at one level of selection will interact with and drive processes at other levels. In these examples, it is clear that using purely external ecological forces to explain the relative performance of different genotypes fails to provide an adequate explanation of their evolution.

Evolutionary studies of physiology and biochemistry have, for the most part, been comparative descriptions of the wonders of adaptation. In the final two essays, the issue of adaptation is explored more closely and more refined evolutionary questions are posed. Albert Bennett and Raymond Huey summarize their studies of physiological performance in lizards, and May Berenbaum examines the evolution of supposedly defensive plant compounds, exemplified especially by her studies of parsnips and webworms. Physiologists and biochemists seldom have examined selection, genetic variation, and the costs and limits of adaptation; these chapters show that such studies are both feasible and rewarding.

D.J.F.
J.A.
November 1990

Contents

Contributors

A. ATLAN, INRA Montpellier, Station de Genetiques et Ameliorations des Plantes, 34130 Mauguio, France

JOHN C. AVISE, Department of Genetics, University of Georgia, Athens, Georgia 30602, USA

R. MARTIN BALL, Department of Genetics, University of Georgia, Athens, Georgia 30602, USA

E. BELHASSEN, INRA Montpellier, Station de Genetiques et Ameliorations des Plantes, 34130 Mauguio, France

ALBERT F. BENNETT, Department of Ecology and Evolutionary Biology, University of California, Irvine, California 92717, USA

MAY R. BERENBAUM, Department of Entomology, 320 Morrill Hall, University of Illinois, 505 S. Goodwin, Urbana, Illinois 61801-3795, USA

HAMPTON L. CARSON, Department of Genetics, University of Hawaii, Honolulu, Hawaii 96822, USA

D. COUVET, Centre Emberger, Centre National de la Recherche Scientifique, Route de Mende, BP 5051, 34033 Montpellier Cedex, France

C. GLIDDON, School of Biological Sciences, University College of North Wales, Bangor, UK

P.H. GOUYON, ESV, Bâtiment 362, Université Paris-Sud, 91405 Orsay Cedex, France

RICHARD G. HARRISON, Section of Ecology and Systematics, Corson Hall, Cornell University, Ithaca, New York 14853, USA

G.M. HEWITT, School of Biological Sciences, University of East Anglia, Norwich NR4 7TJ, UK

RICHARD R. HUDSON, Department of Ecology and Evolutionary Biology, University of California, Irvine, California 92717, USA

RAYMOND B. HUEY, Department of Zoology NJ-15, University of Washington, Seattle, Washington, 98195, USA

F. KJELLBERG, Centre Emberger, Centre National de la Recherche Scientifique, Route de Mende, BP 5051, 34033 Montpellier Cedex, France

MICHAEL J. RYAN, Department of Zoology, University of Texas, Austin, Texas 78712, USA

M.W. SHAW, Department of Agricultural Botany, School of Plant Sciences, University of Reading, Whiteknights, Reading RG6 2AU, UK

OXFORD SURVEYS IN EVOLUTIONARY BIOLOGY

Volume 7
1990

Gene genealogies and the coalescent process

RICHARD R. HUDSON

1. INTRODUCTION

When a collection of homologous DNA sequences are compared, the pattern of similarities between the different sequences typically contains information about the evolutionary history of those sequences. Under a wide variety of circumstances, sequence data provide information about which sequences are most closely related to each other, and about how far back in time the most recent common ancestors of different sequences occurred. If the sequences were obtained from distinct species, then the information is frequently extracted and displayed in the form of an inferred phylogenetic tree, which may represent the evolutionary relationships of the species from which the sequences were sampled. If, instead of being from different species, the sequences are from different individuals of the same population, the information is genealogical, and in this case gene trees can sometimes be inferred. A gene tree shows which sampled sequences are most closely related to each other and perhaps the times when the most recent common ancestors of different sequences occurred. A hypothetical gene tree, or genealogy, of five sampled sequences is shown in Fig. 1. In the absence of recombination, each sequence has a single ancestor in the previous generation. (It is important to distinguish a gene tree of sampled sequences from the pedigree of a sample of diploid individuals, in which the number of ancestors grows as one proceeds back in time, because each diploid individual has two parents.) The possibility of obtaining detailed information about the genealogy of sampled genes dramatically changes the situation for molecular population geneticists.

Before the DNA era, molecular polymorphism data were primarily in the form of frequencies of electromorphs, alleles distinguished by their mobility on electrophoretic gels. With protein electrophoresis, two homologous copies of a gene could be classified as being the same or different. If they were different, one could not measure how different; if the two copies were the same, one could not with confidence distinguish whether

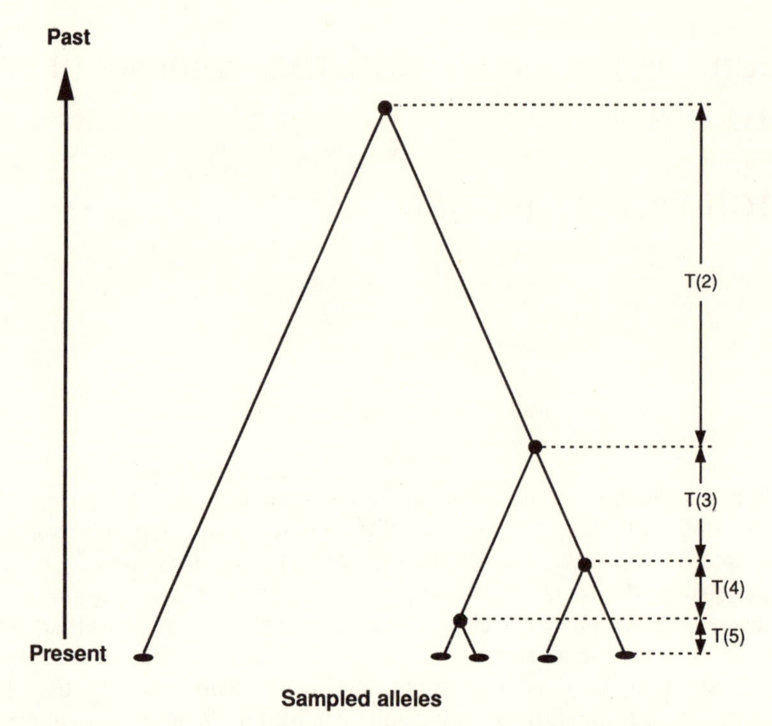

Fig. 1. An example of a genealogy of a sample of five alleles, showing the time intervals between coalescent events. In this figure, the intervals, $T(i)$, are shown with lengths proportional to their expected values as given by eqn (5).

they were really the same or simply convergent in certain physical properties leading to similar electrophoretic mobility. Thus detailed information about the genealogies of genes could not be extracted from data on electromorph frequencies. With modern DNA techniques, sequences of homologous regions of many individuals are obtainable and detailed information about the genealogy of sampled genes will be obtained. Examples of genealogies inferred from sampled alleles are given in Stephens and Nei (1985), Aquadro *et al.* (1986), Bermingham and Avise (1986), Avise *et al.* (1987) and Cann *et al.* (1987).

The obvious challenge for molecular population geneticists is: How can we utilize this information to increase our understanding of the forces acting on molecular variation in natural populations? From the theory side, we can begin by examining the properties of genealogies that arise under a variety of population genetic models. It is important to ask: Are genealogies expected to be very different under different competing models? Can we devise statistical tests that take advantage of the different genealogies expected? To proceed with this task, one needs to examine the statistical properties of genealogies of sampled genes under different models.

In the following, I will describe a variety of circumstances in which properties of genealogies can be derived analytically or by computer simulation. This will not constitute a comprehensive review of gene genealogy theory, but rather a very personal view that concentrates on the infinite-site model. Some properties of genealogies will be described under selectively neutral models, with and without recombination, and with and without geographic structure. The effects of some forms of selection will also be described. I will indicate some applications of this genealogical approach for carrying out statistical tests or estimating parameters or simply allowing an 'eye-ball' test of the fit of observations to data. I will also indicate how simulations based on the coalescent process can be constructed and used to investigate a variety of models.

This will not be a rigorous mathematical treatment. Those interested in a more precise analysis should consult the seminal work of Kingman (1980, 1982*a*,*b*) and the review by Tavaré (1984). Much of the very elegant and useful work of Griffiths (1980), Watterson (1984) and Padmadisastra (1987, 1988) on coalescents and lines of descent that focus on the infinite allele model will not be covered. This includes a large body of work on the ages of alleles (Donnelly 1986; Donelly and Tavaré 1986; Tavaré *et al.* (1989) that is reviewed by Ewens (1989). The infinite-allele models and the infinite-site models are very closely related, as will be described later, and results from one can often be used immediately to answer questions about the other. However, the questions asked and the parameter values considered are often quite distinct for the two models. In this chapter, I will concentrate on results that directly concern infinite-site models, which I feel are most useful in the interpretation of nucleotide variation in populations.

I will focus on properties of relatively small samples of alleles. The work on properties of genealogies of entire populations, including fixation times, will not be considered (Donnelly and Tavaré 1987; Watterson 1982*a*, 1982*b*). Also, the important work on the relationship between gene trees and species trees will not be discussed (Hudson 1983*b*; Neigel and Avise 1986; Pamilo and Nei 1988; Takahata 1989).

Statistical properties of genealogies depend very strongly on the kind of sampling that occurs to produce one generation from the last. In this chapter, only the Wright-Fisher (W-F) model will be considered. The sampling that produces one generation from the last under this model is described briefly in the next section. A range of alternative neutral models have been found that have essentially the same genealogical properties as the W-F model, with only a change of time-scale (Kingman 1982*a*,*b*; Watterson 1975; see also the reviews by Tavaré, 1984, and Ewens, 1989).

2. SEPARATING THE GENEALOGICAL PROCESS FROM THE NEUTRAL MUTATION PROCESS

As will be discussed in great detail in the following pages, the statistical properties of genealogies depend on such factors as population size, geographic structure and the presence of selectively maintained alleles. That properties of genealogies should depend on these demographic properties is obvious, because actual genealogies depend on who had offspring and who did not, who migrated and to where, and whose offspring bore selectively important mutations. It should also be clear that strictly neutral mutations – mutations that have not and will not affect fitness – should have no affect on the genealogies of random samples. This is because, by definition, neutral mutations do not affect the number of offspring or tendency to migrate of individuals bearing those mutations. That being the case, we can study the properties of genealogies without regard to a specific mutation model for *neutral* variants. So, for example, the statistical properties of genealogies do not depend on whether neutral mutations are more frequently transitions than tranversions or whether an infinite-site, finite-site or infinite-allele model is most appropriate. Of course, the statistical properties of our inferences about the genealogical process are likely to depend strongly on the mutation process. For example, if the neutral mutation rate is very low, all the sequences in a sample may be identical and we could get no information about the genealogy of the sample.

With the neutral mutation process that we will consider, each offspring differs from its parent at the locus under consideration by a Poisson distributed number of mutations. The mean number of mutations, μ, will be assumed constant, independent of genotype, population size and time. The mutations are assumed to occur independently in different individuals and different generations. This mutation model will be referred to as the constant-rate neutral mutation process. This is the standard neutral mutation model (Kimura 1983; Watterson 1975). Under these assumptions, mutations accumulate *along lineages* in an inexorable fashion independent of, for example, population size or selection events at linked loci. Given t, the number of generations since the most recent common ancestor of two sampled homologous sequences, S, the number of mutations that have occurred in the descent to the two descendent sequences, is Poisson distributed with mean $2\mu t$. When t is a random quantity, the mean and the variance – in fact all the moments of S – are determined by the moments of t assuming the constant-rate neutral mutation process.

To emphasize this point, consider a population that at time 0 is completely homozygous at a locus at which only neutral mutations occur. After t generations of evolution, one examines the sequence at the locus in a single randomly selected individual. Under the mutation scheme we have described in the previous paragraph, the number of mutations that will have occurred to distinguish our randomly sampled individual from the individuals in the population at time 0, is just the number of mutations that have occurred along a particular lineage of length t. This number of mutations is Poisson distributed with mean μt. It does not matter what the population size has been, whether selection has been occurring at linked loci, or whether there is population subdivision. This is the basis for the results of Birky and Walsh (1988) concerning the rate of accumulation of neutral mutations when selection is occurring at linked loci. In the example above, the number of mutations that have fixed in the entire population between time 0 and time t will depend on these demographic aspects of the population. Similarly, the amount of polymorphism in the population at time t will depend on population size and other demographic factors, but the number of mutations that will have occurred along individual lineages in the past t generations, that distinguish a sampled sequence from their ancestors t generations back, is Poisson distributed with mean μt, regardless of these other factors.

This property of the constant-rate neutral mutation process will be exploited in the following way. Let T_{tot} denote the sum of the lengths of the branches of the genealogy of a sample. As discussed in the previous paragraph, S, the number of mutations on the genealogy, given T_{tot}, is Poisson distributed with mean μT_{tot}. Once the distribution of T_{tot} is determined under a particular model, the distribution of S can easily be obtained. For example, if the first two moments of T_{tot} are determined, then the first two moments of S can be calculated using properties of compound distributions as:

$$E(S) = \mu E(T_{tot}) \tag{1}$$

and

$$\mathrm{Var}(S) = \mu E(T_{tot}) + \mu^2 \mathrm{Var}(T_{tot}) \tag{2}$$

Reiterating, under the models that we will consider, the properties of genealogies do not depend on the neutral mutation process, and therefore can be studied without precise specification of the neutral mutation process. For example, we can study the statistical properties of T_{tot} without specifying the rate or pattern of neutral mutation. Furthermore, statistical properties of neutral variation in samples are completely determined by the statistical properties of the genealogies and the neutral mutation process. In other words, if two different models make the same assumptions about the neutral mutation process and if the two different models lead to the same distribution of genealogies, then the pattern of neutral variation will be the same for the two models. For example, if the neutral mutation process is as we have described above, the mean value of S is completely determined by the mean value of T_{tot}. Two different models that lead to the same mean value of T_{tot} will have the same mean value of S.

Throughout this chapter, we will consider an ideal W-F model, with either N haploids or N diploids. Briefly, this is a discrete generation model in which, for the haploid version, the N haploids of an offspring generation are obtained by sampling (and replicating possibly with mutation) N times with replacement from the parent generation. In the selectively neutral version, all parents are equally likely as parents of each of the N haploid offspring. A detailed description of this model is contained in Ewens (1979). We will assume that N is large and constant, in which case individuals have approximately Poisson distributed numbers of offspring. Most of the results concerning this model will be approximate, ignoring terms of order $(1/N^2)$ relative to $(1/N)$. This corresponds to the usual assumptions made for using diffusion approximations and will be referred to as the diffusion approximation. In contrast to the W-F model, exact results can often be obtained for the Moran model (see, for example, Watterson 1975). The Moran model will not be considered here.

3. THE SIMPLEST CASE: NO SELECTION AND NO RECOMBINATION

Although genealogical processes are implicit in much of the work on identity coefficients that has been carried on for many years, it was the knowledge of the nature of the genetic material and the possibility of obtaining sequence data (or restriction map data) that stimulated some of the earliest work that considers the genealogical process directly. Watterson's (1975) remarkable paper describes the basic properties of genealogies under neutral models and marks the beginning of modern coalescent theory. The following description of the no-recombination genealogy under the W-F neutral model draws heavily from the work of Watterson (1975), Kingman (1980, 1982*a*,*b*) Griffiths (1980) and Tajima (1983).

To begin, we consider an ideal haploid species without recombination, without geographic subdivision and without selection – a typical garden-variety haploid species. We wish to examine properties of the genealogy of a random sample of n individuals from this population. Let us label the population from which the sample was drawn, generation 0. The ancestral population t generations back in time will be referred to as generation t.

The basic property of a sample drawn from such a population, upon which much of the following is based, concerns the probability, $P(n)$, that all the n sampled individuals have separate distinct ancestors in the preceding generation. Consider first a sample of two individuals. The probability that the second individual sampled has the same parent as the first is $1/N$, as under the W-F neutral model each individual of the previous generation is equally likely to be the parent of any individual of the current generation. Thus $P(2)$ is $1-1/N$. If three individuals are sampled, the probability that all three have distinct ancestors in the previous generation, is the probability that the first two have distinct parents $\times$ the probability that the parent of the third individual drawn is distinct from the first two parents. As there are $N-2$ individuals that are distinct from the parents of the first two sampled individuals, the probability that the third individual has a distinct parent from the first two, given that the first two have distinct parents, is $(N-2)/N = 1-2/N$. In general, the probability that n sampled individuals have n distinct parents in the previous generation is:

$$P(n) = \prod_{i=1}^{n-1} (1-i/N) \approx 1 - \frac{\binom{n}{2}}{N} \tag{3}$$

We can ask the same question about these n distinct ancestors: What is the probability that they have n distinct ancestors one generation earlier? Clearly, this is also $P(n)$. This means that the probability that the n sampled individuals have n distinct ancestors in each of the preceding t generations, and that in the $t + 1$ generation back in time, two or more of the sampled individuals have common ancestors is:

$$P(n)^t\,[1-P(n)] \approx \frac{\binom{n}{2}}{N}\, e^{-\frac{\binom{n}{2}}{N}t} \tag{4}$$

In words, the time back until the first occurrence of a common ancestor is geometrically distributed and will be approximated by an exponential distribution with mean $N/\binom{n}{2}$. For large N and small n, as we will assume throughout, the probability that more than two individuals of our sample have common ancestors in a single generation is very small and will be ignored. Thus with high probability, the recent history of our sample consists of t generations in which n distinct lineages exist, and then at generation $t + 1$, a single pair of lineages 'coalesce' at the most recent common ancestor of two of the sampled individuals. Each of the $\binom{n}{2}$ possible pairs of lineages are equally likely to form the coalescing pair. To continue tracing the history of our sample back in time, we note that in the generations preceding the first coalescence, there are $n - 1$ ancestors or lineages to follow. The probability – each generation -- that all of these ancestors have distinct ancestors in the preceding generation is $P(n-1)$. So the time to the next coalescence is approximately exponentially distributed with mean $N/\binom{n-1}{2}$. At this coalescence, each of the $\binom{n-1}{2}$ possible pairs of lineages are equally likely to coalesce at this node. Note that one of these $(n-1)$ lineages has two descendants in our original sample, the other lineages having a single descendant in the sample. We can continue in this way until all the lineages have coalesced into a single lineage, the common ancestor of the entire sample of n individuals.

A genealogy of five sampled alleles is shown in Fig 1. The stochastic process that generates a genealogy, referred to as the coalescent process, can be summarized very briefly. The time, $T(j)$, during which there are j distinct lineages is approximately exponentially distributed, and if time is measured in units of N generations, the mean of $T(j)$ is:

$$E[T(j)] = 1/\binom{j}{2} \tag{5}$$

The two lineages that coalesce at a node in the genealogy, say in generation $t + 1$, are two lineages randomly chosen from the lineages present in generation t. Notice that we have not had to concern ourselves with lineages other than those that are ancestral to our sample. Also note that the intervals between coalescences, the $T(j)$'s, are statistically independent of each other. Also, it is important to note that the older parts of the genealogy (the upper parts of the genealogy in Fig. 1), are identical in statistical properties to the genealogies of smaller samples. For example, the part of the genealogy above the most recent coalescent event in the history of a sample of size n, is distributed exactly as the genealogy of a sample of size $n - 1$. Generating such genealogies on a computer is trivial (an example of a program is given in the Appendix).

These properties of genealogies apply to mitochondrial genomes as well as to garden-variety haploid organisms. If mitochondrial inheritance is strictly maternal and polymorphism within individual females is negligible, then N is the number of females.

For a large population of N diploids, under the W-F model with random mating, no recombination and no selection, the results are also the same, except that N is replaced by $2N$. The genealogy in this case should be thought of as the genealogy for a specific locus within which no recombination occurs. The locus might consist of a single nucleotide site or, if the recombination rate is sufficiently low, of many contiguous nucleotide sites that can be considered completely linked. For the model being considered, sufficiently low means that $Nr \ll 1$, where r is the recombination rate per generation between the ends of the region being considered. If time is measured in units of N generations for haploid models, and in units of $2N$ generations for diploid models, the results are exactly the same for haploids and diploids, i.e. the mean of $T(j)$ is given by eqn (5).

Unlinked loci in large populations are essentially independent and will have their own independent genealogies. Linked loci, which have correlated genealogies, will be considered later.

4. ADDING NEUTRAL MUTATIONS TO THE GENEALOGY

Given the properties of the genealogies just described, we can predict properties of samples under various mutation schemes. As discussed in the previous section, we will assume a constant-rate neutral mutation process, in which each offspring gamete differs from its parent by an average of μ mutations. In addition, we will assume an infinite-site model (Kimura 1969). Under this model, the locus is composed of many sites, so that no more than one mutation occurs at any site in the genealogy of our sample. The oft-employed infinite-allele model (Kimura and Crow 1964) is similar, assuming that each mutation produces a new allele, not present anywhere else in the genealogy of the sample. For our purposes, the infinite-site model and the infinite-allele model are essentially the same but under the infinite-allele model one ignores how many mutations distinguish alleles and notes only whether alleles are the same or different.

The first properties to be considered concern the distribution of the number of mutations that occur on the branches of the genealogy of a sample. Under the infinite-site model, this number of mutations is identical to the number of nucleotide sites that would be polymorphic in the sample. The number of polymorphic sites in the sample, denoted S, is often referred to as the number of segregating sites in the sample. First, we consider the expected value of S.

From eqn (1) we can calculate the expectation of S from the expectation of T_{tot}, the total length of the genealogy. It follows easily from the definition of $T(j)$, that the sum of the lengths of the branches of the genealogy is $\sum_{i=2}^{n} iT(i)$. Therefore, from eqn (5), now measuring time in units of $2N$ generations, it follows that

$$E(S) = \frac{\theta}{2}\sum_{i=2}^{n} iE(T(i)) = \theta \sum_{i=1}^{n-1} 1/i \qquad (6)$$

where $\theta = 4N\mu$ (Watterson 1975). The variance of the total time is also easily obtained, and using eqns (2) and (6), one obtains (Watterson 1975):

$$\mathrm{Var}(S) = \theta \sum_{i=1}^{n-1} 1/i + \theta^2 \sum_{i=1}^{n-1} 1/i^2 \tag{7}$$

In fact, any moment of S can be expressed in terms of the moments of the T_i. Watterson also showed that the number of segregating sizes is approximately normally distributed in samples of sufficient size.

We can obtain the entire distribution of S, but first we consider the probability that $S = 0$, for a sample of size 2. This is equivalent to the expected homozygosity, $E(F)$, or the probability that two sampled alleles are identical. This probability will be derived in two ways. For two sampled alleles to be identical under the infinite-site model (or the infinite-allele model), it must be the case that no mutations have occurred on the lineages that descend to them from their most recent common ancester (denoted MRCA). Given t, the number of generations back to their MRCA, the probability that no mutations have occurred in the descent to the sampled alleles is $e^{-2\mu t}$. This follows from our Poisson assumption about mutation. Therefore, if we take the expectation of $e^{-2\mu t}$, over the distribution of t, which is exponential with mean $2N$ in the diploid model, we find:

$$E(F) = E(e^{-2\mu t}) = \int_0^\infty \frac{e^{-t/2N}}{2N} e^{-2\mu t}\, dt = \frac{1}{1+\theta} \tag{8}$$

This is a classic result (Kimura and Crow 1964) that can, of course, be derived from recursions, but here one gets a sense of its connection to the genealogy.

Equation (8) also illustrates a general connection between the infinite-allele model and the coalescent process. For any model of the population process, which determines the genealogical process, if the mutation process is the infinite-allele constant-rate neutral mutation process that we have been assuming, then the probability that two randomly sampled alleles are identical is $C(\theta) = E(e^{-\theta t})$, where this expectation is with respect to the distribution of t, the time back to the most recent common ancestor of two random alleles measured in units of $2N$ generations. The identity coefficient with $-\theta$ as argument, $C(-\theta)$, is also the moment-generating function of t. The moments of t, and consequently moments of S, are easily obtained from $C(\theta)$ by standard methods. For example, $E(t)$ is $-C'(0)$ and $E(S)$ is $-\theta C'(0)$, where $C'(0)$ represents the derivative of $C(\theta)$ with respect to θ evaluated at $\theta = 0$. This is quite general. For example, in models of gene conversion in multigene families, identity coefficients have been obtained for pairs of alleles sampled in various ways (Nagylaki and Petes 1982). The moments of the number of sites that would distinguish these alleles under an infinite-site model, can be calculated as just described by taking derivatives of the identity coefficients.

An alternative derivation of eqn (8) involves tracing the history of the two sample alleles back in time, until either the MRCA of the alleles is found or a mutation on one of the lineages is found. In each generation, the probability, P_{CA}, that the MRCA occurs is $1/2N$. Also, in each generation, the probability, P_{mut}, that one or the other of the two lineages experiences a mutation is 2μ. The two alleles can be identical if, and only if, the first event encountered is a common ancestor event. Given that one or the other event has occurred, and ignoring the possibility that both occur in the same generation, the probability that the first event encountered is the common ancestor event is:

$$E(F) \approx \frac{P_{CA}}{P_{CA} + P_{mut}} = \frac{1/2N}{1/2N + 2\mu} = \frac{1}{1+\theta} \tag{9}$$

In a similar fashion, one can derive the entire distribution of the number of mutations that have occurred since the MRCA of the sample of size 2. The probability, $P_2(j)$, of j mutations occurring on the lineages since the MRCA, is the probability that the first j events, as we trace backwards in time, are mutations and the $(j + 1)^{st}$ event is a common ancestor event. Thus, we have (Watterson, 1975):

$$P_2(j) = \left(\frac{\theta}{1+\theta}\right)^j \frac{1}{1+\theta} \tag{10}$$

Using a similar argument, we can obtain the probability, $Q_n(j)$, that j mutations occur in the time in which there are n ancestral lineages. To get j mutations during this time, the first j events, during the time there are n lineages, must be mutations, and the $(j + 1)^{st}$ event must be a common ancestor event. Hence, this probability is

$$Q_n(j) = \left(\frac{n\mu}{n\mu + \frac{\binom{n}{2}}{2N}}\right)^j \frac{\frac{\binom{n}{2}}{2N}}{n\mu + \frac{\binom{n}{2}}{2N}}$$

$$= \left(\frac{\theta}{\theta+n-1}\right)^j \frac{n-1}{\theta+n-1} \tag{11}$$

The number of segregating sites in a sample of size n is the sum of the number that occur while there are n lineages, and the number during the rest of the genealogy distributed just like the number in a sample of size $n-1$. It follows that $P_n(j)$, the probability of j segregating sites in a sample of size n, can be written as:

$$P_n(j) = \sum_{i=0}^{j} P_{n-1}(j-i)Q_n(i) \tag{12}$$

The distribution of the number of segregating sites can quickly be calculated using this recursion. Tavaré (1984) obtained an explicit expression for $P_n(j)$. The distribution of S is shown in Fig. 2 for $\theta = 5$ and $n = 20$.

The use of eqn (12) is illustrated by the following example. Recent surveys of polymorphism in the yellow-achaete-scute region of *Drosophila melanogaster* revealed 9 polymorphic sites in 2112 nucleotide sites in 64 chromosomes examined (Aguadé *et al.* 1989). Estimates of θ per base pair from other regions of the *D. melanogaster* genome have averaged about 0.005. Aguadé *et al.* wanted to determine if the observation of 9 polymorphic sites was consistent with the hypothesis that θ per base pair in the yellow-achaete-scute region is 0.005. Using eqn (12), we can calculate that the probability of 9 or fewer polymorphisms, in a sample of 64

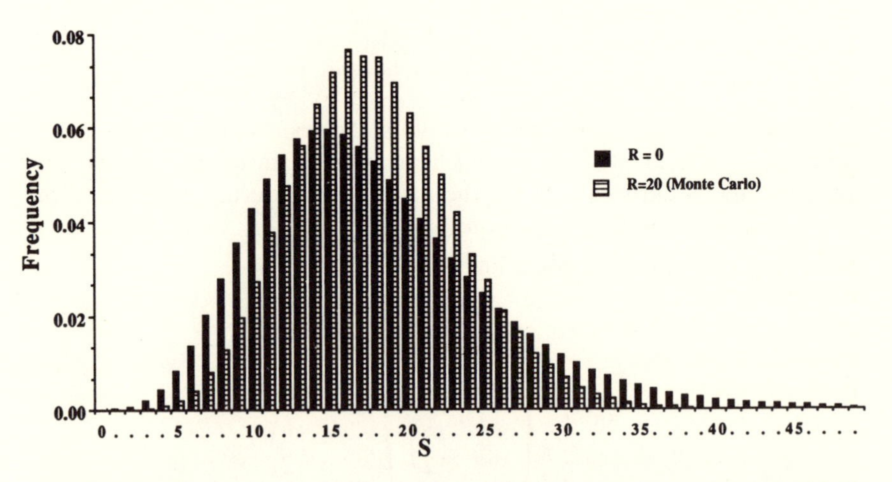

Fig. 2. The distribution of S, the number of segregating sites, in a sample of 20 alleles with θ ($=4N\mu$) = 20. The no-recombination distribution (R = $4N\mu$ = 0) was calculated with eqn (12). For R = 20, the distribution is an estimate obtained by generating 100 000 replicates by a Monte Carlo method described in the text. The expected value of S for both distributions is 17.7, which can be calculated using eqn (6).

with $\theta = 2112(0.005) = 10.6$, is approximately 2×10^{-6}. Assuming the equilibrium neutral model is correct, one must reject a value of 0.005 as the per base pair mutation parameter for this region. If one assumed that some recombination occurs in this region, the probability of 9 or fewer polymorphic sites is even smaller.

5. RECOMBINATION

Let us consider first two loci. It is assumed that no recombination occurs within each locus but, between the two loci, the probability of recombination is r per generation per offspring produced. If $r = 0$, the two loci will always have the same genealogy. If r is large, in a large random mating population, the genealogies of the two loci will be essentially independent (see eqn 13). The difficult case is with intermediate levels of recombination, when the genealogies at the two loci are correlated. Clearly, the marginal distribution of genealogies for each locus under a neutral model, is the single locus no-recombination distribution described above. The only effect of linkage is to produce a correlation between the genealogies for the two loci.

Let us begin by describing how one might simulate on a computer the genealogy of a sample of two gametes, denoted $\mathbf{a}_1(0)\mathbf{b}_1(0)$ and $\mathbf{a}_2(0)\mathbf{b}_2(0)$. We proceed, as before, backward in time. We trace the two lineages back until either a coalescent occurs (probability $1/2N$ per generation) or a recombination event occurs (probability $2r$ per generation). The time back until one of these events is exponentially distributed with mean $2N/(1+R)$, where R is $4Nr$. The probability that the first event is a coalescent event is $1/(1+R)$. In this case, both loci have their MRCA at this time and the genealogies are complete. The other possibility is that the first event is a recombination event. The first event is a recombination event with probability $R/(1+R)$. In this case, one of the two lineages splits in two as illustrated by the genealogy in Fig. 3. In this example, the first event, as one traces backward in time, is a recombination event that occurs in generation t_1. In this example, the ancestral gamete, $\mathbf{a}_2(t_1-1)\mathbf{b}_2(t_1-1)$, is the recombinant descendant of two individuals in generation t_1, which are denoted $\mathbf{a}_2(t_1)$- and -$\mathbf{b}_2(t_1)$. At this point, there are three lineages to follow back in time from the three ancestral gametes in generation t_1. One ancestral gamete, denoted $\mathbf{a}_1(t_1)\mathbf{b}_1(t_1)$, is an ancestor at both loci to one of the sampled gametes. One of the ancestral gametes, denoted $\mathbf{a}_2(t_1)$-, is an ancestor of the $\mathbf{a}_2$ allele in the sample, but the $\mathbf{b}$ allele of this ancestral gamete, indicated by a hyphen, has no descendant in the sample. The history of this allele represented by the hyphen is of no direct interest. The third ancestral gamete, -$\mathbf{b}_2(t_1)$, is the ancester at the $\mathbf{b}$ locus of the $\mathbf{b}_2$ allele in the sample. We continue back in time until the next event,

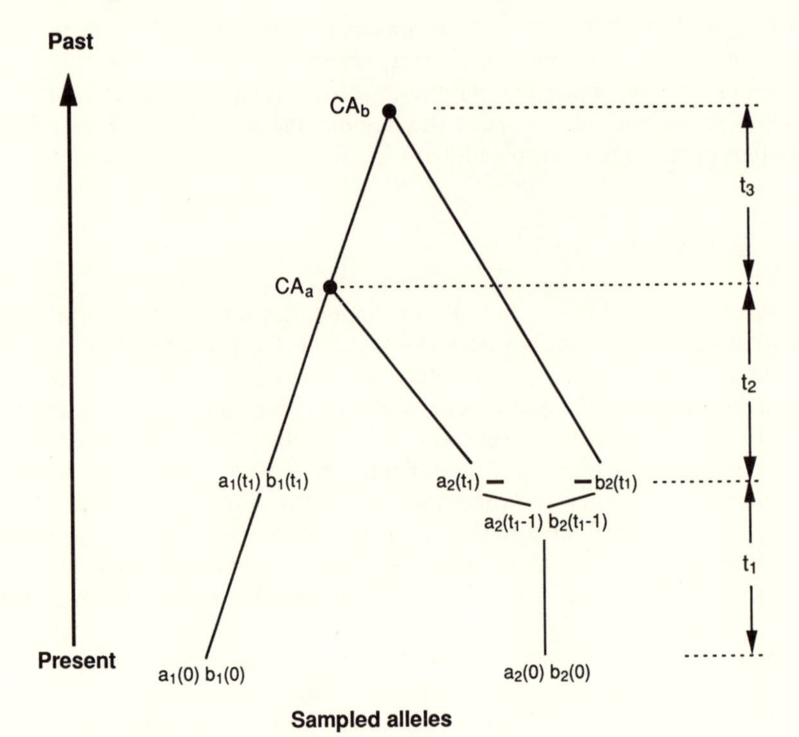

Fig. 3. An example two-locus genealogy for a sample of size 2. In this case, the first event, which occurs in generation t_1, is a recombination event such that the ancestor gamete $\mathbf{a}_2(t_1-1)\mathbf{b}_2(t_1-1)$ is the recombinant descendant of the two gametes $\mathbf{a}_2(t_1)$- and -$\mathbf{b}_2(t_1)$. The second event is a common ancestor event, labeled $CA_{\mathbf{a}}$, at which time, the lineages of $\mathbf{a}_1(0)\mathbf{b}_1(0)$ and $\mathbf{a}_2(t_1)$- coalesce. It is at this point in time, t_1+t_2 generations ago, that the most recent common ancestor of the sampled '**a**' locus alleles occurred. The next event is a common ancestor event, labeled $CA_{\mathbf{b}}$. At this time, $t_1+t_2+t_3$ generations ago, the most recent common ancestor of the sampled '**b**' alleles occurred.

Fig. 4. An example genealogy for an infinite-site recombination model. The two samples gametes, labeled 1 and 2, are represented by the hatched and dotted bars. Recombination events can occur anywhere along the bars. There are five events in this genealogy, designated RE_1, RE_2, CA_1, CA_2 and CA_3, in order from most recent to most ancient. The most recent event, RE_1, is a recombination event that brought two segments together to form the ancestor of gamete 2. Following lineages backward in time, as usual, the result of RE_1 is the splitting of the lineage of gamete 2 into two parts, one being the lineage of the left end of the gamete, and the other being the lineage of the right part of the gamete. The next event back in the genealogy, labeled RE_2, is also a recombination event

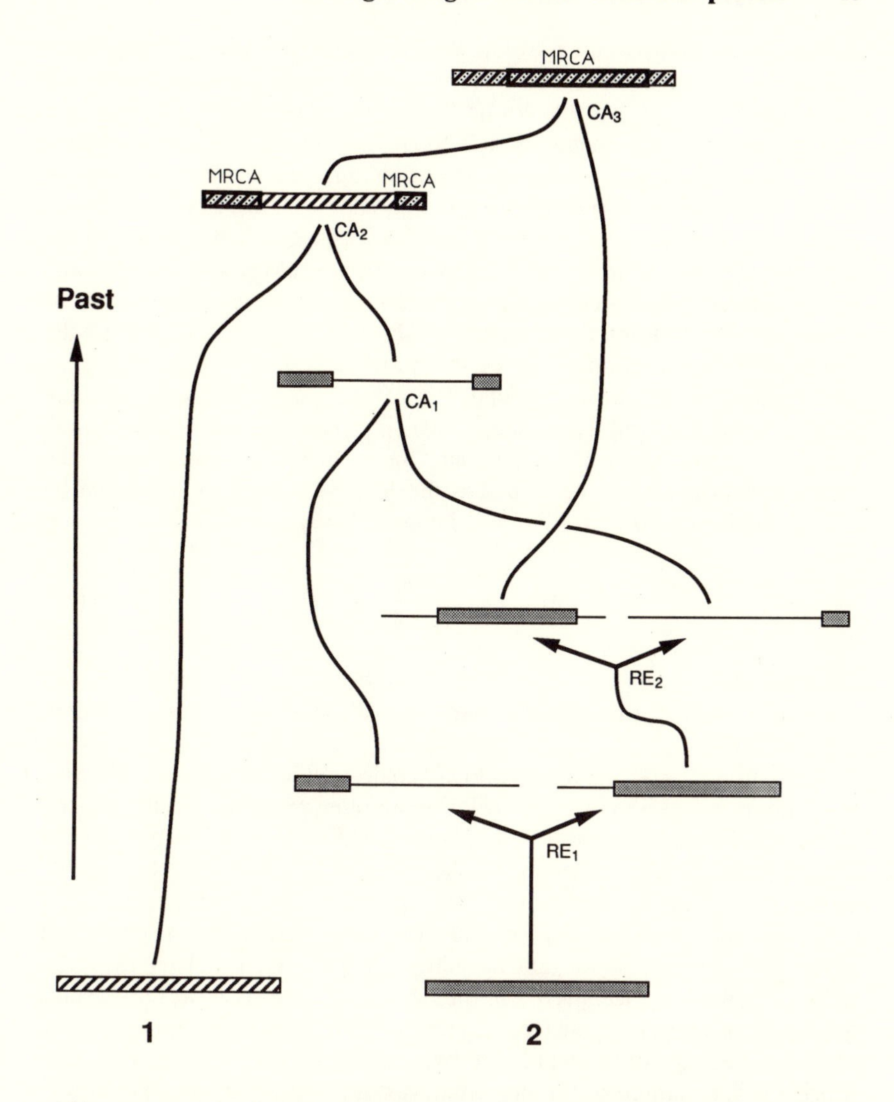

with a crossover in the right-hand segment of an ancestor of gamete 2. At this point in time, there are three distinct ancestors of gamete 2, each being an ancestor of a different part of gamete 2. In contrast, gamete 1 still has a single ancestor. The next event, CA_1, is a common ancestor event involving two ancestors of gamete 2. At this point, one of the two ancestors of gamete 2 is an ancestor for two non-contiguous portions of gamete 2. The next event, CA_2, is a common ancestor event where finally the most recent common ancestor of parts of gametes 1 and 2 occur. The segments with most recent common ancestor at this point are the left end, marked MRCA, and the right end also marked MRCA. The last event, is a common ancestor event where the most recent common ancestor of the sample gametes for the middle segment occurred.

either a coalescent event between any of the three lineages (probability $\binom{3}{2}/2N$ per generation) or a recombination event (probability r per generation). Note that, during this part of the genealogy, only recombinations involving the lineage of $\mathbf{a}_1(t_1)\mathbf{b}_1(t_1)$ are relevant. Recombinations in the lineage of $\mathbf{a}_2(t_1)$- do not result in any change in the state of the process and are irrelevant to the genealogy of the sampled alleles. Eventually, the two alleles at the **a** locus will coalesce and the two alleles at the **b** locus will coalesce, and the two-locus genealogy will be complete.

By consideration of this two-locus process, it is possible to derive various properties of the joint distributions of the times, $t_\mathbf{a}$ and $t_\mathbf{b}$, back to the most recent common ancestors of the **a** and **b** alleles, respectively.

Griffiths (1981*a*) derived properties of the joint distribution of the number of segregating sites at each locus in samples of size 2, when each locus is assumed to be an infinite-site locus. From Griffiths' result, the correlation of $t_\mathbf{a}$ and $t_\mathbf{b}$, the times to the MRCA at locus **a** and **b** can be found (Hudson 1983*a*; Kaplan and Hudson 1985):

$$\mathrm{Cor}(t_\mathbf{a},t_\mathbf{b}) = \frac{R+18}{R^2+13R+18} \qquad (13)$$

Consideration of this two-locus coalescent shows that the probability that $t_\mathbf{a} = t_\mathbf{b}$ is exactly the same as the correlation of $t_\mathbf{a}$ and $t_\mathbf{b}$ (Hudson, unpublished).

Simulations based on the two-locus coalescent were used by Hedrick and Thomson (1986) to study two-locus sampling properties of the neutral model. Kaplan and Hudson (1985) considered the coalescent process for several linked loci to calculate the homozygosity at a global locus made up of several sub-loci between which recombination could occur.

Hudson (1983*a*) and Kaplan and Hudson (1985) also considered an infinite-site version of the above coalescent process, in which recombination could take place anywhere on a continuous interval that represents a contiguous stretch of nucleotide sites. Figure 4 shows a representation of the genealogy of a sample of two gametes under this model. The process is very similar to the preceding two-locus case, except that recombination takes place at random positions along the continuous interval that represents the sequence. In this case, small contiguous segments are likely to have similar genealogies, but the segments farther apart would be likely to have quite different genealogies. The details of how to carry out such a simulation are described in Hudson (1983*a*).

In the genealogy in Fig. 4, the MRCA of the segment of DNA in the middle occurs farther back in time than the MRCA of the end segments. In this sense, the size of the genealogy is larger for the middle segment than for the end segments,and assuming that the neutral mutation rate is the same all along the segment, we would expect the number of neutral mutations per unit length to be greater in the middle segment. In Fig. 5, the outcome of a single realization of this genealogical process is shown for a large contiguous chunk of DNA for a sample size 10. This figure indicates how much the size of the genealogy, as measured by T_{tot}, can vary from one segment to the next. The size of the segment of DNA considered in Fig. 5 is such that $4Nr$ equals 100, where r is the recombination rate per generation between the ends of the region. Although estimates are very rough, this has been estimated to correspond to approximately 5000 base pairs in *D. melanogaster*. (This number can be obtained from estimates of per base pair recombination rate 0.5×10^{-8} and effective population size 10^6: (Hudson and Kaplan 1988; Hudson 1987.)

As before, the total number of segregating sites in a sample, S, conditional on the genealogies of all the segments, is Poisson distributed with mean $\theta T/2$, where in this case T is an average of the sizes of the genealogies of each of the segments weighted by their lengths and θ is $4N$ times the mutation rate for the entire sequence. As the recombination rate increases, the weighted average, T, is made up of greater numbers of

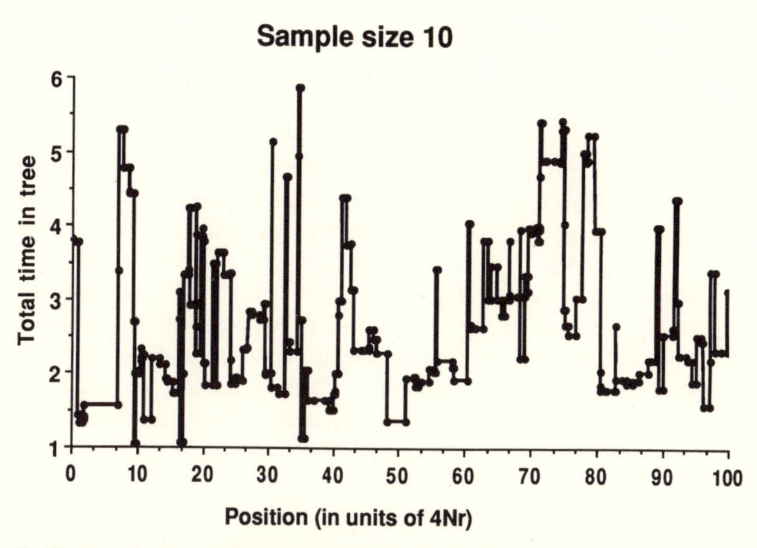

Fig. 5. The total time in the genealogy of the sample, T_{tot}, measured in units of $4N$ generations, plotted as a function of position, for a single realization of the coalescent process for a neutral infinite-site recombination model. The total length of the region of DNA considered is such that the $4Nr = 100$, where r is the recombination rate between the ends of the region. The horizontal axis is the nucleotide position, as measured by the product of the $4N$ and the recombination rate between the site and the left end of the region considered. Evidently, T_{tot} varies considerably from site to site, over a region this size.

relatively smaller segments that have less correlated genealogies. The result is that the variance of T tends to zero, and S becomes Poisson as the recombination parameter (R) tends to infinity (see Ewens 1979, p. 276). Kaplan and Hudson (1985) showed that the variance of S is

$$\mathrm{Var}(S) \approx \theta \left(\sum_{i=1}^{n-1} 1/i \right) + \theta^2 \, \mathrm{Var}(T) \tag{14}$$

and that

$$\mathrm{Var}(T) \approx \frac{2\left(\sum_{i=1}^{n-1} 1/i^2\right)}{R^2} \left(-R + \frac{23R+101}{2\sqrt{97}} \log \left(\frac{2R+13-\sqrt{97}}{2R+13+\sqrt{97}} \frac{13+\sqrt{97}}{13-\sqrt{97}} \right) + \frac{R-5}{2} \log \left(\frac{R^2+13R+18}{18} \right) \right) \tag{15}$$

For sample size 2, the approximation for Var(T) was based on the usual 'diffusion approximations', but for larger sample sizes there is no theoretical justification for the approximation, except that Monte Carlo simulations indicated that it works quite well in the cases examined, namely with small to moderate values of R (Kaplan and Hudson 1985). The number of recombination events in the genealogy of a sample has been examined by Hudson and Kaplan (1985), and an estimator of R based on inferred numbers of events was investigated. A recombination event was inferred to have occurred between two polymorphic sites when all four possible gametic types (haplotypes) involving the two sites were present in the sample.

The distribution of S in a sample of size 20 for $\theta = 5$ and with $R = 0$ and $R = 20$ are shown in Fig. 2. The mean of S does not depend on R, but this figure shows clearly how recombination can reduce the variance in S. The distribution shown for $R = 20$ is based on 100 000 samples generated by the algorithm described above. The variance of S in the Monte Carlo samples was 28.04, whereas the variance calculated with eqns (14) and (15) is 28.28.

6. ESTIMATING θ OR N

One can use S to estimate θ or, if the neutral mutation rate (μ) is known, the population size N. The two commonly used methods are moment estimators. Because the expected number of differences between two alleles is θ, an obvious estimator of θ is $\tilde{\theta}$, the average pairwise number of differences between alleles in a sample (see Nei 1987, eqn 10.6). This is an unbiased estimator of θ. Tajima (1983) showed that under the W-F model with no recombination, the variance of this estimator is (see also Nei 1987, eqn 10.9):

$$\mathrm{Var}(\tilde{\theta}) = \frac{n+1}{3(n-1)}\,\theta + \frac{2(n^2+n+3)}{9n(n-1)}\,\theta^2 \tag{16}$$

Watterson (1975) suggested an estimator based on eqn (6), namely:

$$\hat{\theta} = \frac{S}{\sum_{i=1}^{n-1} 1/i} \tag{17}$$

This estimator is clearly unbiased. Under the no-recombination model, the variance of this estimator can easily be calculated using eqn (7), because:

$$\mathrm{Var}(\hat{\theta}) = \frac{\mathrm{Var}(S)}{\left(\sum_{i=1}^{n-1} 1/i\right)^2} \tag{18}$$

The variance of θ is always less than the variance of $\tilde{\theta}$. With recombination, both of these estimators have substantially reduced variance. The variance of $\hat{\theta}$ in the presence of recombination can be estimated using eqns (14), (15) and (18).

In some circumstances, the reduced variance of S in the presence of recombination may be justification for considering nuclear genes instead of mitochondrial genes for certain problems. For example, recent studies (Avise *et al.* 1988) of mitochondrial genes were used to estimate effective population sizes, using prior estimates of μ. Although practical considerations concerning the relative ease of isolation of mtDNA compared to nuclear DNA may mitigate against the use of nuclear DNA, more precise estimates might be obtained with nuclear data.

For the no-recombination model, maximum likelihood estimates of θ based on S can be obtained, and it has been shown that the maximum likelihood estimates always exceed $\hat{\theta}$ (Tavaré 1984). I have examined a small number of cases and always found that the mean square error of the maximum likelihood estimate exceeds the mean square error of $\hat{\theta}$.

7. MIGRATION AND GEOGRAPHIC STRUCTURE

A number of authors have utilized the genealogical approach to consider properties of samples when there is geographic structure (Griffiths 1981*b*;

Slatkin 1987, 1989; Strobeck 1987; Tajima 1989, Takahata 1988). To illustrate the concepts, let us consider a two-population symmetric island model. Each subpopulation consists of N diploids. Each generation, a small fraction m of each subpopulation is made up of migrants from the other subpopulation. In other words, each individual's parent was resident in the same population with probability $1-m$, and in the other subpopulation with probability m. As with the panmictic model, the probability that two alleles from the same subpopulation have a common ancestor in the previous generation is $1/2N$. Two alleles from different subpopulations have negligible probability of having a common ancestor in the previous generation. Putting these properties together, we can describe the genealogical process for a sample of alleles, n_1 from subpopulation 1 and n_2 from subpopulation 2. We denote the state of the ancestral lineages by an ordered pair, (i,j), indicating that i ancestors reside in subpopulation 1 and j reside in subpopulation 2. As usual, we trace the lineages back in time, in this case until either a common ancestor occurs or one of the lineages changes residence. This time is exponentially distributed with mean

$$\frac{1}{\left(\binom{n_1}{2} + \binom{n_2}{2} + (n_1+n_2)\frac{M}{2}\right)}$$

measuring time in units of $2N$ generations and where $M = 4Nm$. Given that one of the two events occurs, the probability that it is a common ancestor event among the n_i lineages in subpopulation i is:

$$\frac{\binom{n_i}{2}}{\left(\binom{n_1}{2} + \binom{n_2}{2} + (n_1+n_2)\frac{M}{2}\right)}$$

If the common ancestor event occurs in subpopulation 1, the state of the ancestral lineages changes to (n_1-1, n_2). The probability that the event is a change of residence of a lineage in subpopulation i is:

$$\frac{n_i\frac{M}{2}}{\left(\binom{n_2}{2} + \binom{n_2}{2} + (n_1+n_2)\frac{M}{2}\right)}$$

If a lineage changes from subpopulation 1 to subpopulation 2, working backward in time, then the state of the ancestral lineages changes to (n_1-1, n_2+1). And the process continues.

As described, the process is amenable to implementation as a Monte Carlo simulation. Strobeck (1987), Tajima (1989) and Slatkin and Maddison (1989) have carried out Monte Carlo simulations based on this approach.

To illustrate how analytical results can be obtained by this approach, we calculate the probability of identity of two alleles sampled from the same subpopulation, $P_s(\theta)$, and the probability of identity of two alleles from different subpopulations, $P_d(\theta)$. As noted earlier, we can calculate the moments of S once these identity coefficients are obtained. We assume a symmetric island model, as above, except with n subpopulations. We trace backward in time in the genealogy of two alleles from the same subpopulation, until either a coalescent, mutation or a migration event occurs. If the first event is a coalescent event, probability $1/(1+\theta+M)$, the two alleles are identical. If the first event is a mutation, probability $\theta/(1+\theta+M)$, the two alleles are not the same. If the first event is a migration, then the probability of identity of the two alleles is $P_d(\theta)$. This leads to the following equation for $P_s(\theta)$:

$$P_s(\theta) = \frac{1}{1+\theta+M}\cdot 1 + \frac{\theta}{1+\theta+M}\cdot 0 + \frac{M}{1+\theta+M}P_d(\theta) \tag{19}$$

For two alleles from two distinct populations, only mutations and migration events that bring the two lineages into the same subpopulation need to be considered. If the first event is a mutation event, probability $\theta/(\theta+M/n)$, the two alleles are different. If the first event is a migration event that takes one of the lineages into the subpopulation of the other, probability $(M/n)/(\theta+M/n)$, the probability of identity is $P_s(\theta)$. This leads to the following equation for $P_d(\theta)$:

$$P_d(\theta) = \frac{\frac{M}{n}}{\theta+\frac{M}{n}}P_s(\theta) \tag{20}$$

Solving eqns (19) and (20),

$$P_s(\theta) = \frac{(n-1)\theta+M}{(n-1)\theta^2+\theta(n-1+Mn)+M} \tag{21}$$

and

$$P_d(\theta) = \frac{M}{(n-1)\theta^2+\theta(n-1+Mn)+M} \tag{22}$$

These results are not new, having been obtained by several others without consideration of the coalescent process (see Crow and Aoki, 1984, and references therein). To obtain the expectation of the times to the common ancestor, t_s and t_d, for two alleles from the same subpopulation and different subpopulations, respectively, we can use the method described earlier in Section 4. Treating the identity coefficients $P_s(\theta)$ and $P_d(\theta)$ as moment-generating functions, the expectations of t_s and t_d are:

$$E(t_s) = P_s'(0) = n \tag{23}$$

and

$$E(t_d) = -P'_d(0) = n + \frac{n-1}{M} \tag{24}$$

The expected number of differences between two alleles from the same subpopulation is $\theta E(t_s) = n\theta$, and for two alleles from different subpopulations $\theta E(t_d) = n\theta + (n-1)\theta/M$ (Li 1976; Slatkin 1987; Strobeck 1987). Therefore, the expected time to the common ancestor of two alleles sampled from one subpopulation, as well as the expected number of differences, is independent of migration rate. If M is small, the expected time to the common ancestor of two alleles from different populations is relatively large, as is their divergence. This is consistent with our intuition that if the migration rate is low, the two subpopulations will be substantially differentiated. This is illustrated by the genealogies in Fig. 6. Tajima (1989) has used the coalescent approach to study the expected number of segregating sites in samples larger than 2.

Although the mean number of differences between two alleles from the same subpopulation does not depend on the migration rate, other aspects of the distribution do depend on the migration rate. In Fig. 7, the distribution of the average pairwise difference between 10 alleles sampled from the same subpopulation is shown. In this case, there were a total of three subpopulations and $M = 4Nm = 0.2$, and $\theta = 5.0$. Also shown is the distribution of the same statistic when $M = \infty$, i.e. a panmictic population with $\theta = 15.0$, and for a panmictic population with $\theta = 5.0$. The distributions with $M = 0.2$ and $M = \infty$ have the same mean, but otherwise the distributions are quite different. The $M = 0.2$ case has its mode and much of its mass around 5, with a very long tail. Except for the long tail, the distribution looks much like the distribution for a panmictic population with $\theta = 5.0$. This is because with the small migration rate, most of the time coalescent events occur within the subpopulation without any migration, and therefore the sample is like a sample from a single population with parameter $\theta = 5.0$. In contrast, the $M = \infty$ case has its mode around 15.

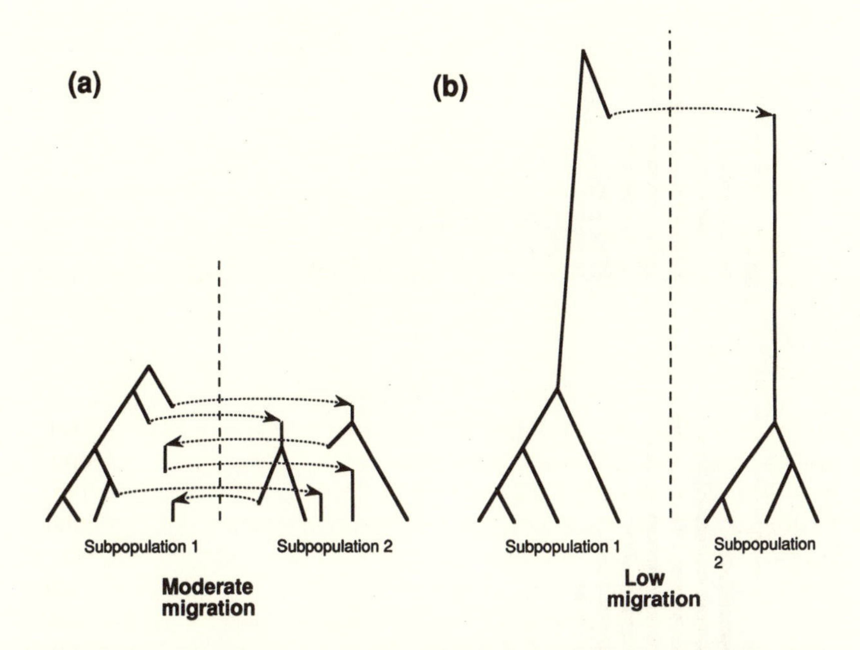

Fig. 6. (a) An example of a genealogy for a sample of size 8, 4 from each of 2 subpopulations, when the migration rate is moderately high. Each migration event is indicated by a dotted line with an arrow that indicates the actual direction of movement of an individual migrant. In this case, there would be relatively little differentiation of the two subpopulations. (b) An example genealogy with low migration rate. In this genealogy there is a single migration event. Alleles from within a subpopulation will be much more similar than alleles from different subpopulations.

These genealogies can also be interpreted as genealogies of gametes bearing different selected alleles (see Section 8). Subpopulation 1 would represent the pool of **S**-bearing gametes, and subpopulation 2 would represent the pool of **F**-bearing gametes. In this case, the dotted lines with arrows indicate mutations making an **F** allele into an **S** allele, and vice versa. If the mutation rate between the selected alleles is high, sequences bearing different alleles will be no more diverged than alleles bearing the same allele. If the mutation rate between **F** and **S** is low, **S**- and **F**-bearing gametes will be relatively diverged from each other. The genealogies could also represent the genealogy of a site linked to the selected locus. In this case, the dotted lines with arrows would represent mutations between the selected alleles and/or recombination events between the site and the selected locus.

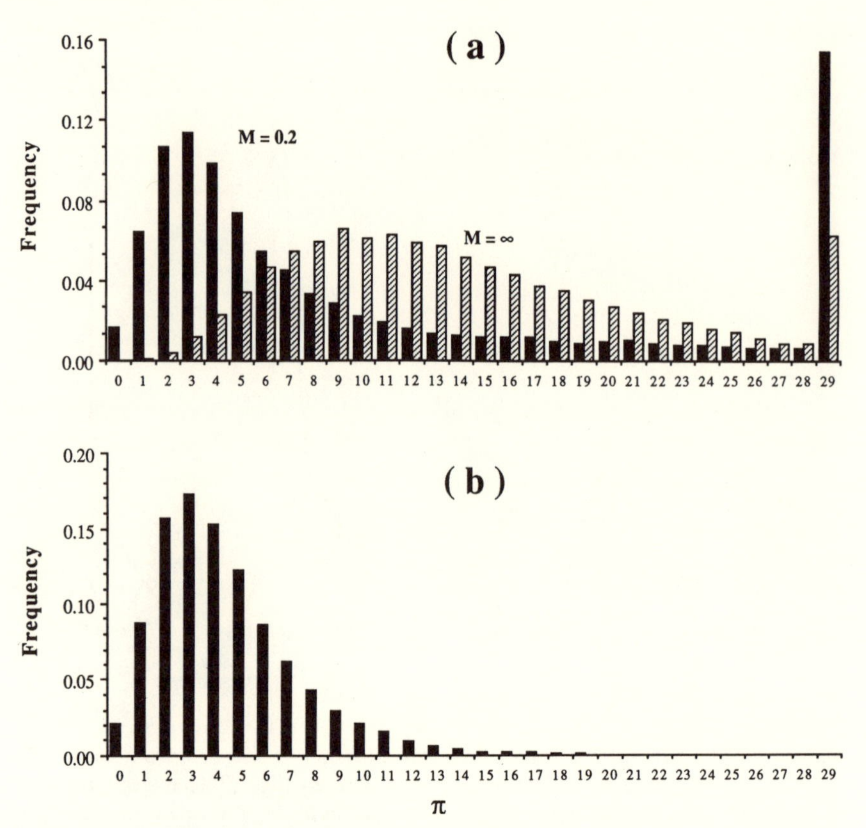

Fig. 7. (a) The distribution of π, the average pairwise number of differences between alleles in a sample of size 10 from a single subpopulation. The population is made up of three subpopulations, each of diploid size N, with $\theta = 4N\mu = 5$, and with $M = 4Nm = 0.2$ (solid bars) and $M = \infty$ (hatched bars). The mean for both distributions is approximately 15. (b) The distribution of the same quantity, for a single panmictic population, with $\theta = 4N\mu = 5$. Note the similarity with the low migration case in (a).

8. BALANCING SELECTION

Kaplan *et al.* (1988) have shown how the coalescent process can be analyzed under models with certain forms of selection. They focus primarily on the case where some form of balancing selection maintains a two-allele polymorphism at a particular nucleotide site, the 'selected site'. It is assumed that recurrent mutation between the two 'selected' alleles, designated **F** and **S**, occurs at rate v per replication. The analysis addresses the question: For sites completely linked to the selected site, how is the genealogy different from a genealogy of a neutral site isolated from any selection? When selection is weak and the frequency of the alleles at the selected site can drift considerably, numerical results can be obtained with some pain (Darden *et al.* 1989). Results are fairly simple when selection is strong and unchanging, so that the frequencies of the selected alleles, **S** and **F**, remain constant.

In the case of strong and constant selection, the coalescent process of sampled alleles is analogous to the coalescent process for the subdivided population model considered earlier, except that migration is no longer symmetric. If the frequencies of **S** and **F** are p and q, respectively, then one can consider the population to be subdivided into two subpopulations of size $2Np$ and $2Nq$. Mutation plays the role of migration. Each generation, an average of $2Nqv$ **F** alleles mutate (migrate) to the **S** allele (subpopulation) and $2Npv$ alleles mutate in the other direction. This means that a fraction, $2Nqv/2Np$, of the **S** alleles in each generation, approximately, are descendants of **F** alleles of the previous generation. In other words, an **S** allele of one generation has as parent an **F** allele with probability qv/p. If one is considering n_1 **S** alleles, the probability, P_{SF}, that one of them has as parent an **F** allele is, approximately:

$$P_{SF} = n_1 \frac{qv}{p}$$

Similarly, the probability, P_{FS}, that one of n_2 **F** alleles has an **S** allele as parent in the previous generation is:

$$P_{FS} = n_2 \frac{pv}{q}$$

The quantities pv/q and qv/p are the analogues of migration in the subdivided population model. In this case, 'migration' is not symmetric and the sizes of the two 'subpopulations' are not equal.

The probability of coalescent events are functions of the size of each subpopulation of alleles. For example, the probability that two **S** gametes have a common ancestor in the previous generation is approximately $1/2Np$, and the corresponding probability for two **F** gametes if $1/2Nq$. More generally, the probability, $P_{CA,S}$, that for n_1 **S** alleles some pair will have a common ancestor in the previous generation is:

$$P_{CA,S} = \frac{\binom{n_1}{2}}{2Np}$$

Similarly, for n_2 **F** alleles, the probability, $P_{CA,S}$, that some pair of the alleles will have a common ancestor in the previous generation is:

$$P_{\mathrm{CA,F}} = \frac{\binom{n_2}{2}}{2Nq}$$

Common ancestor events involving alleles of different type, **F** and **S**, would require both a mutation and a common ancestor event in the same generation. With low mutation rates and large population size, this is very unlikely, and we ignore the possibility.

Putting the elements of the process together, for a sample of n_1 **S** alleles and n_2 **F** alleles, the total rate per generation of events is:

$$P_{\mathrm{tot}} = P_{\mathrm{CA,S}} + P_{\mathrm{CA,F}} + P_{\mathrm{FS}} + P_{\mathrm{SF}}$$

Therefore, the time back until some event is exponentially distributed with mean $1/P_{\mathrm{tot}}$. Given that an event has occurred, the probability of a particular event is the rate of occurrence of that event divided by the total rate of events. For example, the probability that the first event is an **S** allele becoming an **F** allele, as we trace the lineages back in time, is $P_{\mathrm{SF}}/P_{\mathrm{tot}}$, in which case the ancestors on our sample in the generation of the first event would consist of n_1-1 **S** alleles and n_2+1 **F** alleles. Eventually, as with the other models, coalescent events will lead to the most recent common ancestor of the entire sample.

Analysis of this coalescent process shows that, if the mutation rate between **S** and **F** are small, the mean time to the common ancestor of a random **F** and a random **S** allele is large, as we would expect by analogy with the migration model. Kaplan *et al.* (1988) also show that if n alleles are chosen at random, without regard to which allele is present at the selected site, the mean total time in the genealogy of the samples can be considerably larger than under the strictly neutral model if $4Nv$ is small. If $4Nv$ is large, then the mean time to the common ancestor is essentially unaffected by the selection. The genealogy shown in Fig. 6b illustrates a typical genealogy under this model with a low mutation rate. One difference between the geographic subdivision model and the balancing selection model is that with geographic subdivision, all loci should be affected in the same way, whereas with selection, only sites tightly linked to the selected site would show the large genealogy.

Hudson and Kaplan (1988) showed that incorporating recombination in the selection model described above is straightforward. Consider the coalescent process for a locus, perhaps a single nucleotide site, which is linked to the selected site. Let r denote the recombination rate per generation between the locus, call it locus A, and the selected site. Due to recombination, the genealogy of a sample at locus A is not necessarily the same as the genealogy at the selected site. The genealogical process is, however, very similar. Consider a sampled gamete that has the **S** allele at the selected site. We refer to this as an **S**-bearing gamete. The parent (at the A locus) of this gamete is most likely also linked to an **S** allele, but because of the possibility that mutation or recombination occurred in producing our sampled gamete, the parent at the A locus might be linked to an **F** allele. Hudson and Kaplan show that the probability per generation of this change (from being linked to an **S** to being linked to an **F** allele) is $qv/p + 2pqr/p = (v + pr)q/p$. Similarly, the probability that the parent at the A locus of an **F**-bearing gamete is an **S**-bearing gamete is $(v + qr)p/q$. Thus, the coalescent process for linked sites is just like the process for completely linked sites, except that v is replaced by $v + pr$, in some cases, and by $v + qr$ in other cases. Assuming that p, r, v and N remain constant, the time between events is exponentially distributed. The mean time between events and the relative probabilities of the different possible changes in state depend on $2Nv$, $2Nr$ and p.

As one would expect, if a site is tightly linked to the selected site ($2Nr$ small), and $2Nv$ is small, the expected time to the common ancestor at the site of an **S**-bearing and an **F**-bearing gamete is large compared to the strictly neutral case. In other words, sites tightly linked to the selected site will be relatively highly diverged when sequences bearing the different selective alleles are compared. Loosely linked sites will be relatively little affected by the balancing selection. Thus, in comparisons of sequences bearing different selected alleles, there is expected to be a peak in sequence divergence centered on the selected site.

A single nucleotide site with a selectively maintained polymorphism can raise the level of neutral polymorphism at linked sites sufficiently to be detectable in samples. There exists, therefore, the potential for detecting selectively maintained variation by looking for regions of the genome with unusually high levels of polymorphism. One problem with this approach is that selective constraint is expected to vary from site to site and region to region, so that high levels of polymorphism in particular regions might be plausibly explained by an assumption of lowered constraint. It is possible, however, to test the hypothesis of lowered constraint by making a comparison of sequences between closely related species. This is because, under a strictly neutral model, the level of constraint not only determines the expected level of polymorphism within species, but also the level of divergence between species. Thus, if a region has a relatively high level of polymorphism within a species because of lowered constraint, it ought to show a relatively high level of divergence between species. In other words, one should see a strong correlation between the level of divergence and the level of polymorphism as one examines different regions of the genome.

The presence of a polymorphism maintained by balancing selection can cause large deviations from this pattern of correlation between divergence between species and polymorphism with species. A polymorphism maintained by balancing selection and which arose since the divergence of the two species, will have no affect on the accumulation of neutral mutations that differentiate two species, as discussed in Section 2. Balanced polymorphisms that arose before the divergence of the species and that have been maintained continuously since their origin, can result in greater between-species neutral divergence at linked sites than at unlinked sites. The size of this effect diminishes with time since divergence of the two species. Thus, depending on when the balanced polymorphism arose, the balancing selection may have little affect on the divergence between species but, as argued above, will greatly increase the level of neutral polymorphism within species at tightly linked sites.

A statistical test of neutrality was devised to test whether between-species divergence and within-species polymorphism show the correspondence expected under neutrality (Hudson *et al.* 1987; Kreitman and Aguadé 1986). Application of this test to data from the alcohol dehydrogenase (*Adh*) region of *Drosophila melanogaster* and *D. sechellia* resulted in a rejection of neutrality. The departure of the data from the neutral expectations was consistent with the existence of a balanced polymorphism in the coding region of *Adh*. There is a great deal of independent evidence suggesting the importance of selection in the maintenance of the **F/S** polymorphism of *Adh* (Oakeshott *et al.* 1982).

If this departure of the data from the expectations of the neutral model is due to balancing selection acting on the **F/S** protein polymorphism of *Adh*, the model with balancing selection and recombination described above should be applicable. We should be able to predict quantitatively the increased level of divergence at tightly linked sites that result from a large genealogy induced by selection. The **F/S** polymorphism of *Adh* is produced by a nucleotide polymorphism at codon 192 of the *Adh* gene. To predict the level of polymorphism at sites linked to codon 192 of the *Adh* gene under the selection model, one must assign values to a number of parameters. The parameters needed are: p, the frequency of the **S** allele; $2Nv$, where v is the mutation rate between the selected alleles; θ, the neutral mutation rate per base pair at linked sites; and $R = 4Nr$, where r is the recombination rate per base pair. For all of these parameters, some prior information was available to permit us to assign approximate values to these parameters and then make comparisons between expected levels of divergence between sequences and observed levels in the sequence data of Kreitman (1983).

To display the observed and predicted divergence between sequences, as a function of position along the sequence, a sliding window method was used. In this method, a window is sequentially slid along the aligned sequences. At each position of the window, the level of polymorphism is noted for the collection of contiguous sites in the window. In this way, polymorphism as a function of nucleotide position can be displayed. The level of polymorphism was measured by the average number of pairwise differences between sequences in the window. The protein coding constraints were also incorporated by varying the window width so that the number of silent changes possible remained constant. A reasonable fit of observed and expected was achieved for **F/S** comparisons, using our prior estimates of the parameters, except for R. Our prior estimate of R was 24.0, but the best fit was obtained with $R = 4.0$. However, there was considerable discrepancy between the expected and the observed divergence between sequences bearing the **S** allele. The interested reader should consult Hudson and Kaplan (1988) for more details.

Hudson and Kaplan's (1988) predictions were based on the assumption that all sites that do not affect the amino acid sequence of the *Adh* enzyme, that is non-coding sites and silent sites of the coding exons, have the same neutral mutation rate parameter, θ. It was also assumed that sites at which zero, one or two silent nucleotide changes are possible, have neutral mutation parameters of 0, $\theta/3$ and $2\theta/3$, respectively. With these assumptions, only one neutral mutation parameter was needed. It is quite plausible, however, that the level of constraint, and hence the neutral mutation parameter, is not the same for all non-amino acid changing sites.

A more detailed analysis is possible using sequence data from a closely related species, *Drosophila simulans*, to estimate the mutation parameter at different sites. To use the between-species data to estimate mutation parameters, we can proceed as follows. Let us suppose that the neutral mutation parameter varies from site to site, denoting the neutral mutation parameter for site i by θ_i. Let t denote the average time since the most recent common ancestor of two sequences, one from *D. melanogaster* and one from *D. simulans*. If $\theta_i t$ is small enough, the probability that the two sequences from the species are different at site i is approximately $\theta_i t$. Consequently, the average value of θ for a small region, consisting of l nucleotides, can be estimated by d/tl, where d is the number of sites differing between the two species in the region l nucleotides long. Such estimates were used as follows.

A window (the same that will be used to display the observations and predictions of within-species polymorphism) is slid along an aligned pair of sequences, one from *D. melanogaster* and one from *D. simulans*. At each position of the window, the number of differences between the sequences in the window is counted and divided by the product of t and the width of the window. This quantity is taken as an estimate of θ for the site in the middle of the window. This is clearly a smoothing procedure, which assigns a value of θ to a site that is determined by the variation of a collection of sites surrounding the site. The value of t was chosen to produce a good fit, that is, to give an average level of polymorphism that fits the observations. With these θ_i's in hand, one can predict the level of polymorphism at any collection of sites linked to the selected site.

Applying this method to the polymorphism data of Kreitman (1983), supplemented by additional 5′ and 3′ sequence data from *D. melanogaster* and a sequence from *D. simulans* (all kindly provided by Martin Kreitman, (pers. com.), the observed and predicted levels of polymorphism are shown in Fig. 8. Assuming complete neutrality, ignoring the **F/S** polymorphism entirely, the expected number of differences between two sampled sequences in a region is the sum of the θ_i's for the sites in the region. This prediction is shown in Fig. 8a. Because the θ_i's are proportional to the divergence between the two species, the predicted level of polymorphism in Fig. 8a is proportional to the level of divergence between *D. melanogaster* and *D. simulans*. There appear to be regions of higher and lower constraint. Note especially the low predicted and low observed level of polymorphism in the 5′ flanking region, numbered approximately 400–450 in Fig. 8. Note also that the silent sites of the third coding exon of *Adh* are predicted to have a relatively high level of polymorphism. This is due to a relatively high level of divergence between species for these sites. And yet the observed level of polymorphism for these sites is still much higher than the predicted. If a balanced polymorphism is assumed at the **F/S** site in codon 192, this high level of polymorphism can be accounted for quite easily, as shown in Fig. 8b. The recombination parameter that produces this fit is R = 12.0, only a factor of 2 lower than the *a priori* estimate given by Hudson and Kaplan (1988). Expected and observed levels of polymorphism between **S** alleles are shown in Fig. 9. The fit of observed and expected shown in Fig. 9, obtained using the θ_i's estimated from between-species data, is considerably better than the fit obtained by Hudson and Kaplan using a constant value of θ. The observed and predicted divergence between **F** alleles are both low and are not shown.

Analyses utilizing both within-species polymorphism and between-species divergence may be extremely powerful for detecting the action of natural selection, both balancing as indicated here and also recent fixations of advantageous mutants as described in the next section. Population geneticists have for more than 20 years debated about whether selection plays a significant role in maintaining the electrophoretically detectable polymorphisms of soluble enzyme loci. It seems that significant headway could be made if 20–30 such loci could be examined in the same detail as *Adh* in *D. melanogaster*. With modern polymerase chain reaction methods, such an undertaking does not seem out of the question. Many situations where selection is already strongly indicated, such as *Ldh* in *Fundulus* (Powers *et al.* 1983), *Lap* in *Mytilis* (Hilbish and Koehn 1985), *Gpt* in *Tigriopus* (Burton and Feldman 1983), *Gpdh* in *D. melanogaster* (Barnes and Laurie-Ahlberg 1986; Oakeshott *et al.* 1984), and others described

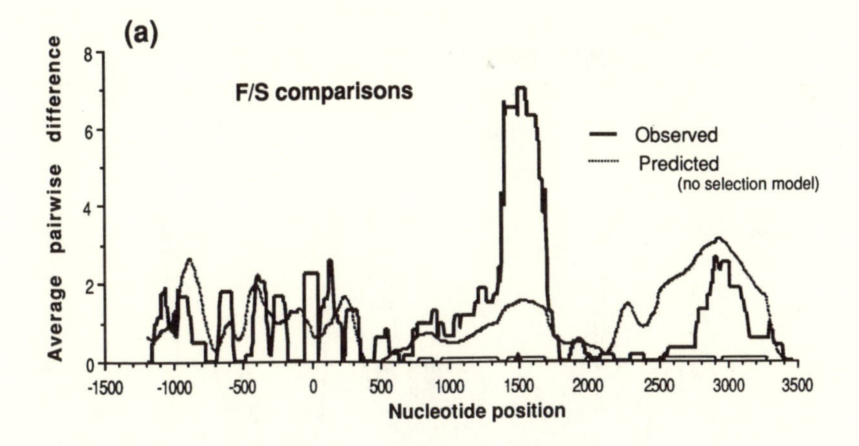

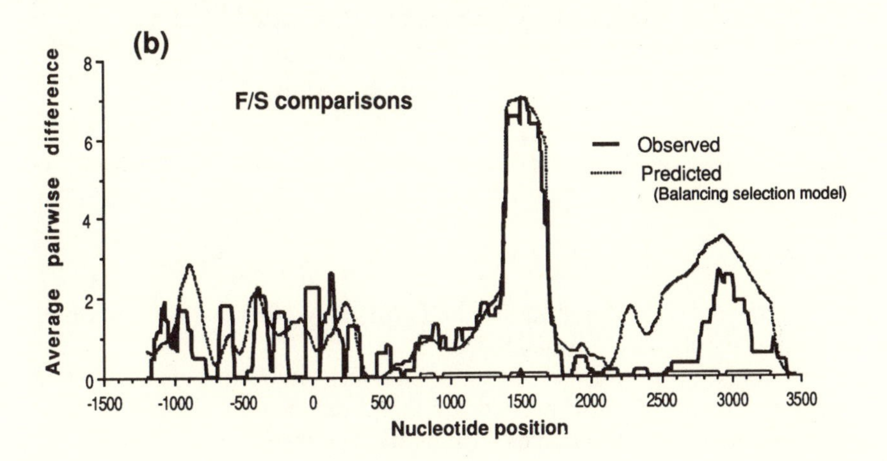

Fig. 8. The predicted and observed number of differences between **F**- and **S**-bearing sequences in a 'sliding window' plotted as a function of nucleotide position. The coding exons of *Adh* and the duplicate locus, are shown by the low rectangles on the position axis. The site of the **F/S** polymorphism, codon 192 of *Adh*, is indicated by the black triangle. The width of the window was adjusted so that there were always 300 possible silent changes in the window.

(a) The predicted curve based on the strict neutral model without balancing selection. The between-species comparison of sequences from *D. melanogaster* and *D. simulans* was used to estimate θ for each site as described in the text. The predicted pairwise difference under the neutral model is simply the sum of the θ's for the sites in the window. The value assumed for t, the time since divergence of the species in units of $2N$ generations, was 5. (b) The predicted curve is based on the balancing selection model with parameter values: $\beta = 0.001$, $R = 4Nr = 12.0$, $t = 5$, and $p = 0.7$ (see the text for an explanation of these parameters).

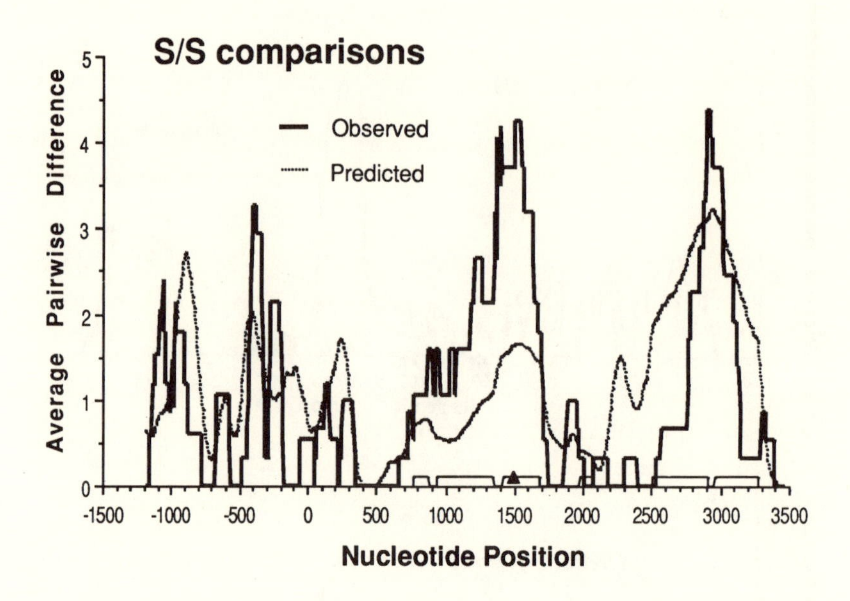

Fig. 9. The predicted (under the balancing selection model) and observed number of differences between **S**-bearing sequences in a 'sliding window' plotted as a function of nucleotide position. All parameters are as in Fig. 8b.

by Koehn *et al.* (1983) and Zera *et al.* (1985) may provide good test cases, if sequence data can be obtained.

The statistical power of Hudson *et al.*'s test (1987) to detect selection needs investigation. Also needed are measures of the goodness of fit of observed and predicted curves such as those shown in Figs 8 and 9.

9. HITCHHIKING

Using coalescent methods, Kaplan *et al.* (1989) have analysed a model in which rare advantageous variants sweep through a population. They reanalyzed the 'hitchhiking' effect (Maynard Smith and Haigh 1974) of these advantageous variants on selectively neutral variation at linked sites. The process is very similar to the balancing selection model, except that the frequency of the selected alleles change through time and therefore the probability of coalescent events, as well as other events, change through time. Consequently, the time intervals between events are not exponentially distributed. If the frequency of the advantageous allele can be approximated by a deterministic function, results can be obtained by straightforward numerical methods. In this way, one can assess how neutral variation at linked nucleotide sites is reduced by the rapid fixation of a favored variant.

A genealogy is shown in Fig. 10 that illustrates how a gene tree can be very different in shape after a recent hitchhiking event compared to genealogies under the equilibrium neutral model (Fig. 1). Most of the genealogy in Fig. 10 consists of lineages that descend without branching to a single sampled gamete. With this form of genealogy, a 'star' genealogy, most neutral mutations would result in a polymorphism such that, at each polymorphic site, the mutant nucleotide is present only once and the non-mutant nucleotide is present in all the remaining sequences. That is, most polymorphic nucleotide sites would have a low-frequency allele, present once or twice in the sample, and a very high-frequency allele. No particular gamete would carry a large number of the rare mutations, as would be expected if there was one highly diverged lineage. This pattern

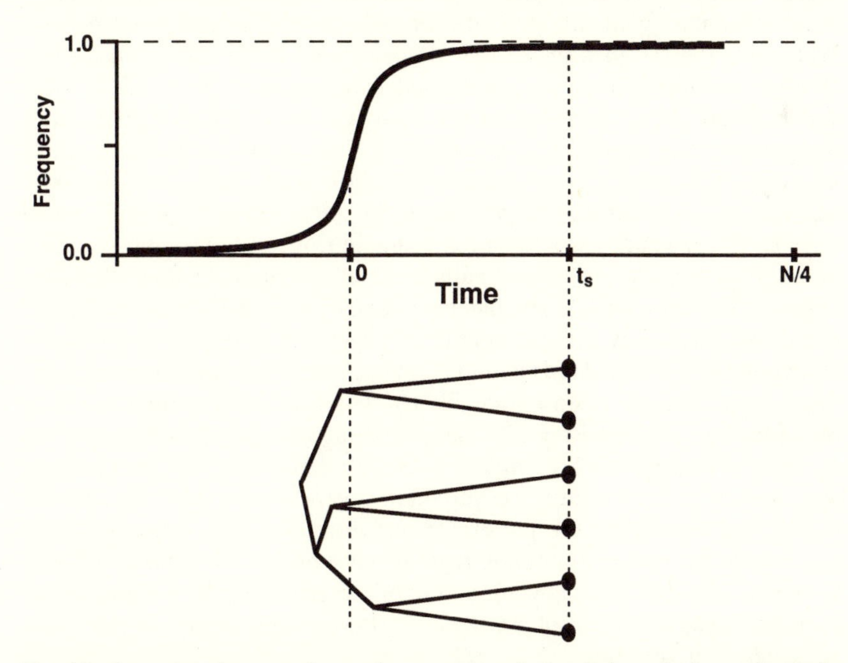

Fig. 10. An example genealogy of a sample of six alleles when a selectively advantageous mutant has recently fixed. The bold curve shows the frequency of the advantageous mutant as a function of time. The genealogy of the sample, taken at time t_s, is shown superimposed on the plot. No recombination is depicted in this example. If t_s is considerably less than N, most coalescent events will occur close to the time marked 0. In this case, compared to the genealogy without selection, the overall size of the genealogy will be small, and so there will be relatively little variation. Also, the shape of the genealogy is different than the neutral equilibrium genealogy (see text).

of mostly low-frequency polymorphisms would also be characteristic of populations after a recent very severe bottleneck, which results in most lineages coalescing in a relatively short period of time. In contrast to the hitchhiking model, after a population bottleneck, all loci are expected to have the distinctive genealogy. The frequency spectrum of polymorphisms may prove to be an extremely useful characteristic of nucleotide variation.

10. CONCLUSIONS

Focusing on the coalescent process can be a useful way of thinking about sampling properties in a variety of circumstances. We have seen that the coalescent process can be relatively simply described in a variety of circumstances. In addition to the situations described in the preceding sections, the coalescent process has been used to examine the variation of alleles in multigene families with conversion (Kaplan and Hudson 1987; Watterson 1989*b*), in transposable element families (Hudson and Kaplan 1986*b*), and highly repetitive dispersed elements such as ALU (Kaplan and Hudson 1987). In many of the analyses described here, the entire complexity of genealogies was reduced to one or two statistics, namely how big is the genealogy or, equivalently, how many mutations, S, would be expected to occur on the genealogy. In some cases, it was also asked how big the genealogy would be for a sample from within a subpopulation or within a certain selectively maintained allelic class, or how divergent different classes or alleles would be. (Results are also available for the size of the genealogy of a subsample of sequences all bearing a particular electrophoretic allele: Hudson and Kaplan 1986*a*.) It should be noted that these properties are often obtainable by classic methods, without explicit reference to the genealogy of a sample, using one-generation recursions or diffusion approaches (Watterson 1989*a*). However, the genealogical approach is broad and intuitive, providing a general approach for deriving results and visualizing the pattern of variation expected under a variety of models. For non-recombining regions, genealogies will be a useful way to summarize complex information, both empirical and theoretical, as exemplified by the work of Avise *et al.* (1987) and Cann *et al.* (1987) on mitochondrial variation.

Methods that exploit actual details of the genealogy of samples, as opposed to just S, may in some cases be extremely powerful. Noteworthy in this regard is the work of Golding (1987; Golding *et al.* 1986) and Iizuka (1989), which examines the distribution of different classes of mutations on the genealogy in order to detect selection. Slatkin (1989) and Slatkin and Maddison (1989) have begun to explore methods of inferring geographic structure using detailed genealogical information. Estimation of mutation parameters using detailed genealogical information in the form of an unrooted tree has been examined (Strobeck 1983). A great deal of information resides in the genealogy of sampled alleles. Additional methods of exploiting this information are needed.

For regions with high levels of recombination, detailed genealogies may be impossible to infer with any reliability. With nuclear genes, genealogies can differ substantially for sites a few hundred base pairs apart. Our ability to make inferences about the genealogies of segments a few hundred base pairs long is very limited. The genealogy for a sample of sequences of moderate length will frequently be an incredibly complex network of lineages, coalescing and breaking apart.

Although some kinds of genealogical information may resist at least moderate levels of recombination (Golding 1987), summary statistics, rather than detailed genealogical reconstruction, may be essential when analysing data from highly recombining regions. Possible useful quantities include the level of polymorphism (measure by S or the average number of pairwise differences), the frequency spectrum of variation, and contrasts of these quantities between different classes of variants, e.g. silent vs. coding, insertion/deletion vs. substitution, or between variants linked to different electrophoretic alleles. Sawyer *et al.* (1987) used such an approach. Comparisons of patterns seen within species to patterns observed between species will be very informative (Kreitman 1987), suggesting sites where balancing selection, hitchhiking events, or drastic changes in mutation rates or constraints have occurred. Ten or fifteen other loci with enzyme polymorphisms might usefully be examined in the same fashion as *Adh* has. The coalescent process can be useful in deriving some statistical properties of these quantities under a variety of models. No-recombination analyses apply, after all, to genealogies of single sites regardless of how much recombination occurs. (If mutation rates are sufficiently low per nucleotide site, infinite-site or infinite-allele models can still be accurate even for single nucleotide sites. For higher mutation rates, finite-allele models may be required.). However, adequately taking into account the non-independence of linked sites may prove extremely difficult analytically by any approach, classical or coalescent. For this reason, simulations based on the coalescent process will play an important role in the investigation of statistical methods, testing and estimation.

Intuition from consideration of no-recombination models, together with Monte Carlo simulation, may be an important route for the development of methods of analysis of molecular variation within species. Simulations based on the coalescent process with recombination can be orders of magnitude faster than the analogous simulations implemented the old-fashioned way with entire generations represented in the computer and the time-consuming sampling to produce large numbers of successive generations. The method of simulation can straightforwardly incorporate, simultaneously, recombination, geographic structure, population size variation and non-equilibrium situations, as well as circular genomes and some forms of selection, and a variety of mutation schemes. Such simulations will play a role in establishing confidence intervals of estimates, significance points for test statistics, and the power of tests against a variety of alternative models.

Despite the difficulties of analyzing data from recombining DNA, such data may be preferred in some circumstances. Consider a sample of mitochondria which, because it lacks recombination, has a single genealogy for the entire molecule. With the large number of sites that constitute a mitochondrial genome, one can hope to make accurate inferences about that genealogy. But the genealogy one gets is a single realization of the stochastic process that governs the evolution of that molecule in the population from which the sample is drawn. The confidence one can have in inferences about the stochastic process is limited from one observation. With nuclear genes, we may frequently be unable to infer a genealogy accurately, but the information that we do gather comes from many segments, with different degrees of statistical independence. Each segment will have experienced the same history of population subdivision and population size. If there are statistical properties of such samples that can help us estimate parameters or test hypotheses, then the possibility of accumulation of information from distinct loci makes the nuclear genome potentially much more informative. Estimation of migration rates and mutation parameters, for example, may be much more precise using nuclear data. The type of analysis used to examine variation in and near the *Adh* locus could not be carried out with mitochondrial data alone. One could, however, employ Hudson *et al.*'s (1987) test using mtDNA as one locus and a nuclear gene for the other locus.

The genealogies that are expected under neutral models and under certain models with constant selection coefficients have been characterized to some extent. An important class of models for which predictions about genealogies are not available are random environment models. It is important to know if these models predict genealogies that are very different in some way from the genealogies predicted from neutral models.

DNA polymorphisms within populations have already shown their usefulness in addressing longstanding questions about genetic variation within populations. As more data are collected, it is clear that genealogical analysis will play a large part in understanding the patterns that are revealed.

ACKNOWLEDGMENTS

I am very grateful to Norm Kaplan for many useful hours of discussion over a period of several years. Thanks also to Janis Antonovics, Norm

Kaplan, Marty Kreitman and Marcie McClure for comments on the manuscript.

REFERENCES

Aguadé, M., Miyashita, N. and Langley, C. H. (1989). Reduced variation in the yellow-achaete-scute region in natural populations of *Drosophila melanogaster*. *Genetics* **122,** 607–615.

Aquadro, C. F., Deese, M. M., Bland, C. H., Langley, C. H. and Laurie-Ahlberg, C. C. (1986). Molecular population genetics of the alcohol dehydrogenase gene region of *Drosophila melanogaster*. *Genetics* **114,** 1165–90.

Avise, J. C., Arnold, J., Ball, R. M., Bermingham, E., Lamb, T., Neigel, I. E., Reeb, C. A. and Saunders, N. C. (1987). Intraspecific phylogeography: The mitochondrial DNA bridge between population genetics and systematics. *Ann. Rev. Ecol. Syst.* **18,** 489–522.

——, Ball, R. M. and Arnold, J. (1988). Current versus historical population sizes in vertebrate species with high gene flow: A comparison based on mitochondrial DNA lineages and inbreeding theory for neutral mutations. *Mol. Biol. Evol.* **5,** 331–44.

Barnes, P. T. and Laurie-Ahlberg, C. C. (1986). Genetic variability of flight metabolism in *Drosophila melanogaster*. III. Effects of GPDH allozymes and environmental temperature on power output. *Genetics* **113,** 267–94.

Bermingham, E. and Avise, J. C. (1986). Molecular zoogeography freshwater fishes in southeastern United States. *Genetics* **113,** 939–65.

Birky, C. W. and Walsh, J. B. (1988). Effects of linkage on rates of molecular evolution. *Proc. Natl Acad. Sci. USA* **85,** 6414–18.

Burton, R. S. and Feldman, M. W. (1983). Physiological effects of an allozyme polymorphism: Glutamate-pyruvate transaminase and response to hyperosmotic stress in the copepod *Tigriopus californicus*. *Biochem. Genet.* **21,** 239–51.

Cann, R. L., Stoneking, M. and Wilson, A. C. (1987) Mitochondrial DNA and human evolution. *Nature* **325,** 31–6.

Crow, J. F. and Aoki, K. (1984). Group selection for a polygenic behavioral trait: Estimating the degree of population subdivision. *Proc. Natl Acad. Sci. USA* **81,** 6073–7.

Darden, T., Kaplan, N. L. and Hudson, R. R. (1989). A numerical method for calculating moments of coalescent times in finite populations with selection. *J. Math. Biol.* **27,** 355–68.

Donnelly, P. (1986). Partition structures,, *Polya urns*, the Ewens sampling formula, and the age of alleles. *Theoret. Popul. Biol.* **30,** 271–88.

—— and Tavaré, S. (1986). The ages of alleles and a coalescent. *Adv. Appl. Prob.* **18,** 1–19.

—— and Tavaré, S. (1987). The population genealogy of the infinitely-many neutral alleles model. *J. Math. Biol.* **25,** 381–91.

Ewens, W. J. (1972). The sampling theory of selectively neutral alleles. *Theoret. Popul. Biol.* **3,** 87–112.

—— (1979). *Mathematical population genetics*. Springer-Verlag, New York.

—— (1989). Population genetics theory – the past and the future. In *Mathematical and statistical problems of evolutionary theory* (ed. S. Lessard), pp. Kluwer Academic, Dordrecht.

Golding, G.B. (1987). The detection of deleterious selection using ancestors inferred from a phylogenetic history. *Genet. Res. Camb.* **49,** 71–82.

——, Aquadro, C. F. and Langley, C. H. (1986). Sequence evolution within populations under multiple types of mutation. *Proc. Natl Acad. Sci. USA* **83,** 427–31.

Griffiths, R. C. (1980). Lines of descent in the diffusion approximation of neutral Wright-Fisher models. *Theoret. Popul. Biol.* **17,** 37–50.

—— (1981*a*). Neutral two-locus multiple allele models with recombination. *Theoret. Popul. Biol.* **19,** 169–86.

—— (1981*b*). The number of heterozygous loci between two randomly chosen completely linked sequences of loci in two subdivided population models. *J. Math. Biol.* **12,** 251–61.

Hedrick, P. W. and Thomson, G. (1986). A two-locus neutrality test: Applications to humans, *E. coli* and lodgepole pine. *Genetics* **112,** 135–56.

Hilbish , T. J. and Koehn, R. K. (1985). The physiological basis of natural selection at the *Lap* locus. *Evolution* **39,** 1302–1317.

Hudson, R. R. (1983*a*). Properties of a neutral allele model with intragenic recombination. *Theoret. Popul. Biol.* **23,** 183–201.

—— (1983*b*). Testing the constant-rate neutral model with protein sequence data. *Evolution* **37,** 203–217.

—— (1987). Estimating the recombination parameter of a finite population model without selection. *Genet. Res. Camb.* **50,** 245–50.

—— and Kaplan, N. L. (1985). Statistical properties of the number of recombination events in the history of a sample of DNA sequences. *Genetics* **111,** 147–64.

—— and Kaplan, N. L. (1986*a*). On the divergence of alleles in nested subsamples from finite populations. *Genetics* **113,** 1057–76.

—— and Kaplan, N. L. (1986*b*).On the divergence of members of a transposable element family. *J. Math. Biol.* **24,** 207–215.

—— and Kaplan, N. L. (1988). The coalescent process in models with selection and recombination. *Genetics* **120,** 831–40.

——, Kreitman, M. and Aguadé, M. (1987). A test of neutral molecular evolution based on nucleotide data. *Genetics* **116,** 153–9.

Iizuka, M. (1989). Population genetical model for sequence evolution under multiple types of mutation. *Genet. Res. Camb.* **54**, 231–7.

Kaplan, N. L. and Hudson, R. R. (1985). The use of sample genealogies for studying a selectively neutral M-loci model with recombination. *Theoret. Popul. Biol.* **28,** 382–96.

—— and Hudson, R. R. (1987). On the divergence of genes in multigene families. *Theoret. Popul. Biol.* **31,** 178–94.

——, Darden, T. and Hudson, R. R. (1988). The coalescent process in models with selection. *Genetics* **120,** 819–29.

——, Hudson, R. R. and Langley, C. H. (1989). The 'hitchhiking effect' revisited. *Genetics* **123**, 887–99.

Kimura, M. (1969). The number of heterozygous nucleotide sites maintained in a finite population due to steady flux of mutations. *Genetics* **61,** 893–903.

—— (1983). *The neutral theory of molecular evolution*. Cambridge University Press, Cambridge.

—— and Crow, J. F. (1964). The number of alleles that can be maintained in a finite population. *Genetics* **49,** 725–38.

Kingman, J. F. C. (1980). *Mathematics of genetic diversity*. CBMS-NSF Regional Conference Series in Applied Mathematics, No. 34. Society for Industrial and Applied Mathematics, Philadelphia.

—— (1982*a*). The coalescent. *Stochast. Proc. Appl.* **13,** 235–48.

—— (1982*b*). On the genealogy of large populations. *J. Appl. Prob.* **19A,** 27–43.

Koehn, R. K., Zera, A. J. and Hall, J. G. (1983). Enzyme polymorphism and natural selection. In *Evolution of genes and proteins* (ed. M. Nei and R. K. Koehn), pp. 115–36. Sinauer, Sunderland, Mass.

Kreitman, M. (1983). Nucleotide polymorphism at the alcohol dehydrogenase locus of *Drosophila melanogaster. Nature* **304,** 412–17.

—— (1987). Molecular population genetics. In *Oxford surveys in evolutionary biology* (ed. P. H. Harvey and L. Partridge), Vol. 4, pp. 38–60. Oxford University Press, Oxford.

—— and Aquadé, M. (1986). Excess polymorphism at the *Adh* locus in *Drosophila melanogaster. Genetics* **114,** 93–110.

Li, W.-H. (1976). Distribution of nucleotide differences between two randomly chosen cistrons in a subdivided population: The finite island model. *Theoret. Popul. Biol.* **10,** 303–308.

Maynard Smith, J. and Haigh, J. (1974). The hitchhiking effect of a favorable gene. *Genet. Res. Camb.* **23,** 23–35.

Nagylaki, T. and Petes, T. D. (1982). Intrachromosomal gene conversion and the maintenance of sequence homogeneity among repeated genes. *Genetics* **100,** 315–37.

Nei, M. (1987). *Molecular evolutionary genetics*. Columbia University Press, New York.

Neigel, J. E. and Avise, J. C. (1986). Phylogenetic relationships of mitochondrial DNA under various demographic models of speciation. In *Evolutionary processes and theory* (ed. E. Nevo and S. Karlin), pp. 513–34. Academic Press, London.

Oakeshott, J. G., Gibson, J.B., Anderson, P. R., Knibb, W. R., Anderson, D.G. and Chambers, G. K. (1982). Alcohol dehydrogenase and glycerol-3-phosphate dehydrogenase clines in *Drosophila melanogaster* on three continents. *Evolution* **36,** 86–96.

——, McKechnie, S. W. and Chambers, G. K. (1984). Population genetics of the metabolically related *Adh*, *Gpdh* and *Tpi* polymorphisms in *Drosophila melanogaster*. I. Geographic variation in *Gpdh* and *Tpi* allele frequencies in different continents. *Genetica* **63,** 21–9.

Padmadisastra, S. (1987). The genetic divergence of three populations. *Theoret. Popul. Biol.* **32,** 347–65.

—— (1988). Estimating divergence times. *Theoret. Popul. Biol.* **34,** 297–319.

Pamilo, P. and Nei, M. (1988). Relationships between gene trees and species trees. *Mol. Biol. Evol.* **5,** 568–83.

Powers, D. A., DiMichele, L. and Place, A. R. (1983). The use of enzyme kinetics to predict differences in cellular metabolism, developmental rate, and swimming performance between *Ldh-B* genotypes of the fish, *Fundulus heteroclitus*. In *Isozymes: Current topics in biological and medical research, Vol. 10: Genetics and evolution* (ed. M. C. Rattazzi, J. G. Scandalios and G. S. Whitt), pp. 147–70. Alan R. Liss, New York.

Press, W. H., Flannery, B. P., Teukolsky, S. A. and Vetterling, W. T. (1988). *Numerical recipes in C.* Cambridge University Press, Cambridge.

Sawyer, S. A., Dykhuizen, D. E. and Hartl, D. L. (1987). Confidence interval for the number of selectively neutral amino acid polymorphisms. *Proc. Natl Acad. Sci. USA* **84,** 6225–8.

Slatkin, M. (1987). The average number of sites separating DNA sequences drawn from a subdivided population. *Theoret. Popul. Biol.* **32,** 42–9.

—— (1989). Detecting small amounts of gene flow from phylogenies of alleles. *Genetics* **121,** 609–612.

—— and Maddison, W. P. (1989). A cladistic measure of gene flow inferred from the phylogenies of alleles. *Genetics* **123,** 603–613.

Stephens, J. C. and Nei, M. (1985). Phylogenetic analysis of polymorphic DNA sequences at the *Adh* locus in *Drosophila melanogaster* and its sibling species. *J. Mol. Evol.* **22,** 289–300.

Strobeck, C. (1983). Estimation of the neutral mutation rate in a finite population from DNA sequence data. *Theoret. Popul. Biol.* **24,** 160–72.

—— (1987). Average number of nucleotide differences in a sample from a single subpopulation: A test for population subdivision. *Genetics* **117,** 149–53.

Tajima, F. (1983). Evolutionary relationship of DNA sequences in finite populations. *Genetics* **105,** 437–60.

—— (1989). DNA polymorphism in a subdivided population: The expected number of segregating sites in the two-subpopulation model. *Genetics* **123,** 229–40.

Takahata, N. (1988). The coalescent in two partially isolated diffusion populations. *Genet. Res. Camb.* **52,** 213–22.

—— (1989). Gene genealogy in three related populations: Consistency probability betweeen gene and population trees. *Genetics* **122,** 957–66.

Tavaré, S. (1984). Line-of-descent and genealogical processes, and their applications in population genetic models. *Theoret. Popul. Biol.* **26,** 119–64.

——, Ewens, W. J. and Joyce, P. (1989). Is knowing the age-order of alleles in a sample useful in testing for selective neutrality? *Genetics* **122,** 705–711.

Watterson, G. A. (1975). On the number of segregating sites in genetical models without recombination. *Theoret. Popul. Biol.* **10,** 256–76.

—— (1982*a*). Mutant substitutions at linked nucleotide sites. *Adv. Appl. Prob.* **14,** 206–224.

—— (1982*b*). Substitution times for mutant nucleotides. *J. Appl. Prob.* **19A,** 59–70.

—— (1984). Lines of descent and the coalescent. *Theoret. Popul. Biol.* **26,** 77–92.

—— (1989*a*). Allele frequencies in multigene families. I. Diffusion equation approach. *Theoret. Popul. Biol.* **35,** 142–60.

—— (1989*b*). Allele frequencies in multigene families. II. Coalescent approach. *Theoret. Popul. Biol.* **35,** 161–80.

Zera, A. J., Koehn, R. K. and Hall, J. G. (1985). Allozymes and biochemical adaptation. In *Comprehensive insect physiology, biochemistry and pharmacology* (ed. G. A. Kerkut and L. I. Gilbert), Vol. 10, pp. 633–74. Pergamon Press, New York.

APPENDIX

Monte Carlo simulations based on the coalescent process are an efficient way to investigate properties of samples of alleles under a variety of models. In contrast to standard simulation methods, where entire populations are represented in the computer and many generations of sampling are required to reach equilibrium, with the coalescent approach only the lineages of sampled alleles need be represented. When times are measured in units of $2N$ generations and one is willing to use 'diffusion approximations', the population size does not enter as a separate parameter. Instead, all other parameters enter as products with the population size, such as $4Nr$, $4N\mu$ and $4Nm$.

It is frequently useful to generate the genealogy first, then add mutations to produce gametes. The subroutine shown below is based on the simplest neutral model with no recombination and illustrates the basic ideas. As with most of the models considered in this chapter, the coalescent process consists of a random series of events, separated by exponentially distributed time intervals. For this neutral model, there are only common ancestor events. The times of all the common ancestor events can be generated prior to determining the topology, that is, which lineages coalesce at which time. The time intervals are the $T(i)$ referred to in the main text, and have means given by eqn (5).

The following subroutine, make_tree, was written in C and generates a genealogy of a sample of alleles. Times are measured in units of $2N$ generations. The genealogy is represented by an array of nodes, designated 'tree' in this routine. Let n denote the sample size, which is called sample_size in the program. The first n nodes represent the sampled alleles. The next $N-1$ nodes represent the actual nodes of the genealogy. Each node in the genealogy is represented in the computer by a structure, which records the time of the node, the node that is ancestral to the node and the nodes that are descendants of the node. The arguments to make_tree are a pointer at the first node and the sample size. The function ran1 is a subroutine that returns a random variable that is uniformly distributed on the open interval (0, 1).

```
struct node {
    double time;
    struct node *desc1;
    struct node *desc2;
    struct node *ancestor;
    } ;

make_tree(tree, sample_size)
    struct node *tree;
    int sample_size;
{
    int in, pick ;
    double t, ran1(), x ;
    struct node **list ;

/* Initialize things */
    list = (struct node **)malloc( sample_size*sizeof(struct node *) );
```

```
    for(in=0; in<sample_size; in++) {
         tree[in].time = 0. ;
         tree[in].desc1 = tree[in].desc2=0 ;
         list[in] = tree + in ;
    }

/* Generate the times of the nodes  */

    t = 0. ;
    for(in = sample_size ; in>1; in--){
        t += -2.0 * log(1.-ran1()) / ( ((double)in)*(in-1) ) ;
        tree[2*sample_size - in].time = t ;
     }

/* Generate the topology of the tree  */

    for( in=sample_size; in>1; in--){
         pick = in*ran1();
         list[pick]->ancestor = tree + 2 * sample_size - in ;
         tree[2 * sample_size - in].desc1 = list[pick] ;
         list[pick]=list[in-1];
         pick = (in-1)*ran1();
         list[pick]->ancestor = tree + 2 * sample_size - in;
         tree[2 * sample_size - in].desc2 = list[pick];
         list[pick] = tree + 2 * sample_size - in ;
    }

    free(list);
}
```

The following program, which illustrates the use of make_tree( ), can be used to study the frequency spectrum of variation under the neutral model at equilibrium. This program can also be used to study how the frequency spectrum will be affected by a single change in population size at some time in the past. The input to the program consists of θ, the sample size, the number of independent samples to generate, the time since the population changed, and the factor by which the population size before the change differs from the current population size.

For each sample generated, the following steps are performed. A genealogy is generated (make_tree). The tree is distorted by changing the times of nodes before the population size change by the 'factor' (bottleneck). (Because the only effect of population size changes is to change the time-scale, all that needs to be done is change the times of the nodes.) The lineage above each node is assigned a Poisson distributed number of mutations, with mean $\theta t/2$, where t is the duration of the branch above the node. The number of descendants of the node is counted (count_desc). The number of descendants of the node is the frequency in the sample of the variants produced by mutations that occur in the lineage above the node under consideration. The numbers are tabulated. After all the samples have been generated, the fraction of all mutations that produce variants of each possible frequency are output.

The reader must supply his or her own version of the subroutine Poisson, which returns a Poisson distributed random variable with mean equal to the argument (see, for example, Press *et al.* 1988).

This program can be used to see how low-frequency variants are more common after a severe bottleneck. When the size factor equals 1, there is no change in population size, and the program output results for the equilibrium neutral model.

In this case, the results can be checked against the Ewens (1972) sampling distribution.

```
#include <stdio.h>

struct node {
    double time;
    struct node *desc1;
    struct node *desc2;
    struct node *ancestor;
    } ;

main()
{
    struct node *tree;
    int sample_size, number_samples, in, *spectrum, nmuts, ndes, node;
    double theta, time, total_muts, time_of_size_change, factor ;

    scanf(" %lf",&theta);                /*  θ               */
    scanf(" %d",&sample_size);         /* sample size      */
    scanf(" %d",&number_samples);      /* number of samples to generate */
    scanf(" %lf", &time_of_size_change); /* in units of 2N generations */
    scanf(" %lf", &factor);              /* the factor by which the
            population size differed before population size changed.  */

    tree = (struct node *) malloc( 2*sample_size*sizeof(struct node ));
    spectrum = (int *)malloc( sample_size*sizeof(int) );  /* for storing
                                                              the results */
    for(in=0;in<sample_size;in++) spectrum[in] = 0 ;
    total_muts = 0.0 ;


    for(in=0;in<number_samples; in++) {
        make_tree(tree, sample_size);
        bottleneck(tree,sample_size,time_of_size_change,factor);
        for( node = 0; node< sample_size*2 -2 ; node++) {
            time = (tree[node].ancestor->time) - tree[node].time ;
                /*  time is the length of the branch above the node  */
            nmuts = poisso(time*theta/2.);  /* returns a poisson deviate
                                                  with  mean  θt/2  */
            if( nmuts > 0 ) {
                ndes = count_desc(tree+node);
                spectrum[ndes] += nmuts ;
                total_muts += nmuts ;
            }
        }
    }

    printf(" Average number of mut's per sample: %lf\n\n",total_muts/number_samples);
    printf("freq of mut     fraction of mutations\n\n");
    for( in=1; in<sample_size; in++)
        printf("%d                      %lf\n",in,spectrum[in]/total_muts);
}

 /* a recursive method for counting the number of descendents of a node */
    int
 count_desc(node)
    struct node *node;
 {
    int sum=0 ;

    if( node->desc1 == NULL) return( 1 );
    sum += count_desc( node->desc1 ) ;
    sum += count_desc( node->desc2 );
    return(sum);
 }
```

```
/*  When the population size prior to a certain "time" differs from the current
population size by a certain "factor", pass the tree through this routine.   */
    int
bottleneck(tree, sample_size,time,factor)
    struct node *tree;
    int sample_size;
    double time, factor;
{
    int in;

    for(in=sample_size; in<2*sample_size-1 ; in++)
         if( tree[in].time > time ) tree[in].time = factor*(tree[in].time - time ) +
time ;
}
```

Principles of genealogical concordance in species concepts and biological taxonomy

JOHN C. AVISE and
R. MARTIN BALL, JR.

1. INTRODUCTION

For more than 50 years, the 'biological species concept' (BSC) has been a major theoretical framework orienting research on the origins of evolutionary diversity (Dobzhansky 1937). Under the BSC, species are 'groups of interbreeding natural populations that are reproductively isolated from other such groups' (Mayr 1970, p. 12). Numerous authors have expressed sentiments on the BSC similar to those of Ayala (1976, p. 18):

> among cladogenetic processes, the most decisive one is speciation – the process by which one species splits into two or more. . . . Species are, therefore, independent evolutionary units. Adaptive changes occurring in an individual or population may be extended to all members of the species by natural selection; they cannot, however, be passed on to different species.

Thus under the BSC, species are perceived as biological and evolutionary entities that are more meaningful and less arbitrary than other taxonomic categories such as subspecies or genera (Dobzhansky 1970). The BSC has served to focus attention on questions concerning the evolution of intrinsic reproductive barriers (RBs), including: What genetic changes produced RBs, and hence new species? What morphological, developmental or behavioral traits are involved? What ecological, demographic or evolutionary conditions favor RB evolution?

As judged by its continued widespread employment in textbooks and as a guide to research strategy (eg. Coyne and Orr 1989), the BSC appears to have survived a variety of criticisms (both philosophical and operational) leveled against it over the last 30 years (Ehrlich 1961; Ehrlich and Raven 1969; Levin 1979; Raven 1976; Sokal and Crovello 1970; Sokal 1973; Wiley 1978). Recently, another serious challenge has come from some systematically-oriented evolutionists who argue that the BSC lacks a sufficient phylogenetic perspective, and hence provides an inappropriate guide to the origins and products of evolutionary diversification (Cracraft 1983; de Queiroz and Donoghue 1988; Donoghue 1985; Eldredge and

Cracraft 1980; McKitrick and Zink 1988; Mishler and Donoghue 1982; Nelson and Platnick 1981; Rosen 1979). Many of these critics of the BSC argue that 'reproductive isolation should not be a part of species concepts' (McKitrick and Zinc 1988, p. 3). This has led to another call for abandonment of the BSC, and its replacement by the 'phylogenetic species concept' (PSC).

We believe there is much of value in the PSC, but that some of its proponents go too far in suggesting a total abandonment of the BSC. The purpose of this chapter is to introduce another conceptual approach for taxonomic and species recognition based on principles of genealogical concordance defined below. These principles derive most easily from theories and observations in molecular evolution, but can also be applied to hereditary morphological, behavioral and other phenotypic attributes traditionally studied by systematists. Concepts of genealogical concordance in taxonomic recognition combine what we perceive to be the better elements of the PSC and BSC.

2. SYNOPSIS AND CRITIQUE OF THE PHYLOGENETIC SPECIES CONCEPT

Under the PSC, a species has been defined as a monophyletic group composed of 'the smallest diagnosable cluster of individual organisms within which there is a parental pattern of ancestry and descent' (Cracraft 1983, p. 170). Cracraft intentionally avoids explicit reference to reproductive disjunctions, and instead focuses directly on the distributions of diagnostic, heritable trait(s). As emphasized by Cracraft (1983, p. 170), assemblages constituting a phylogenetic species 'simply must be diagnosable from all other species'. Each successful diagnosis is taken to indicate a phylogenetic separation – the key concept underlying the phylogenetic species. The PSC has several suggested advantages over the BSC, including the following (Cracraft 1983): (a) assemblages diagnosed under the PSC are phylogenetic units, and as such provide more informative subjects for evolutionary and ecological study; (b) the PSC eliminates the need for direct concern with reproductive compatibilities (that in nature are often difficult to observe directly, and normally cannot be assessed among allopatric forms); and (c) attention is focused on the geographic and genealogical histories of populations.

We are sympathetic to the general goal of the PSC of calling for greater attention to historical relationships of populations (Avise 1989*a*; Avise *et al*. 1987*a*). However, current formulations of the PSC (e.g. Cracraft 1983; McKitrick and Zink 1988) have limitations as a replacement for the BSC:

1. *The number of species recognized under the PSC depends on the*

resolving power of the analytical tools available. The first clause of the PSC definition, 'the smallest diagnosable cluster of individual organisms', indicates that any diagnostic trait uniting an array of individuals to the exclusion of others is sufficient to define a phylogenetic species (provided also that the trait is inherited and monophyletic). Some authors (e.g. Cracraft 1983) suggest that the diagnostic trait may be either primitive or derived, whereas others (McKitrick and Zink 1988) restrict consideration to derived traits only. However, there is considerable sentiment that a single diagnostic trait, no matter how 'trivial', is sufficient for clade definition (Wiley 1981), and hence for species recognition under the PSC. For example, McKitrick and Zink (1988, pp. 9–10) contemplate the prospect of finding a group of birds with one extra hooklet on the barb of the seventh primary feather, and conclude 'There is no theory to suggest that a trait must be of a certain quality or magnitude to provide historical information or to delimit species. . . . Thus, no character is potentially more or less useful as a tool to reconstruct patterns of speciation (sensu the PSC).'

What would be the consequences of recognizing a distinct species for each diagnostic trait? Cracraft (1983, p. 173) suspects that within ornithology 'the phylogenetic species concept should not increase the number of taxa already recognized' (although many subspecies would be elevated to species status). McKitrick and Zink (1988, p. 9) suspect that closer scrutiny of biochemical, morphological and other characters on a microgeographic scale will likely reveal the existence of many more phylogenetic avian taxa, but that 'The notion that there should be an upper limit to the number of species described does not appear to have any value, heuristic or otherwise.'

We believe that such statements concerning the PSC overlook the huge size and extreme variability of eukaryotic genomes. Evidence from molecular biology demonstrates enormous genetic polymorphism within most taxa. For example, nucleotide diversity (a measure of heterozygosity at the nucleotide level) ranges from about 0.002 to 0.019 for various loci (Nei 1987, p. 267). Because a typical gene consists of several thousand nucleotide pairs, randomly drawn haplotypes from conspecific individuals will normally differ in nucleotide sequence. Even cursory assays of restriction-fragment-length-polymorphisms (RFLPs) of particular genes such as mitochondrial DNA (mtDNA) or minisatellite nuclear DNA have revealed extensive intraspecific genetic heterogeneity. Most recognized biological species are already divisible into large numbers of diagnosable subunits (often geographically subdivided: Avise *et al.* 1987*a*), and in some taxa nearly every organism can be distinguished with the limited genetic information already at hand (Avise *et al.* 1989; Burke 1989). The data from direct nucleotide sequencing, and from multiple loci, will make such levels of genetic distinction commonplace (e.g. Jeffreys *et al.* 1985;

Kocher *et al.* 1989; Lander 1989). If each individual organism is genetically unique at a high level of resolution, then the grouping of individuals requires that we ignore distinctions that occur below some arbitrary threshold. The evolutionary significance of any such threshold must surely be questionable.

2. *Unless persistent extrinsic (geographic) or intrinsic RBs are present, different gene genealogies will usually disagree in the boundaries for 'species' under the PSC.* A strict application of the PSC definition is also difficult to apply given the vast numbers of gene genealogies and their expected idiosyncratic distributions within and among population pedigrees. The nuclear genome of most birds and mammals, for example, consists of about 2–3 billion nucleotide pairs. Under a conventional mutation rate of $\mu = 10^{-9}$ per nucleotide site per generation (Nei 1987), a typical individual is likely to carry 2–3 newly arisen mutations, and a species composed of even a few million animals would be expected to carry several million new mutations every generation. Under a reasonable population demography (Poisson distribution of surviving progeny with mean 2 per family), nearly two-thirds of new neutral mutants are expected to survive for at least one generation, and nearly 2 per cent will likely survive beyond 100 generations (Spiess 1977, pp. 376–7). Each new mutation that survives (a derived trait) will have its own particular geographic and population distribution, depending on such factors as its place of origin and age, the fitness conferred on its bearers, and the historical gene flow regime of the species (Fig. 1). Except in the special evolutionary circumstances discussed beyond (which involve intrinsic or extrinsic RBs), little or no concordance should exist among the assemblages of individuals diagnosable with independently-derived mutations or their derivatives: non-overlaps, partial overlaps, or nested arrangements in group membership should typically characterize various organismal assemblages diagnosed by independent genetic traits (Fig. 1). Due to inherently stochastic aspects of the hereditary process – mutational origins, allelic segregation during meiosis and the vagaries of population demography – alleles at each genetic locus that have 'trickled' through an organismal pedigree will exhibit a unique phylogenetic tracing (Ball *et al.* 1990).

3. *Shared ancestry in sexually reproducing organisms implies historical membership in a reproductive community.* The PSC could clearly be applied to the identification of 'species' in asexually reproducing organisms. However, the second clause of Cracraft's (1983) PSC definition – parental pattern of ancestry and descent – was intended to extend the PSC to sexually reproducing organisms. Species then constitute a phylogenetic 'lineage' within which matings and successful reproduction have taken place (McKitrick and Zink 1988). Thus when the PSC is applied to

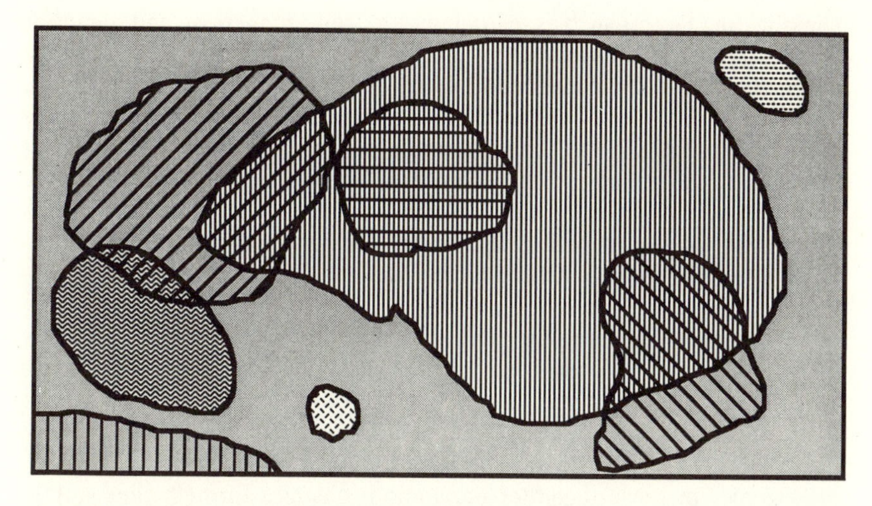

Fig. 1. Examples of possible geographic distributions of various derived mutations in a continuously distributed species with limited dispersal capability (compared to the geographic range occupied).

sexually reproducing forms, historical reproductive communication and continuity implicitly accounts for shared possessions of heritable diagnostic traits. In other words, 'species' under the PSC are recognizable precisely because their members derive from reproductive communities of individuals (a general philosophy that the BSC has always emphasized).

Thus overall, if a broader framework of the PSC is to contribute to a significant advance in systematic practice (and we believe it can), a shift from issues of diagnosability to those of magnitudes and patterns of phylogenetic differentiation (and of the historical and reproductive reasons for such patterns) will be required. A powerful approach to these issues should involve reference to the following principles of genealogical concordance.

3. GENEALOGICAL CONCORDANCE

The extant haplotypes (DNA sequences) present in any 'species' represent the gene lineages that have survived through an organismal pedigree. Within any pedigree, such lineage tracings (gene phylogenies) can differ greatly from locus to locus (Ball *et al.* 1990), due to the vagaries of meiotic segregation, mating pattern and the reproductive success of individuals through which the alleles were transmitted. Such differences among gene phylogenies within an organismal pedigree arise inevitably, even when all loci experience nucleotide substitutions at the same rate, and when complicating factors such as recombination, gene conversion and 'sampling error' due to the idiosyncratic origins of the particular mutations assayed are neglected. A clear distinction must therefore be drawn between gene phylogenies and organism phylogenies (Avise 1989*a*; Nei 1987; Takahata 1989; Wilson *et al.* 1985), and hence between phylogenetic diagnoses based on single genetic traits versus those based on broader trends in the information content from multiple loci.

Suppose, as shown in Fig. 2, that an ancestral, random mating population is sundered into two daughter populations, through either a geographic or genetic barrier to gene flow. Immediately following this separation, at any locus some haplotypes in daughter population 1 will likely be genealogically closer to some haplotypes in daughter population 2 than they are to other haplotypes in population 1 (and vice versa). Phylogenetic partitions based on traits encoded by a single gene would therefore be discordant with the population subdivision. The status of particular gene phylogenies in these daughter populations changes through time due to demographically based processes of lineage sorting, until eventually all remaining haplotyes in population 1 are genealogically closer to others in population 1 than to any haplotypes in population 2, and vice versa (Neigel and Avise 1986; Pamilo and Nei 1988; Tajima 1983). The rate of the process is demography-dependent, but commonly takes about $2N_e$–$4N_e$ generations, where N_e is the effective size of the daughter populations. In other words, in terms of any gene phylogeny, two isolated daughter populations are expected to evolve from conditions of initial polyphyly, through paraphyly, and eventually to a state of reciprocal monophyly (Neigel and Avise 1986), at which point the major phylogenetic subdivisions in the gene genealogy become coincident with the major population-level subdivisions (as defined by the intrinsic or extrinsic barriers to reproduction). Secondary admixture and introgressive hybridization between the two populations could, of course, blur the evidence for this historical separation.

Thus in taxonomic recognition, the operational challenge involves an assessment of when the arrays of individuals grouped by particular genetic trait(s) coincide with the historical, organismal-level partitions that are seldom observable directly. Under what biological or evolutionary conditions should the biotic partitions registered by various genetic traits faithfully mirror the phylogenetic separations of the taxa that we might wish to recognize formally?

3.1 Gene–gene phylogenetic concordances

One important consideration must be whether many independent gene phylogenies (those from unlinked and non-epistatic loci) provide *concordant* support for the organismal assemblages identified. As shown below,

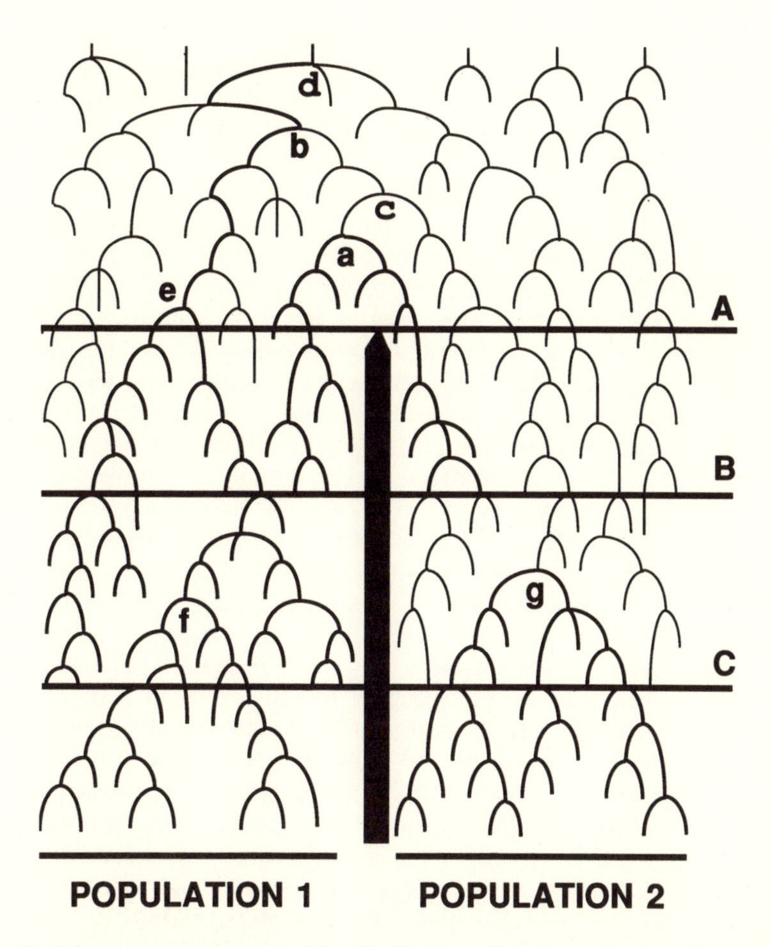

Fig. 2. Schematic presentation of the distribution of haplotype lineages (at a single gene) through an ancestral population subdivided (at time A) by a geographic or other barrier to reproduction. With respect to this particular gene genealogy, between levels A and B the daughter populations 1 and 2 are polyphyletic, i.e. at any point in that time interval, some individuals in population 1 are genealogically closer to some individuals in population 2 than to other individuals in population 1, and vice versa (see nodes a, b, c and d). Between levels B and C, the populations exhibit a paraphyletic relationship in the gene tree, i.e. at any point in that time interval, all individuals in population 1 form a monophyletic subset (tracing to node e) within the more ancient gene tree of population 2, some of whose extant members diverged at nodes c and d. Below time level C, populations 1 and 2 are reciprocally monophyletic in the gene tree, tracing to nodes at f and g, respectively.

such concordances are likely to arise only when populations have been reproductively separated from one another (either by intrinsic or extrinsic RBs) for reasonably long periods of time. The following examples from computer simulations are intended to provide a heuristic introduction to concepts of genealogical concordance. They represent a preliminary extension of prior single-gene treatments (Neigel and Avise 1986; Ball *et al.* (1990) to multiple loci considered together.

The current programs are slight modifications of those described by Ball *et al.* (1990), which should be consulted for assumptions, demographic conditions and other details. The original programs were designed to: (a) produce a random-mating population pedigree; (b) choose for analysis haplotypes at particular loci that in effect have 'trickled' through the organismal pedigree under Mendelian rules of segregation; and (c) find the times to the most recent common ancestor for randomly drawn pairs of haplotypes from those loci. In the present application and extension of these simulations, we considered again random mating ancestral populations (in this case, each composed of $N \cong 100$ individuals) that become split into two daughter populations (each $N \cong 50$) by a geographic or genetic barrier to successful reproduction.

Ten simulated populations were created and allowed to run for 1000 generations with no barriers to gene flow. At the end of the 1000th generation, 100 gene genealogies were chosen from each population and a barrier to gene flow was erected. The 10 pairs of daughter populations were allowed to continue for another 500 generations with 100 additional gene genealogies being chosen for analysis for each pair of populations every 10 generations. The 51 000 gene genealogies so created (100 genes $\times$ 10 simulated populations $\times$ 51 time points) were then analyzed by picking one of the daughter populations at random and determining the size of the largest monophyletic group containing only individuals from that daughter population. By this process, we hoped to show the development of concordant gene patterns through time. The results are shown in Fig. 3.

The results from these preliminary simulations support our intuitive expectations that populations isolated from one another for increasing lengths of time should be genealogically differentiable by increasing numbers of loci. Thus, if the phylogenetic histories of many independent genes in an empirical survey were to separate individuals into concordant arrays, such a finding would be consistent with long-term reproductive separation of those arrays. Furthermore, concordance among the gene arrays is highly unlikely if there were no reproductive barrier. Therefore, we suggest that such population subdivisions concordantly identified by multiple independent genetic traits should constitute the population units worthy of recognition as phylogenetic taxa (see below).

Our simulations involve monitoring actual times to common ancestry

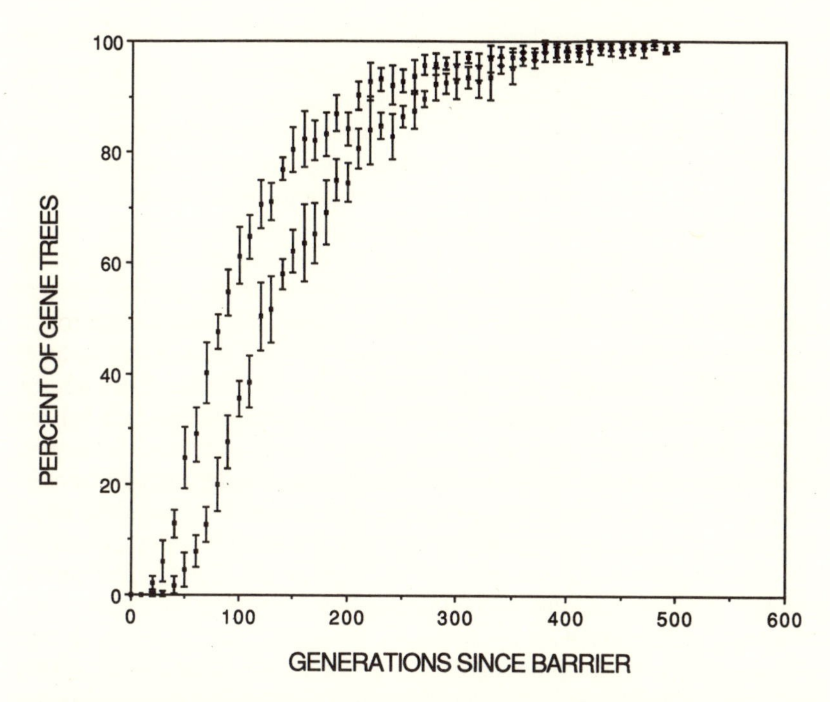

Fig. 3. The time-course of the approach to monophyly in the gene genealogies of daughter populations separated by an absolute barrier to reproduction at time zero. In computer simulations, for each of 10 independent pedigrees, genealogies of 100 genes were monitored per pedigree every 10th generation. The vertical axis is the proportion of outcomes (± 0.5 S.D.) in which at least 75 per cent (upper curve) (or 100 per cent, lower curve) of the individuals were part of the largest monophyletic group (in a gene tree) in the daughter population (see the text for additional details).

in the pedigree (and thus ignore stochastic errors in the gene tree arising from the mutational process). In reality, many loci will evolve too slowly to provide markers for identifying recently separated populations and, in addition, the sensitivities of assay methods employed may be inadequate to detect all genetic differences that exist. Thus in practice, it is unrealistic to suppose that most or all loci should contribute to the concordance distinguishing historically separated population units. None the less, unless multiple genetic traits distinguish arrays of organisms, those arrays cannot necessarily be assumed to reflect significant phylogenetic population subdivisions.

It might be argued that the requirement of multiple gene concordance for taxonomic recognition is overly restrictive. After all, genealogical concordance will not develop until some time after reproductive isolation has occurred, and in the interim, the isolated groups are not acknowledged. The problem is based on more than the simple failure to diagnose reproductively isolated groups. It revolves rather on the problem of determining what a significant period of reproductive isolation is. There must be innumerable instances of temporary population isolation within any 'species' exhibiting limited dispersal capabilities. Most of these instances will be of such short duration that the populations cannot develop independent evolutionary histories. It is only when reproductive separations among such populations extend to longer periods of time that genetic differences accumulate, and genealogical concordances appear.

In theory, the times to reciprocal monophyly in the gene genealogies distinguishing two isolated populations are also directly related to effective population size (Neigel and Avise 1986): small populations reach a state of reciprocal monophyly more rapidly than large populations. However, the probability of detecting distinguishing characters over short time-scales is also lower. Thus local populations will seldom exhibit high degrees of concordance among gene genealogies (or prove to be distinguished by large genetic distances), particularly if they are periodically connected to other such populations through gene flow and/or are evolutionarily ephemeral (Avise 1989*a*).

We suspect that the number of phylogenetic population units within most currently recognized 'biological species', as identified by genealogical concordance, will be relatively low (certainly far less than the number of 'local populations' or family units, though often greater than 1 – see below). Such phylogenetic units, supported by concordant distributions of multiple, independent traits (which can be biochemical, morphological, behavioral, etc., provided they have independent genetic bases), should represent the population assemblages that have been isolated from one another by long-term impediments to interbreeding.

A distinction should also be made between the use of multiple genes (or characters) in a discriminatory versus a concordance sense. For example, discriminant function analysis (Sneath and Sokal 1973) is a multivariate statistical approach designed to maximally separate populations based on the accumulated information from many characters, each of which may overlap in distribution between the populations. Although concordance principles are similar in that they also apply to multiple characters, the concern is not whether the populations can be distinguished, but rather with the level of concordant support for such distinctions.

3.2 Gene–geography concordances among taxa

The branches in the phylogenetic trees for particular loci often show strong geographic clustering (Fig. 1; Avise *et al.* 1987*a*). While we have argued that justification for the recognition of the major subdivisions in an organismal phylogeny may ultimately necessitate concordant support from the phylogenetic partitions of multiple genes, such information may seldom be available. Are there any conditions under which concordances between geography and the subdivisions in single-gene genealogies (such as those provided by mtDNA) yield compelling evidence for longstanding, phylogenetic population subdivisions?

We propose that if major phylogenetic distinctions in particular gene trees were geographically concordant across populations of a number of independent taxa, separations in the organismal pedigrees due to historical isolation would be strongly implicated. In a similar fashion, vicariance biogeographers have long emphasized that patterns of geographic congruence in the phylogenies of multiple, unrelated taxa should reflect the historical patterns of disjunctions in environments occupied by those organisms (e.g. Platnick and Nelson 1978; Rosen 1978).

Work in our laboratory over the past decade provides two applications of such comparative methods as applied to the 'intraspecific' gene trees registered in mtDNA. First, within each of five recognized species of freshwater fishes, the earliest branching point observed in the mtDNA phylogenies readily distinguished populations in the eastern versus western portions of the species' ranges in the southeastern USA (Fig. 4; Bermingham and Avise 1986). These concordant patterns in the mtDNA gene trees therefore suggest two major areas of endemism for southeastern fish populations, a result further supported by a conventional biogeographic reconstruction involving concentrations of species' distributional limits in the region (Swift *et al.* 1985). For one of the assayed species (*Lepomis macrochirus*), allozyme data were also available, and they provided strong, independent genetic support (gene–gene concordance) for the phylogenetic units identified by mtDNA (Fig. 4; Avise *et al.* 1984).

A second example of geographic concordance in mtDNA phylogenies across a number of taxa involves coastal marine species in the southeastern USA (Avise *et al.* 1987*b*; Bowen and Avise, unpublished; Lamb and Avise, unpublished; Reeb and Avise, 1990; Saunders *et al.* 1986). Within each of six taxonomically recognized species or species groups, ranging from oysters (*Crassostrea virginica*) and horseshoe crabs (*Limulus polyphemus*) to toadfishes (*Opsanus beta* and *O. tau*), diamondback terrapins (*Malaclemys terrapin*) and seaside sparrows (*Ammodramus maritumus*), the earliest (and most strongly supported) separations in the reconstructed mtDNA phylogenies involved distinction of most Atlantic coast populations from those in the Gulf of Mexico and southeastern Florida (Fig. 5). These assayed species are confined to coastal margins, and are typically associated with saltmarsh and estuarine conditions. Reeb and Avise (1990) discuss the paleoclimatic and geologic evidence for Pliocene/Pleistocene disjunctions in suitable habitats that likely initiated the phylogenetic population separations, as well as the ecologic and

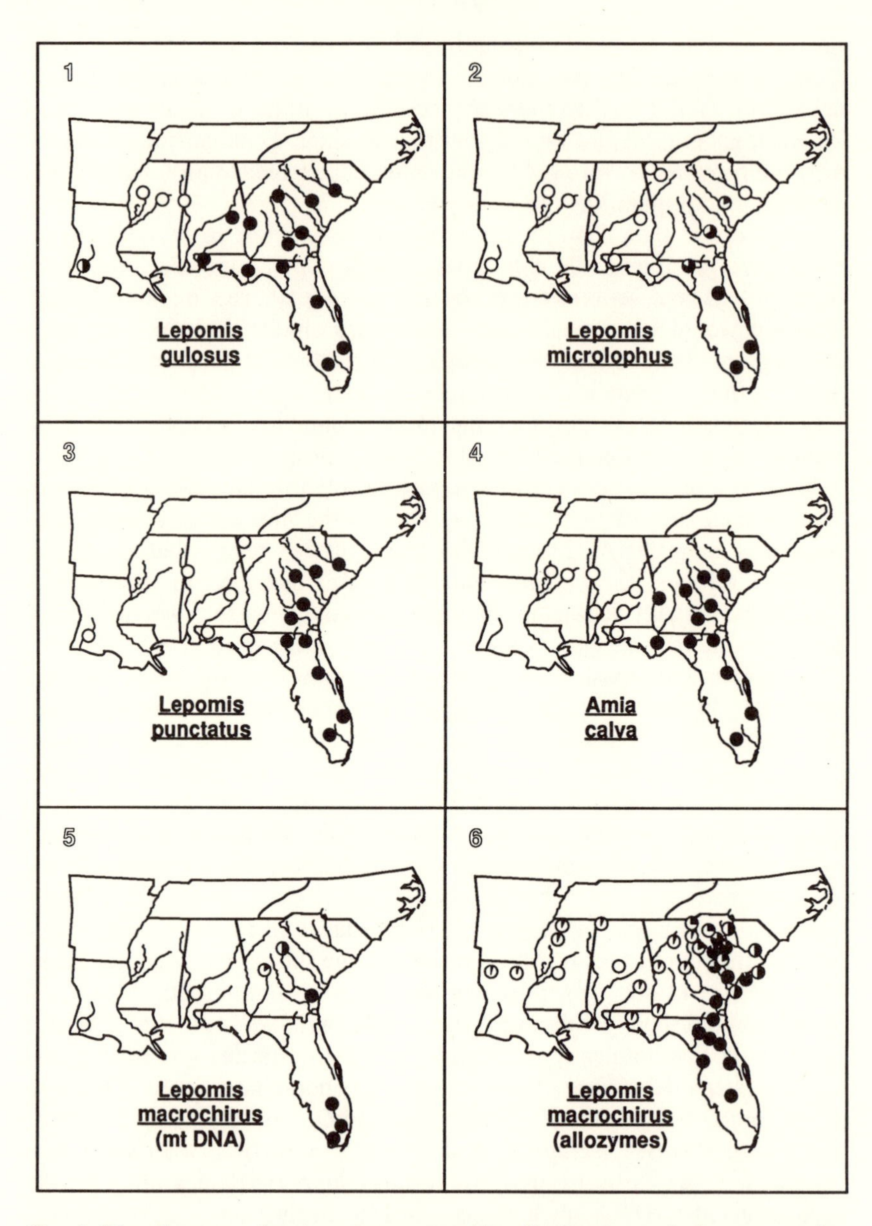

Fig. 4. Pie diagrams showing the geographic distributions of the two major mtDNA phylogenetic branches observed within each of five species of freshwater fishes: 1, warmouth sunfish; 2, redear sunfish; 3, spotted sunfish; 4, bowfin (Bermingham and Avise 1986); 5, bluegill sunfish (Avise *et al.* 1984). Also shown (panel 6) are frequencies in the bluegill sunfish of the two electromorphs at the allozyme locus *Got-2* (which are also very similar to observed geographic distributions at alleles at another nuclear gene, *Es-3*: (Avise and Smith 1974).

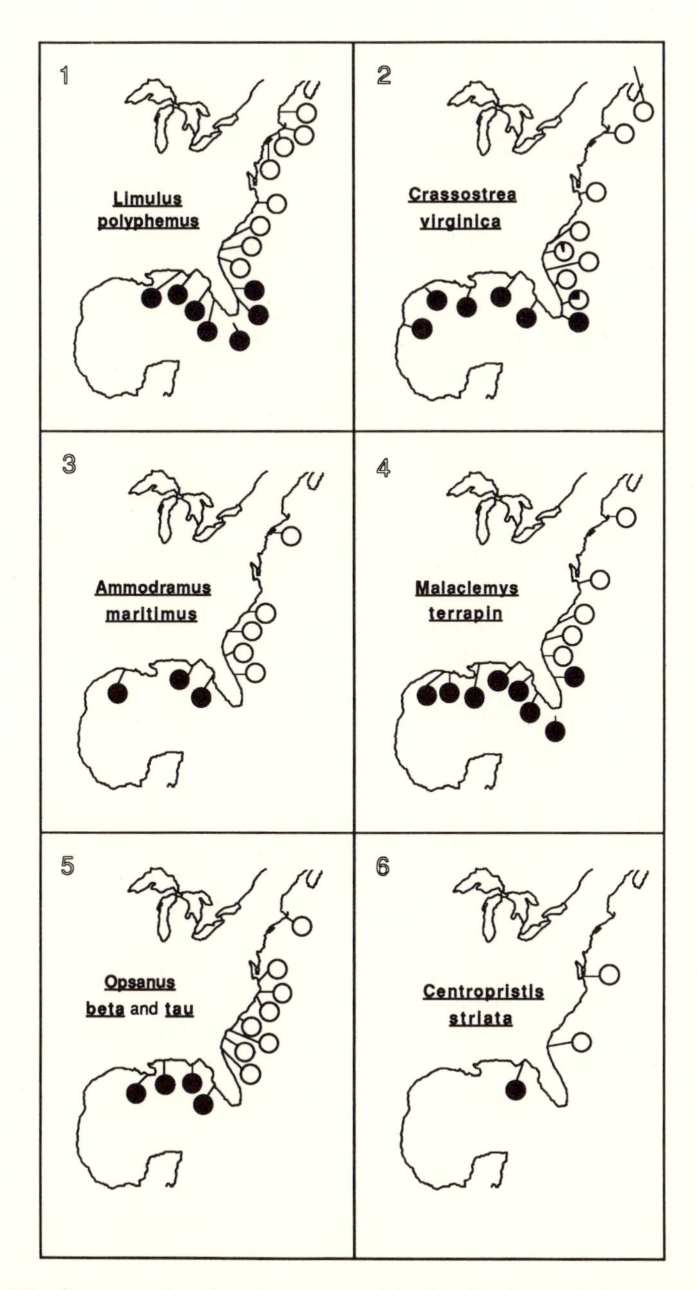

Fig. 5. Pie diagrams showing the geographic distributions of the two major mtDNA phylogenetic branches observed within each of six coastal-restricted marine species: 1, horseshoe crab (Saunders *et al.* 1986); 2, American oyster (Reeb and Avise 1990); 3, seaside sparrow (Avise and Nelson 1989); 4, diamondback terrapin (Lamb and Avise, in prep.); 5, toadfish (Avise *et al.* 1987*b*); 6, black sea bass (Bowen and Avise, in prep.).

hydrologic limits to gene flow in southern Florida that today may help to maintain these historical patterns of population subdivision.

4. SUGGESTIONS FOR TAXONOMIC PRACTICE

Concordances among the genealogical histories of independent loci are likely to arise only under conditions where intrinsic or extrinsic RBs have severed genetic exchange between populations for periods of time considerably longer than the evolutionary effective sizes of those populations. If the RBs are genetically based (intrinsic), the populations exhibiting gene–gene phylogenetic concordances may be either sympatric or allopatric, but in either event would constitute valid species under the conventional BSC. If the RBs are exclusively extrinsic (based solely on longstanding geographic barriers to gene flow), phylogenetic concordances may also be exhibited, either among independent loci within a taxon, or at a given locus among several unrelated taxa that had been concomitantly subdivided by the geographic barrier.

However, there remains a fundamental distinction between concordances in gene genealogies that have arisen from intrinsic versus purely extrinsic isolating barriers. The former are likely to be irreversible evolutionarily, whereas the latter can be ephemeral due to breakdowns of geographic barriers. If intrinsic RBs had not yet evolved in allopatry, secondary population contacts could lead to introgressive hybridization and disintegrations of genealogical concordance (decay of gametic-phase disequilibrium). For this reason, species deemed distinct under the BSC (isolated by intrinsic RBs) will tend to evolve independently and will accumulate genetic differences through time, while the genetic differences between allopatric populations separated solely by extrinsic gene-flow barriers would remain in jeopardy – unless the barriers remained intact for very long periods, in which case intrinsic RBs (and hence biological species) would likely arise as a byproduct of overall genetic divergence.

4.1 Species descriptions: Retention of the BSC

Therefore, we suggest that the biological and taxonomic category 'species' continue to refer to groups of actually or potentially interbreeding populations isolated by *intrinsic* RBs from other such groups. In other words, we favor a retention of the philosophical framework of the BSC. However, the *evidence* for evolutionarily significant RBs (whether intrinsic or extrinsic) will normally be indirect, and will involve concordant genetic differences among the populations involved. We recognize (as have many others) that there are operational difficulties with a strict application of this BSC, particularly as applied to allopatric forms, due to: (1) elements of uncertainty and subjectivity in applying the definition to certain disjunct populations, where the issue of reproductive compatibility in nature is difficult to assess; and (2) the occurrence of intermediate situations where limited hybridization and introgression occur in localized areas. Despite these problems, by objectively examining organisms from the perspective of the gene–gene and gene–geography concordances described in this chapter, assessments of the degree and pattern of phylogenetic separation will be facilitated.

4.2 Subspecies descriptions: Application of a modified PSC

The growing literature of molecular biology has demonstrated that many species (as defined by the BSC) are further subdivided into genetically distinct sets of geographic populations exhibiting considerable historical, phylogenetic separation from one another (e.g. Avise *et al.* 1987; Wilson *et al.* 1985). While such populations may not qualify as distinct species under the BSC (i.e. they are not isolated by intrinsic RBs), their longstanding separations have resulted in the accumulation of considerable genetic differences that in principle and practice are reflected in geographically congruent patterns of divergence at multiple loci.

To illustrate, the bluegill sunfish (*Lepomis macrochirus*) in the southeastern USA consists of two genetically distinct forms distributed east versus west of the Apalachicola River drainage separating Alabama from Georgia (Fig. 4). The two forms differ in nucleotide sequence divergence by $\cong$ 8.5% in mtDNA (Avise *et al.* 1984), exhibit two nearly fixed allozyme differences at assayed nuclear genes (and an overall genetic distance of $D \cong 0.15$: Avise and Smith 1974) and evidence additional differences in morphological and physiological traits (Hubbs and Allen 1944; Hubbs and Lagler 1958). Yet the two forms hybridize extensively in a secondary contact zone in parts of Georgia and the Carolinas. Within particular hybrid populations, alleles at nuclear loci are in gametic phase equilibrium with one another, as well as with the distinct mtDNA genotypes (Avise and Smith 1974; Asmussen *et al.* 1987), thus providing strong evidence for near random-mating. In these freshwater fish, barriers to gene flow between populations in eastern and western river drainages are purely extrinsic. But the barriers have been of sufficiently longstanding duration that these populations are now distinguished by many genetic characteristics. We propose that such sets of populations, concordantly recognizable by many independent genetic differences, reflect the major phylogenetic partitions within a biological species that are worthy of formal subspecies recognition. (The two bluegill forms has indeed been assigned the Latin trinomials *L.m. macrochirus* and *L.m. purpurescens*).

We therefore suggest the following definitional guidelines for the taxonomic category 'subspecies' in sexually reproducing organisms: *subspecies are groups of actually or potentially interbreeding populations phylogenetically distinguishable from, but reproductively compatible with, other such groups. Importantly, the evidence for phylogenetic distinction must normally come from the concordant distributions of multiple, independent, genetically based traits.*

Because the populations constituting distinct subspecies are reproductively compatible, subspecies will normally be allopatric (though some of their populations may meet in secondary hybrid zones), and any significant phylogenetic partitions will be registered by multiple loci exhibiting congruent geographic distributions. The longer the geographic isolation, the greater the opportunity for genetic divergence and also for the accumulation of concordant phylogenetic distinctions at multiple loci. It remains to be seen how often genealogical concordance will be observed among geographic populations within biological species, such that subspecies designations are warranted, but we suspect that many biological species (such as the bluegill sunfish) will prove to have rather fundamental phylogenetic subdivisions resulting from historical population separations.

5. ADVANTAGES AND DIFFICULTIES OF CONCORDANCE PRINCIPLES IN TAXONOMY

5.1 Advantages of concordance principles

1. *The category 'subspecies' will rest on a firmer empirical foundation.* Conventionally, subspecies descriptions have been based on one or a few traits (such as pelage color or size) that allowed distinction of a high percentage of individuals in a geographic region from those in other areas (e.g. Mayr 1969). As noted and criticized by Wilson and Brown (1953, p. 104), 'The tendency in this method has been to delimit races on the basis of one or several of the most obvious characters. . .; the remainder of the geographically variable characters are then ignored, or if they are considered at all, they are analyzed only in terms of the subspecific units previously defined.' We agree with Wilson and Brown's (1953, p. 110) contention that 'geographical variation should be analyzed first in terms of genetically independent characters, which would then be employed synthetically to search for possible racial groupings'.

2. *Major phylogenetic units within biological species will be afforded formal taxonomic recognition.* Sets of long-isolated populations, reflecting historical disjunctions in suitable habitat, probably occur within the boundaries of many biological species (as delimited by intrinsic RBs). Such populations would be afforded subspecies status under concordance principles. The resulting taxonomies can be of great use in reconstructing

historical biogeography (Avise *et al.* 1987), as well as in summarizing the apportionment of intraspecific genetic diversity that should aid the management and preservation efforts of conservation biology (Avise 1989*b*). With the partitions in intraspecific phylogeny properly recognized (by concordance principles), the interpretation of characters at variance with the primary pattern should also be facilitated. For example, characters likely to be under intense selection pressures related to ecological circumstances (such as pelage color or body size) may often be geographically discordant with the phylogenetic subdivisions. Thus under concordance principles (as under the PSC, but not the BSC), explicit attention is focused on phylogenetic histories of populations.

3. *The category 'species' will remain similar to that currently employed, and intrinsic RBs retain primacy as a conceptual guide to species distinctions.* Unlike the PSC, which would require a drastic revision of taxonomic designations at the species level, application of concordance principles should have little effect on current species-level taxonomies. The conventional procedure for distinguishing biological species already rests implicitly on concordance principles – intrinsic reproductive barriers are deduced from the indirect evidence of differences (sometimes requiring close scrutiny) in numerous morphological, behavioral and other assayable traits. Because reproductive assessments seldom can be made directly (see Sokal and Crovello 1970), such character-state surrogates of reproductive unions and disjunctions are likely to remain of prime utility in species descriptions.

To the extent that genetically-based RBs are irreversible evolutionarily, they constitute irrevocable partitions of biotic diversity. The genes directly responsible for RBs, by cementing biotic subdivisions, will have a correlated effect on the phylogenetic partitions of many other loci in the genome (in the presence of RBs, all neutral loci will eventually evolve to a status of concordant reciprocal monophyly). Thus under concordance principles (as under BSC, but not the PSC), intrinsic RBs properly occupy a position of fundamental evolutionary significance.

5.2 Potential difficulties of concordance principles

1. *Some species may be overlooked.* Occasionally, populations may have evolved intrinsic barriers to reproduction so recently that phylogenetic separation and concordance are not yet evident in genes other than those directly responsible for the RBs themselves. While it seems unlikely that RB-producing genetic traits alone would normally distinguish species (except perhaps in very young polyploid assemblages, or in other situations where chromosomal or other genetic changes rapidly and recently precipitated intrinsic reproductive isolation), some 'good' biological species could

be missed under concordance principles (they would likely be overlooked under the PSC and BSC also, unless the critical genetic trait conferring reproductive isolation were examined).

Some events that give species under the BSC may not be immediately recognizable under an implementation of concordance principles. In the special case where a single character change results in the intrinsic reproductive isolation of two groups, there will be no immediate concordance between that character and others, although such concordance will inevitably develop through time. Overlooking valid biological species that were recently isolated is of special concern because such taxa should afford the chance to study the initial stages of speciation. Thus intrinsic RBs should retain primacy as a conceptual basis for species recognition (as under the BSC).

2. *Subjective taxonomic judgments may be required for intermediate levels of genetic congruence and divergence.* Admittedly, there are several gray areas in this heuristic construct for taxonomic recognition under concordance principles. For example, how many gene phylogenies must support the subspecies distinctions? And, how much concordance must characterize the geographic distributions of the gene phylogenies? In principle, levels of genealogical concordance can vary along a continuum (see Fig. 3), so any specific suggestions for implementing concordance methods will necessarily involve an element of arbitrariness. However, all taxonomic schemes that divide the elements of the continuous pattern of evolutionary differentiation into discrete categories (such as subspecies and species) face similar difficulties with intermediate situations.

3. *Obtaining phylogenies for independent genes may be difficult.* Ideally, complete information on the histories of allelic relationships at each of many loci would be most desirable in the search for genealogical concordance. However, apart from the large number of mtDNA phylogenies presented in recent years (Avise *et al.* 1987*a*; Wilson *et al.* 1985), very few gene trees have as yet been generated for any taxa at the microevolutionary scale (Aquadro *et al.* 1986; Avise 1989*a*; Langley *et al.* 1988), despite the increased availability of laboratory methods for restriction site mapping and nucleotide sequencing. In the assay of nuclear genomes, one technical complication involves the isolation and assay of haplotypes from diploid organisms; but a more serious problem involves the likelihood that intragenic recombination or gene conversion will have shuffled nucleotide sequences at a locus, thus confounding reconstruction of gene genealogies (Aguadé *et al.* 1989; Hudson and Kaplan 1988; Stephens 1985; Templeton *et al.* 1987). It remains to be seen whether significant disequilibria involving intragenic restriction sites or nucleotide

sequences (of use to phylogenetic reconstruction) will commonly be found among geographically separated populations.

As complete gene genealogies may remain difficult to obtain for most nuclear loci, frequencies of phylogenetically unordered alleles [such as those provided by allozyme methods, or restriction fragment length polymorphism (RFLP) analyses: Avise 1989*a*] will continue to provide an important source of molecular character states normally used in the search for geographically concordant population subdivisions. Differences in morphological, behavioral and other phenotypic attributes, provided they are genetically based and independent, will of course continue to be important characters for survey in the search for patterns of geographic and genetic concordance.

6. CONCLUSIONS

One important root of the PSC probably traces to Simpson's (1951) paleontological perspective on taxa, summarized in his definition of an evolutionary species as 'a lineage (ancestral-descendant sequence of populations) evolving separately from others and with its own unitary evolutionary role and tendencies' (Simpson 1961, p. 153). Current versions of the PSC, apparently motivated by a perceived lack within the BSC of an adequate emphasis on history and phylogeny, have led some PSC proponents to reject the BSC's emphasis on reproductive isolation. Principles of genealogical concordance provide a compromise or composite stance between the BSC and PSC. Concepts of concordance are far from new in systematics (see, for example, Wilson and Brown 1953) and numerous statements can be found in support of the desirability of concordant information prior to taxonomic recognition. For example, Mayr (1969, p. 192) notes that 'geographic variation in the salamander *Plethodon jordani* is too discordant to justify the recognition of formal subspecies, even though the variation of each individual character shows a definite geographic trend'. Yet such sentiments too seldom have been followed, and many taxa continue to be recognized on the basis of one or a few diagnostic traits. The new generation of systematists may avoid a repeat of such errors by requiring concordance among several independent characters before advocating formal taxonomic recognition of putative population disjunctions. By focusing on the phylogenetic consequences of intrinsic RBs, and by emphasizing that important historical partitions can also be present within biological species because of extrinsic RBs, concordance principles should provide a useful set of philosophical and operational guidelines for the recognition of biotic and taxonomic diversity.

ACKNOWLEDGMENTS

This work was supported by NSF grants to J.C.A., and by a NIH Genetics Training grant to R.M.B.

REFERENCES

Aguadé, M., Miyashita, N. and Langley, C. H. (1989). Restriction-map variation at the *Zeste-tko* region in natural populations of *Drosophila melanogaster*. *Mol. Biol. Evol.* **6,** 123–30.

Aquadro, C.F., Desse, S. F., Bland, M. M., Langley, C. H. and Laurie-Ahlberg, C. C. (1986). Molecular population genetics of the alcohol dehydrogenase gene region of *Drosophila melanogaster*. *Genetics* **114,** 1165–90.

Asmussen, M. A., Arnold, J. and Avise, J. C. (1987). Definition and properties of disequilibrium statistics for associations between nuclear and cytoplasmic genotypes. *Genetics* **115,** 755–68.

Avise, J. C. (1989*a*). Gene trees and organismal histories: A phylogenetic approach to population biology. *Evolution* **43,** 1192–1208.

—— (1989*b*). A role for molecular genetics in the recognition and conservation of endangered species. *Trends Ecol. Evol.* **4,** 279–81.

—— and Nelson, W. S. (1989). Molecular genetic relationships of the extinct Dusky seaside sparrow. *Science* **243,** 646–8.

—— and Smith, M. H. (1974). Biochemical genetics of sunfish. I. Geographic variation and subspecific intergradation in the bluegill, *Lepomis macrochirus*. *Evolution* **28,** 42–56.

——, Bermingham, E., Kessler, L. G. and Saunders, N. C. (1984). Characterization of mitochondrial DNA variability in a hybrid swarm between subspecies of bluegill sunfish (*Lepomis macrochirus*). *Evolution* **38,** 931–41.

——, Arnold, J., Ball, R. M. Jr, Bermingham, E., Lamb, T., Neigel, J. E., Reeb, C. A. and Saunders, N. C. (1987*a*). Intraspecific phylogeography: The mitochondrial DNA bridge between population genetics and systematics. *Ann. Rev. Ecol. Syst.* **18,** 489–522.

——, Reeb, C. A. and Saunders, N. C. (1987*b*). Geographic population structure and species differences in mitochondrial DNA of mouthbrooding marine catfishes (Ariidae) and demersal spawning toadfishes (Batrachoididae). *Evolution* **41,** 991–1002.

——, Bowen, B. W. and Lamb, T. (1989). DNA fingerprints from hypervariable mitochondrial genotypes. *Mol. Biol. Evol.* **6,** 258–69.

Ayala, F. J. (1976). Molecular genetics and evolution. In *Molecular evolution* (ed. F. J. Ayala), pp. 1–20. Sinauer, Sunderland, Mass.

Ball, R. M., Jr, Neigel, J. E. and Avise, J. C. (1990). Gene genealogies within the organismal pedigrees of random mating populations. *Evolution* **44**, 1109–19.

Bermingham, E. and Avise, J. C. (1986). Molecular zoogeography of freshwater fishes in the southeastern United States. *Genetics* **113,** 939–65.

Burke, T. (1989). DNA fingerprinting and other methods for the study of mating success. *Trends Ecol. Evol.* **4,** 139–44.

Coyne, J. A. and Orr, H. A. (1989). Patterns of speciation in *Drosophila. Evolution* **43,** 362–381.

Cracraft, J. (1983). Species concepts and speciation analysis. In *Current ornithology* (ed. R. F. Johnston), Vol. 1, pp. 159–87. Plenum Press, New York.

Dobzhansky, T. (1937). *Genetics and the origin of species*. Columbia University Press, New York.

—— (1970). *Genetics of the evolutionary process*. Columbia University Press, New York.

Donoghue, M. J. (1985). A critique of the biological species concept and recommendations for a phylogenetic alternative. *Bryologist* **88,** 172–81.

Ehrlich, P. R. (1961). Has the biological species concept outlived its usefulness? *Syst. Zool.* **10,** 167–76.

—— and Raven, P. H. (1969). Differentiation of populations. *Science* **165,** 1228–32.

Eldredge, N. and Cracraft, J. (1980). *Phylogenetic patterns and the evolutionary process*. Columbia University Press, New York.

Hubbs, C. L. and Allen, E. R. (1944). Fishes of Silver Springs, Florida. *Proc. Fla. Acad. Sci.* **6,** 110–30.

—— and Lagler, K. F. (1958). *Fishes of the Great Lakes region*. University of Michigan Press, Ann Arbor.

Hudson, R. R. and Kaplan, N. L. (1988). The coalescent process in models with selection and recombination. *Genetics* **120,** 831–40.

Jeffreys, A. J., Wilson, V. and Thein, S. L. (1985). Individual-specific 'fingerprints' of human DNA. *Nature* **316,** 76–9.

Kocher, T. D., Thomas, W. K., Meyer, A., Edwards, S. V., Paabo, S., Villablanca, F. X. and Wilson, A. C. (1989). Dynamics of mitochondrial DNA evolution in animals: Amplification and sequencing with conserved primers. *Proc. Natl Acad. Sci. USA* **86,** 6196–200.

Lander, E. S. (1989). DNA fingerprinting on trial. *Nature* **339,** 501–505.

Langley, C. H., Shrimpton, A. E., Yamazaki, T., Miyashita, N., Matsuo, Y. and Aquadro, C. F. (1988). Naturally occurring variation in the restriction map of the *Amy* region of *Drosophila melanogaster*. *Genetics* **119,** 619–29.

Levin, D. A. (1979). The nature of plant species. *Science* **204,** 381–4.

Mayr, E. (1969). *Principles of systematic zoology*. McGraw-Hill, New York.

—— (1970). *Populations, species, and evolution*. Belknap Press, Harvard University, Cambridge, Mass.

McKitrick, M. C. and Zink, R. M. (1988). Species concepts in ornithology. *The Condor* **90,** 1–14.

Mishler, B. D. and Donoghue, M. J. (1982). Species concepts: A case for pluralism. *Syst. Zool.* **31,** 491–503.

Nei, M. (1987). *Molecular evolutionary genetics*. Columbia University Press, New York.

Neigel, J. E. and Avise, J. C. (1986). Phylogenetic relationships of mitochondrial DNA under various demographic models of speciation. In *Evolutionary processes and theory* (ed. E. Nevo and S. Karlin), pp. 515–34. Academic Press, London.

Nelson, G. J. and Platnick, N. I. (1981). *Systematics and biogeography: Cladistics and vicariance*. Columbia University Press, New York.

Pamilo, P. and Nei, M. (1988). Relationships between gene trees and species trees. *Mol. Biol. Evol.* **3,** 254–9.

Platnick, N. I. and Nelson, G. J. (1978). A model of analysis for historical biogeography. *Syst. Zool.* **27,** 1–16.

de Queiroz, K. and Donoghue, M. J. (1988). Phylogenetic systematics and the species problem. *Cladistics* **4,** 317–38.

Raven, P. H. (1976). Systematics and plant population biology. *Syst. Botan.* **1,** 284–316.

Reeb, C. A. and Avise, J. C. (1990). A genetic discontinuity in a continuously distributed species: Mitochondrial DNA in the American oyster, *Crassostrea virginica. Genetics* **124**, 397–406.

Rosen, D. E. (1978). Vicariant patterns and historical explanation in biogeography. *Syst. Zool.* **27,** 159–88.

—— (1979). Fishes from the uplands and intermontane basins of Gautemala: Revisionary studies and comparative geography. *Bull. Amer. Mus. Nat. Hist.* **162,** 267–376.

Saunders, N. C., Kessler, L. G. and Avise, J. C. (1986). Genetic variation and geographic differentiation in mitochondrial DNA of the horseshoe crab. *Limulus polyphemus. Genetics* **112,** 613–27.

Simpson, G. G. (1951). The species concept. *Evolution* **5,** 285–98.

—— (1961). *Principles of animal taxonomy*. Columbia University Press, New York.

Sneath, P. H. A. and Sokal, R. R. (1973). *Numerical taxonomy*. W. H. Freeman, San Francisco.

Sokal, R. R. (1973). The species problem reconsidered. *Syst. Zool.* **22,** 360–74.

—— and Crovello, T. J. (1970). The biological species concept: A critical evaluation. *Amer. Nat.* **104,** 127–53.

Spiess, E. B. (1977). *Genes in populations*. John Wiley, New York.

Stephens, J. C. (1985). Statistical methods of DNA sequence analysis: Detection of intragenic recombination or gene conversion. *Mol. Biol. Evol.* **2,** 539–56.

Swift, C. C., Gilbert, C. R., Bortone, S. A., Burgess, G. H. and Yerger, R. W. (1985). Zoogeography of the freshwater fishes of the southeastern United States: Savannah River to Lake Ponchartrain. In *Zoogeography of North American freshwater fishes* (ed. C. H. Hocutt and E. O. Wiley), pp. 213–65. John Wiley, New York.

Tajima, F. (1983). Evolutionary relationships of DNA sequences in finite populations. *Genetics* **105,** 437–60.

Takahata, N. (1989). Gene genealogy in three related populations: Consistency probability between gene and population trees. *Genetics* **122,** 957–66.

Templeton, A. R., Boerwinkle, E. and Sing, C. F. (1987). A cladistic analysis of phenotypic associations with haplotypes inferred from restriction endonuclease mapping. I. Basic theory and an analysis of alcohol dehydrogenase activity in *Drosophila. Genetics* **117,** 343–51.

Wiley, E. O. (1978). The evolutionary species concept reconsidered. *Syst. Zool.* **27,** 17–26.

—— (1981). *Phylogenetics: The theory and practice of phylogenetic systematics*. John Wiley, New York.

Wilson, A. C., Cann, R. L., Carr, S. M., George, M. Jr, Gyllensten, U. B.,

Helm-Bychowski, K. M., Higuchi, R. G., Palumbi, S. R., Prager, E. M., Sage, R. D. and Stoneking, M. (1985). Mitochondrial DNA and two perspectives on evolutionary genetics. *Biol. J. Linn. Soc.* **26,** 375–400.
Wilson, E. O. and Brown, W. L. Jr (1953). The subspecies concept and its taxonomic application. *Syst. Zool.* **2,** 97–111.

Hybrid zones: windows on evolutionary process

RICHARD G. HARRISON

1. INTRODUCTION

Spatial patterns of variation and patterns of change over time in genetic or organismic diversity are the fundamental data of evolutionary biology. Describing such patterns is often a straightforward and satisfying task, and modern morphometric and molecular techniques have enormously increased the resolution that can be achieved. One major goal, of course, is to understand evolutionary process. Unfortunately, definitive answers to questions about process have remained remarkably elusive. Although most evolutionary biologists acknowledge that natural selection, random events and historical constraints have all contributed to current patterns of variation, in relatively few cases has a specific role for these evolutionary forces been clearly demonstrated. Questions about speciation have proved particularly frustrating, with debate focusing on the geographic context as well as on the nature of the evolutionary forces involved.

Theory and experiment are essential for defining the realm of the possible, but they cannot reveal the unique series of historical events leading to current patterns of diversity and adaptation. Resolution of debates about evolutionary process must ultimately emerge from careful quantitative analysis of observed patterns. Natural zones of hybridization, especially those that occur as abrupt discontinuities between otherwise relatively homogeneous entities, have always attracted the attention of evolutionary biologists. Hybrid zones have been a focus of study, both because they are windows on evolutionary process and because, as patterns of variation, they do not fit comfortably into the traditional classification schemes of taxonomists. Although there is a vast literature documenting the existence of hybrid zones in both plants and animals and an expanding literature of hybrid zone theory, their origin and fate (including their role in the speciation process) continue to be vigorously debated. Many of the traditional hybrid zone questions remain unanswered and some have been deemed unanswerable. Even those issues that were once thought to be fully resolved have been reopened. At the same time, new approaches (especially the application of biochemical and molecular genetic techniques) have revealed patterns that were not evident from simple morphological analyses and have provided important insights into the dynamics of hybrid zones.

This chapter reviews the contributions of hybrid zone research to our understanding of evolutionary pattern and process. It focuses first on the origin and dynamics of hybrid zones – whether history or current ecology are the primary determinants of observed patterns of differentiation, whether hybrid zones are stable or transient, and what forces act to modify them. I then discuss how genetic and ecological analyses of hybrid zones provide insights into the nature of species, the functioning of barriers to gene exchange, the genetic architecture of these barriers, and the dynamics of the speciation process. The issues raised are obviously among the most important and contentious in evolutionary biology.

2. WHAT ARE HYBRID ZONES?

Before embarking on a discussion of pattern and process in hybrid zones, it is important to clarify concepts and terminology and reach agreement on exactly what is implied by the term 'hybrid zone'. A survey of the recent literature suggests that there is no clear consensus and that definitions vary substantially in their underlying assumptions. It is useful to begin by considering what is meant by 'hybridization'. Mayr (1942, pp. 258–59) summarized the problem as follows:

> It is very difficult to define this term [hybridization], or at least to delimit it against various forms of intraspecific interbreeding. The use of the term hybridization is undoubtedly justified if individuals of different families, genera or good species interbreed. But to what extent can the interbreeding of individuals of different subspecies or merely distinct populations of the same species be called hybridization?

Restriction of the term hybridization to interbreeding between individuals assigned to different taxa is necessarily arbitrary (as the decisions in making the assignment obviously vary among taxa and among taxonomists, especially when subspecies designations are involved). I prefer to apply the term hybridization very broadly and therefore subscribe to a modification of Woodruff's definition of natural hybridization as 'the interbreeding of [individuals from] two populations, or groups of populations, which are distinguishable on the basis of one or more [heritable] characters' (Woodruff 1973, p. 214). I have inserted the words 'individuals from' and 'heritable' because the basic units of hybridization are individual organisms, not populations, and because the character differences must have a genetic basis. Others have arrived at similarly broad definitions for hybridization (e.g. Sibley 1954; Mallet 1986).

Definitions of hybrid zones are nearly as numerous as the evolutionary biologists who have written about them. Furthermore, hybrid zones have frequently been described using other terms (e.g. cline, contact zone, zone of intergradation, tension zone), each of which has implications that are only sometimes made explicit. It will be useful to review a number of possible definitions before arriving at a working definition that will serve as a basis for further discussion in this chapter.

Definitions of hybrid zones may be based simply on observed patterns of variation or they may include assumptions about the evolutionary mechanisms and processes that are responsible for the patterns. Mayr (1942) distinguished zones of hybridization, which involve secondary contact between previously isolated populations, from 'ordinary zones of primary intergradation', in which differentiation arises within a series of populations that are in continuous contact. Several recent reviews (Heiser 1973; Moore 1977; Rising 1983; Littlejohn and Watson 1985) have explicitly restricted the term hybrid zone to zones of secondary contact. The difficulty with this approach is that distinguishing primary integradation from secondary contact based on simple observations of patterns of variation is certainly not trivial and may be exceedingly difficult (see discussion below). Because very different histories can give rise to identical patterns, it would seem unwise to assume or imply a particular history when using the term hybrid zone.

Barton and Hewitt (1981*b*) originally used the term hybrid zone to describe a 'narrow cline maintained by some sort of hybrid unfitness', i.e. their definition included an explicit assumption about the dynamics of the interactions they were describing. Recognizing that other evolutionary biologists were not following their lead, they have abandoned this assumption and, in a more recent review (Barton and Hewitt 1985), they simply equate a hybrid zone with a cline. Following Key (1968), they refer to clines maintained by a balance between dispersal and selection against hybrids (i.e. one class of hybrid zones) as 'tension zones'.

Endler (1977, p. 4) characterizes hybrid zones as regions where 'distinct groups of relatively uniform sets of populations are separated by narrow belts that show greatly increased variability in fitness compared to random mating'. He distinguishes this pattern of variation from other patterns that he terms conjunction (steep clines) or gradation (shallow clines). According to his definition, hybrid zones only form when sufficient differentiation has occurred such that interbreeding results in 'unbalanced gene complexes' in individuals of mixed ancestry. This definition is impractical, both because it carves up a continuum (extent of differentiation or degree of incompatability) into three arbitrary categories and because it depends on estimates of variance in fitness, which are not available.

Hybrid zone definitions have sometimes been based on the spatial distribution of parental types and individuals of mixed ancestry. Thus, Short (1969) restricted the term hybrid zone to regions in which only individuals of mixed ancestry are found. He distinguished these cases from 'zones of overlap and hybridization' in which the two parental types occur together with recombinant types throughout the zone of interaction. Woodruff (1973) also proposed a classification based on the distribution of genotypic (phenotypic) classes. He recognized three categories of natural hybridization – allopatric, parapatric and sympatric – based on the geographic relationships of the parental types. Although the spatial distribution of genotypic classes is important as an indicator of evolutionary process, the divisions proposed by Short and Woodruff are arbitrary, especially when only a limited number of unmapped markers are used to define the parental types.

A definition of hybrid zones that is conceptually sound and also operational must make a minimum of assumptions about the history of the interaction and about the evolutionary forces currently acting. It is best to start with a definition that includes a broad array of superficially similar phenomena, which can subsequently be subdivided on the basis of detailed analyses of pattern and process. In that spirit, I propose the following operational definition. Hybrid zones are interactions between genetically distinct groups of individuals resulting in at least some offspring of mixed ancestry. Pure populations of the two genetically distinct groups are found outside of the zone of interaction.

Hybrid zones vary in a number of characteristics: (1) width (from tens of meters to hundreds of kilometers); (2) nature of the characters used to distinguish the hybridizing taxa; (3) extent of genetic differentiation of the hybridizing taxa; (4) relative fitness of individuals of mixed ancestry; (5) role of environmental heterogeneity in determining structure and position; (6) level of assortative mating within the zone; and (7) origins. The well-characterized hybrid zones summarized in Fig. 1 (pages 74/75) are examples selected to illustrate the diversity of spatial patterns of variation that have been observed.

3. HYBRID ZONE ORIGINS

3.1 Primary intergradation and secondary contact

The origin of hybrid zones has been the subject of debate ever since Mayr (1942) proposed a distinction between zones of primary and secondary intergradation. The former represent clinal variation that develops within a continuous series of populations, whereas the latter are the result of contact between populations that have become differentiated in allopatry. For Mayr (1942) the relatively abrupt discontinuities characteristic of many hybrid zones were indicative of secondary contact; he did not hesitate to draw conclusions about process from observations of patterns of variation. As the most forceful opponent of non-allopatric models of speciation, he naturally found it difficult to accept that changes from one 'character combination' to another over a very short distance could be the product of differentiaton *in situ*.

In fact, the debate about hybrid zone origins parallels the longstanding and often acrimonious debate between those who believe that allopatric speciation is the exlusive mode of speciation and those who argue that speciation can occur in the absence of geographic isolation (see Futuyma and Mayer 1980). In the early 1960s, few evolutionary biologists gave much credence to non-allopatric models of speciation. Hybrid zones were confidently interpreted as zones of secondary contact between forms that had diverged in allopatry but had not reached the point of being 'good' species.

This was the calm before the storm. Models developed by Clarke (1966), Slatkin (1973) and Endler (1973, 1977) revived interest in cline models originally outlined by Fisher (1950) and Haldane (1948). They clearly demonstrated that steep single-locus clines could form in the absence of any barriers to gene flow. The models examined the impact of varying intensities of selection and levels of dispersal and the consequences of postulating additional loci that modify the fitnesses of the three single-locus genotypes (i.e. the consequences of differential coadaptation). The question of hybrid zone origins was discussed at length in Endler's (1977) influential book, *Geographic variation, speciation, and clines*. There he argued that, because primary intergradation and secondary contact can produce identical patterns of variation, it is futile to attempt to infer process from pattern. Only in cases of very recent secondary contact will it be possible to draw conclusions about hybrid zone origins. The clear message was that the previous bias in favor of secondary contact as an explanation for hybrid zones was unwarranted. Evolutionary biologists were once again confronted with ambiguity and uncertainty.

In a later paper, Endler (1982) argued that distinguishing historical from current ecological factors is a general problem in biogeography. The essence of the problem is that any observed distribution pattern could be produced either by vicariance events coupled with dispersal (historical factors) or by environmental heterogeneity resulting in spatially varying selection coefficients. However, biogeographic scenarios do not necessarily attempt to specify what evolutionary forces have led to the origin of variation (either within or between species). For this reason, it is essential to partition the original issue of hybrid zone origins (primary vs. secondary intergradation) into two distinct questions (Barton and Hewitt 1985): (1) Where and how did existing variants arise (in what geographic context and as a result of what evolutionary forces)? (2) What historical and ecological factors account for the current distribution of these variants

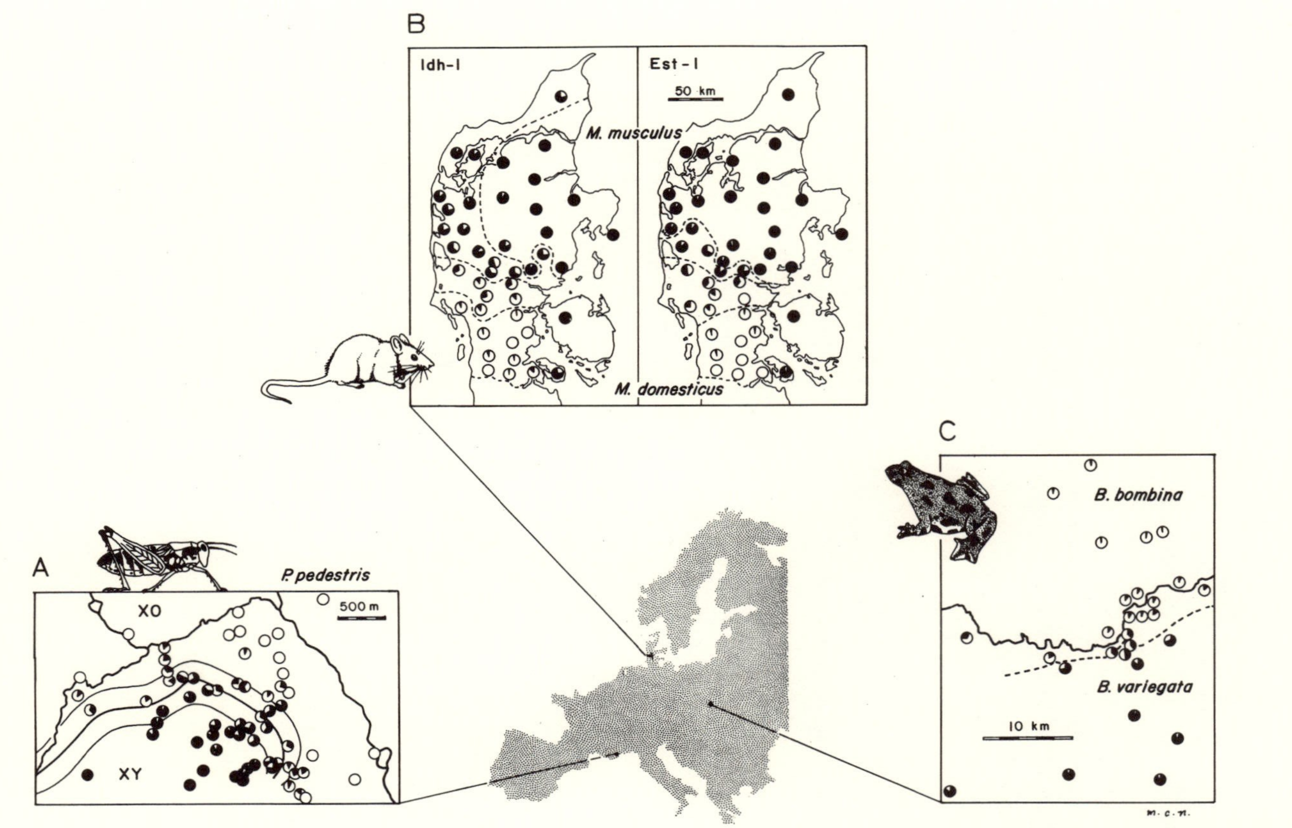

Fig. 1. Patterns of variation in six well-characterized hybrid zones. In all cases except the Northern Flicker hybrid zone, the maps represent only a small part of a more extensive hybrid zone. (a) A steep chromosomal cline in the grasshopper *Podisma pedestris* in the Alpes Maritimes, showing the center of the cline and the 25 and 75 per cent contours (Barton and Hewitt 1981*c*). (b) Alloyzme variation across a hybrid zone between *Mus musculus* and *M. domesticus* in Denmark (Hunt and Selander 1973). (c) Map of a hybrid zone between species of fire-bellied toads (*Bombina bombina* and *B. variegata*) showing average frequency of *variegata* alleles (five enzyme loci) (Szymura and Barton 1986).

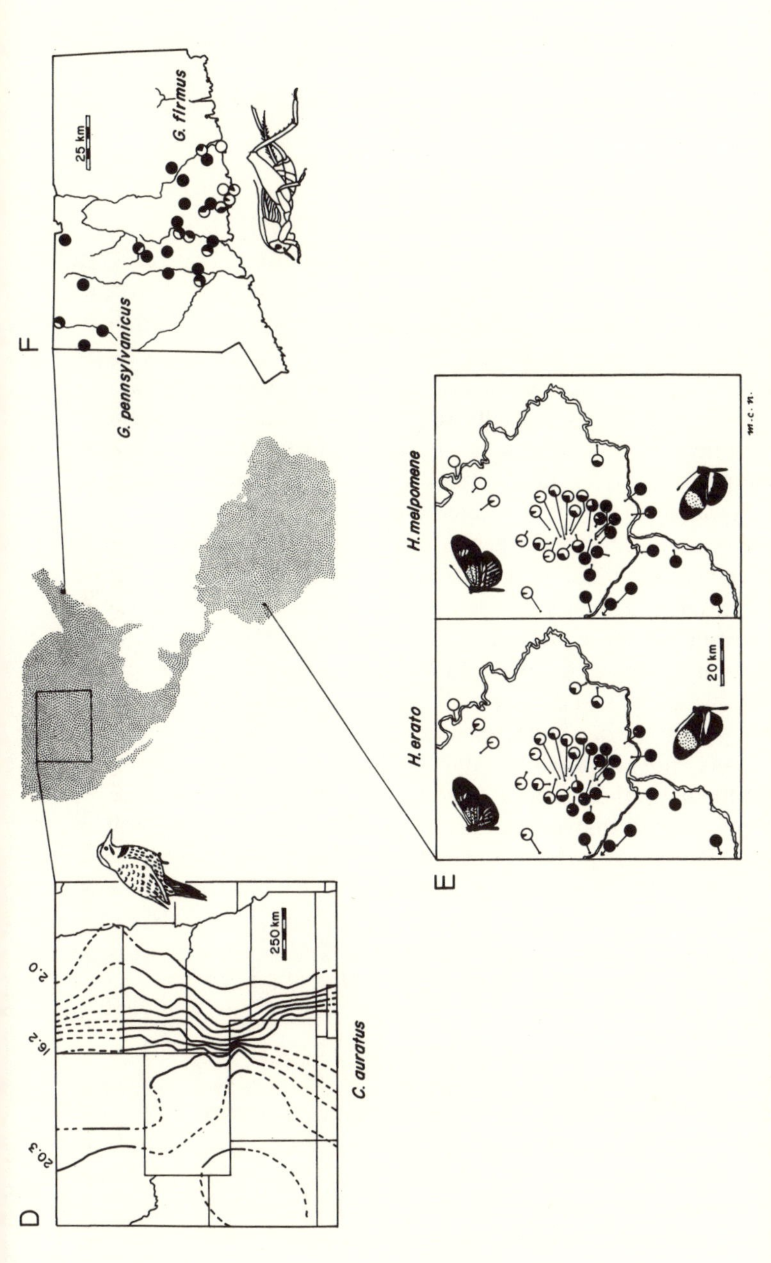

Fig. 1. (continued) (d) Variation in plumage characters (summarized as a single hybrid index score) in the Northern Flicker (*Colaptes auratus*) hybrid zone in the central United States (Moore and Koenig 1986). (e) Variation in allele frequencies (genes controlling color pattern) in parallel hybrid zones of the South American butterflies *Heliconius erato* and *H. melpomene* (Mallet and Barton 1989a). (f) Mitochondrial DNA genotype frequencies across the state of Connecticut in a hybrid zone between the field crickets *Gryllus firmus* and *G. pennsylvanicus* (Harrison *et al.* 1987; Harrison and Rand 1989).

(i.e. the current geographic extent of the hybrid zone and patterns of variation within it)? It is clear that answers to the first question will not emerge simply from observations of patterns of variation in hybrid zones. The process of differentiation (speciation) will often be obscured by subsequent events. However, answers to the second question are not beyond reach.

3.2 Inferences from patterns of variation

The fundamental question is whether a particular pattern (e.g. a steep cline or abrupt change in a particular character) arose *in situ* as the result of spatially varying selection or whether it reflects the coming together of two already differentiated populations. For a single character and a single transect across a hybrid zone, distinguishing the two possible scenarios is (as Endler, 1977, suggested) virtually impossible. Even if the character difference is effectively neutral (e.g. a silent substitution in a DNA sequence comparison), it is always possible to invoke tight linkage to some other character that is under selection.

But what if many different characters change across the zone? Hybrid zones are often characterized by transitions in morphology, behavior, life history, chromosome complement, allozymes and mitochondrial DNA genotype (or some subset of these traits). Selection acting independently on many loci can produce a set of concordant clines. Likewise, if co-adaptation is widespread and if most newly arising alleles are advantageous in only one genetic background, concordant clines can arise *in situ*. Although clearly not impossible, the probability of these scenarios becomes less likely as the number of independent character differences increases. Explanations that invoke selection are notoriously difficult to disprove, as selection can produce virtually any observed pattern of variation. Nevertheless, 'the existence of multiple zones [concordant change in several characters] is one of the stronger arguments in favor of origin by secondary contact' (Hewitt and Barton 1980). The argument has been used frequently, e.g. in wood warblers (Barrowclough 1980), pocket gophers (Haffner *et al.* 1983; Heaney and Timm 1985), field mice (Stangl 1986; Nelson *et al.* 1987) and grasshoppers (Marchant *et al.* 1988).

The same argument can be applied to hybrid zones that are mosaics rather than sets of coincident clines (Briggs 1962; Harrison and Rand 1989). Any scenario for the origin of a mosaic hybrid zone must explain the complex internal structure, with the distribution of multi-locus genotypes often associated with an underlying environmental mosaic. For example, in a hybrid zone between the cricket species *Gryllus pennsylvanicus* and *G. firmus* (Fig. 1f), there is a mosaic population structure determined at least in part by the patchwork of habitat (soil) types (Harrison 1986; Rand and Harrison 1989). At many sites within the zone, consistent differences in morphology, allozyme frequencies and mtDNA genotype frequencies characterize neighboring populations on sand and loam soils. The possibility that differentiation for each of these markers occured *in situ* seems remote, as this would require independent selection on each of the characters within each 'patch'. It is far more likely that this part of the hybrid zone was formed as a result of the sorting of two already differentiated lineages into alternative patch types. The implication is that the observed non-random association of markers (linkage disequilibrium) is a consequence of prior differentiation rather than a product of selection within each of the patches of the environmental mosaic. The same logic can be extended to situations in which identical sets of concordant changes occur across geographically distant sites within the same hybrid zone.

The essence of the approach described above is to construct a set of phylogenies based on samples of many populations from within a mosaic hybrid zone or from either side of a linear zone (Fig. 2). Trees in which all populations from one side (or one patch type) cluster together could reflect recent common ancestry (evolutionary history), current ecology (parallel or convergent evolution) or widespread co-adaptation. If many independent phylogenies (i.e. based on different sets of characters) exhibit the same pattern, the first explanation is far more likely. Thorpe (1984) has devised a tree-building approach for distinguishing primary and secondary intergradation based on deriving historical patterns of range expansion from population phylogenies and then inferring hybrid zone origins from the position of the zone with respect to these postulated expansions. The method depends on having reliable (and concordant) population phylogenies and on being able to estimate accurately branch lengths on a tree.

It is now possible to generate a set of independent allele phylogenies based on comparative DNA sequence data. Selective amplification of specific genes or gene fragments using the polymerase chain reaction (PCR), followed by sequencing of the amplified regions, allows evolutionary biologists to generate data for many gene regions in large numbers of individuals. Because these sequence variants are unlikely to be strongly selected, they are precisely the sort of markers needed to reveal recent history, i.e. to confront the issue of the origin of hybrid zones.

However, several problems remain. First, the traits that distinguish hybridizing taxa may have different histories (and therefore give trees with different branching patterns). Some markers may have differentiated in allopatry, whereas others may have differentiated (or possibly converged) subsequent to secondary contact (e.g. due to environmental heterogeneity – either selection gradients or mosaics). Population histories may be complex, reflecting several episodes of range expansion and contraction (see Hewitt 1989). In such cases, the origin and spread of new

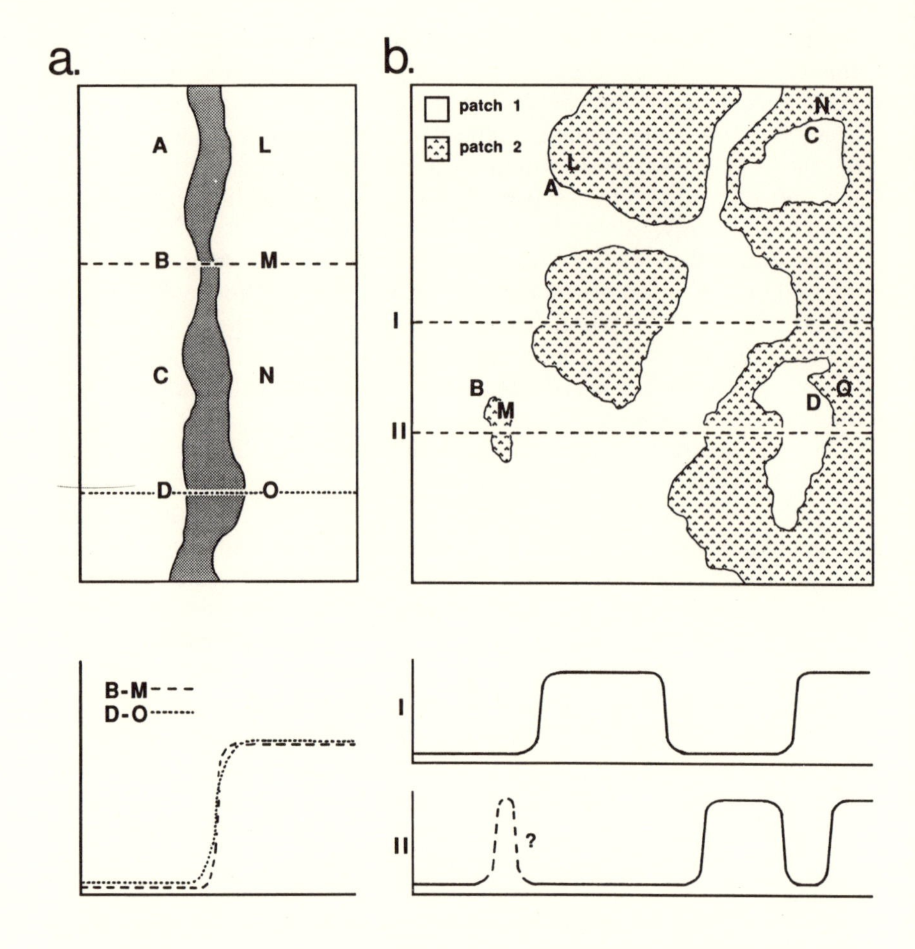

Fig. 2. Hybrid zone structure, population sampling and phylogenetic inference. (a) An extensive but narrow hybrid zone with several (many) characters exhibiting concordant clinal variation. The dashed lines are transects across the zone, with patterns of variation along these transects shown below the map. A–D and L–O are population samples from either side of the hybrid zone. (b) A hybrid zone in which two distinct forms occupy 'patches' in a mosaic environment. The patches represent different environments (e.g. soil types). A–D and L–O are population samples from the two patch types. The dashed lines are transects across the zone, with patterns of variation along these transects shown below. These patterns depend on the location of the transect. The patch from which sample M is taken may be small enough (compared with dispersal distance) that the distribution of genotypes does not reflect its presence (see p. 86).

C.

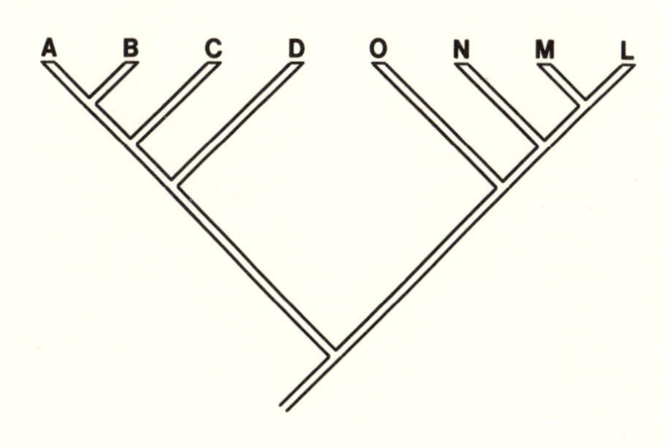

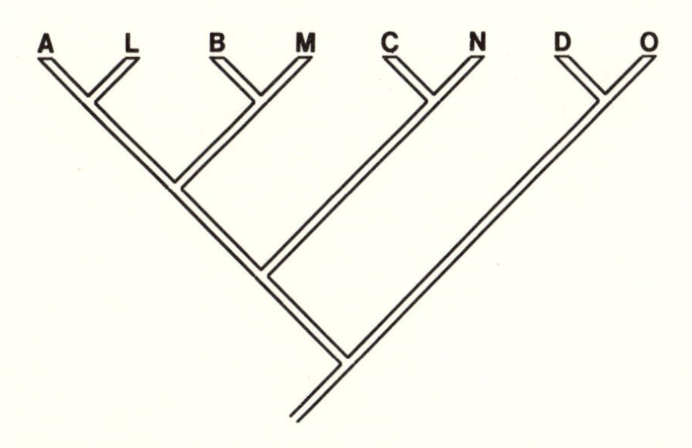

Fig. 2. (c) Two possible allele phylogenies based on comparisons of DNA sequences from the eight populations (DNA sequences *not* involved in initial identification of the hybrid zone). If most gene trees show A–D and L–O as distinct clusters (tree 1), the two lineages probably existed prior to the formation of the hybrid zone (i.e. secondary contact). Tree 2 is expected for at least some characters if differentiation has occurred *in situ*.

variants may have occurred at different times and in different geographic contexts. Furthermore, differential introgression or random lineage sorting in populations derived from a polymorphic ancestor can produce lack of concordance among allele phylogenies (Neigel and Avise 1985; Harrison 1989). Lineage sorting (a consequence of the random extinction of lineages at each generation) is particularly troublesome in phylogenetic inference for recently diverged populations and species – precisely those taxa between which hybrid zones are likely to occur. Because of recombination, genomes are not coherent through time and allele phylogenies are, therefore, not equivalent to organism phylogenies (Nei 1987); discordance among individual markers is to be expected. This 'background noise' will make more difficult the task of sorting out explanations for current patterns of variation.

In a few cases, recent population history seems unambiguous. For example, hybridization between two subspecies of the salamander *Ensatina eschscholtzii* occurs in the foothills of the Sierra Nevada, California (Wake and Yanev 1986; Wake *et al.*, 1989). The two subspecies are genetically very distinct (Nei D of 0.42) and are clearly not sister taxa. One of the hybridizing pair is virtually identical (morphologically and genetically) to salamanders from the region east of San Francisco Bay. It is almost certain that this coastal subspecies has recently crossed the Central Valley (a region currently not inhabited by *Ensatina*) and established an 'outpost' in the Sierran foothills. In general, hybridization between non-sister taxa would seem to signal secondary contact (Rosen 1979). An interesting variant of the argument is used by Heaney and Timm (1985), who favor secondary contact as an explanation for the origin of a pocket gopher (*Geomys*) hybrid zone because the two species harbor parasites that are non-sister species.

In contrast, the observation that hybrid zones involve taxa that have diverged very little can be used to argue for differentiation *in situ*. Detailed analysis of hybrid zones between morphologically distinct populations in the land snail genus *Cerion* suggests that very distinct morphologies arise as a consequence of small changes in early development and do not reflect extensive genetic differentiation (Gould *et al.* 1975; Galler and Gould 1979; Woodruff 1981). The 'traditional' explanation for the numerous *Cerion* hybrid zones – secondary contact due to hurricane transport – is therefore rejected because it is 'unlikely that a random process would have brought such similar ontogenetic types into proximity. . . ' (Galler and Gould 1979).

3.3 Hybrid zone biogeography

Hybrid zones between many different pairs of species or subspecies often appear to coincide geographically. Remington (1968) described a number of such 'suture zones' in the continental USA, which he interpreted as regions of recent secondary contact between previously isolated biota. Perhaps best known are the avian hybrid zones that occur in the Great Plains. A review of these zones (Rising 1983) indicates that distributional limits do not coincide precisely; in fact, several different patterns of east–west replacements can be identified. Unfortunately, there has been little quantitative analysis to determine whether hybrid zones are really non-randomly distributed. Concordant patterns of geographic variation have also been suggested for various faunal elements in the tropical lowland forests (e.g. Haffer 1969; Brown *et al.* 1974). Concordance has been interpreted as evidence for secondary contact between populations that were isolated in forest refugia during cool, dry periods in the Pleistocene. The data are also consistent with the hypothesis that current ecological factors are responsible (Endler 1982). However, there are no clear examples in which observed patterns of variation can be related directly to environmental factors. On balance, evidence from the distribution of hybrid zones is at best equivocal and does not allow discrimination among the alternative explanations for hybrid zone origins.

Many hybrid zones in northern temperature regions must have come to occupy their present positions within the last 10 000–20 000 years (see Hewitt, 1989, for a summary of hybrid zones in Europe and North America that occur in regions that could not have been occupied during the last ice age). Obviously, one explanation for the origin of these zones is that they were established as a result of range expansions from Pleistocene refugia following the retreat of the glaciers, as new areas of suitable habitat opened up. Such scenarios imply that populations were already differentiated when contact was established. However, as emphasized above, questions about how, where and when the differences arose remain unanswered. Molecular data (initially allozymes, now DNA sequences) can potentially provide a framework for estimating times of divergence.

Many European and North American hybrid zones would appear to have developed as a result of post-Pleistocene secondary contact – Barton and Hewitt (1985) say that 37 per cent of the hybrid zones they surveyed are 'clear examples'. Current distribution patterns are often suggestive of Pleistocene disjunction and subsequent secondary contact, e.g. chromosomal races of the grasshopper *Podisma pedestris* in southeastern France (Barton and Hewitt 1981*b*), subspecies of the grasshopper *Chorthippus parallelus* in the Pyrenees (Hewitt 1989), subspecies of grackles in eastern North America (Yang and Selander 1968), many pairs of birds in the Great Plains of the United States (Rising 1983; Grudzien *et al.* 1987), swallowtail butterflies in the *Papilio machaon* group in western North America (Sperling 1987), and species of *Impatiens* in the northwestern USA (Ornduff 1967). Detailed scenarios have been developed where evidence from palynological data provide clues about changes in climate and vegetation type, e.g. species of *Juniperus* in southern Wisconsin (Palma-Otal *et al.* 1983). Information about sea-level changes allows inferences about previous disperal corridors for marine organisms or inundation of coastal areas now above sea level, e.g. stone crabs (*Menippe*) in the Gulf of Mexico and along the Atlantic coast (Bert 1986; Bert and Harrison 1988) and populations of *Plethodon* salamanders along the coastal plain of the mid-Atlantic USA (Highton 1977; Wynn 1986).

Patterns of change in climate and vegetation make it likely that many of these examples are, in fact, classic cases of secondary contact between populations/species that have been geographically isolated for varying lengths of time. However, skeptics and pan-selectionists can still point to other possible scenarios that could account for observed patterns of variation. To a large extent, this is because none of these hybrid zones have been examined in sufficient detail. Careful phylogenetic analysis, involving population samples from throughout the ranges of the interacting taxa and comparisons of many independent DNA sequences, should clarify the recent history of dispersal and colonization and distinguish among alternative hypotheses for origins. Obviously, evolutionary biologists will have to settle on a limited number of well-chosen model systems.

3.4 The impact of recent disturbance

To what extent has recent human disturbance altered the distributions of animals and plants and led to the formation of hybrid zones? Examples of intentional or accidental introductions resulting in hybridization have been documented (e.g. Echelle and Connor 1989), and these obviously represent hybrid zones of very recent origin.

Human influence often involves habitat alteration or fragmentation that breaks down ecological barriers or provides corridors of dispersal. In either case, populations that were previously isolated come into contact. Botanists have been particularly impressed by the consequences of human disturbance. Anderson and Hubricht (1938, p. 399) wrote that 'the more that natural ecological conditions have been upset the greater would be the opportunity for introgressive hybridization'. They argued that, in the genus *Tradescantia*, habitat segregation is the most important barrier to gene exchange between closely related species. Human disturbance removes this barrier and thereby promotes hybridization. Examples of closely related plant species that exhibit different habitat 'preferences' and only appear to hybridize in disturbed situations may be common, e.g. *Liatris aspera* and *L. spicata* (Levin 1967) and *Phlox drummondi* and *P. cuspidata* (Levin 1975).

Parallel cases can be found among the animals. The clearing of natural vegetation for agriculture and lumber has catalyzed hybridization between towhees (*Pipilo*) in Mexico (Sibley 1954). The damming of streams in Texas has increased the extent of hybridization between species of *Rana* (Hillis 1988). For African paradise flycatchers, fragmentation of forest habitat and invasion of these isolated patches by birds of clearings and second growth has apparently led to the breakdown of reproductive isolation (Chapin 1948). For species that occupy disturbed habitats (pastures, old fields, etc.), the clearing of forests and widespread farming and grazing have presumably not only led to greatly increased population sizes but have provided corridors of dispersal between what were once far more isolated populations (McDonnell *et al.* 1978; Harrison and Arnold 1982).

In virtually all populated areas, human disturbance has reduced the effectiveness of ecological barriers and therefore increased the extent of genetic exchange. In many of these cases, hybridization probably occurred (perhaps only to a very limited extent) prior to disturbance. However, habitat alteration has no doubt seriously perturbed many of the interactions, perhaps rendering previously stable zones unstable (Littlejohn and Watson 1985) and certainly altering their configuration.

4. THE DYNAMICS OF HYBRID ZONES

Once formed, hybrid zones may persist unchanged or become modified in various ways. Much of the early interest in hybrid zones focused on their possible role in the speciation process. The model of speciation championed by Dobzhansky (1940, 1941) and many of his disciples is based on the premise that genomes are 'integrated systems' and that recombination breaks up these co-adapted complexes, leading to individuals of reduced fitness. According to this view, hybrid zones are sites where speciation is completed through the evolution of pre-mating barriers in response to selection against individuals of mixed ancestry (a process often referred to as 'reinforcement').

Although speciation via reinforcement represents one scenario for evolution within a hybrid zone, it is clear that other outcomes are possible (even probable). Much of the early literature emphasized that hybrid zones are likely to be transient phenomena, leading either to fusion or speciation (Sibley 1957; Wilson 1965; Remington 1968). More recent reviews have assumed that many hybrid zones are stable and have proposed models for how they are maintained (Moore 1977; Barton and Hewitt 1981*b*, 1985, 1989; Hewitt 1988). Thus, there has been a very obvious shift in emphasis. Here I briefly review certain aspects of hybrid zone dynamics. I focus on how selection can maintain stable hybrid zones, on fusion and extinction as possible outcomes of hybrid zone interactions, and on the influence of population structure and environmental grain on hybrid zone dynamics. I also review the evidence for the long-term stability of hybrid zones. The debate about reinforcement will be discussed in detail in a later section.

4.1 The maintenance of stable hybrid zones

Stable hybrid zones (or clines) can be maintained by selection alone or by a balance between selection and dispersal (Barton and Hewitt 1985). Dispersal-independent models include those in which hybrids (heterozygotes in the case of single-locus models) are more fit than either parental phenotype (homozygote) within a restricted geographic area (Endler 1977; Moore 1977). The 'bounded hybrid superiority' model (Moore 1977; Moore and Koenig 1986) implies that environmental factors are primary determinants of relative fitness, that parental types are adapted to different environments, and that habitats exist in which hybrids (individuals of mixed ancestry) are more successful than either parent. The distribution of these habitats will determine the boundaries of the hybrid zone.

Are there environments in which individuals of mixed ancestry are likely to be at an advantage? The prevailing view (most frequently voiced by botanists) has been that only in unstable, rapidly changing or disturbed habitats will recombinant types be successful (Anderson 1948; Stebbins 1950; Grant 1981; Heiser 1973). This follows from the assumption that parental types 'represent gene combinations so well adapted to the environments they occupy that any new combinations created by hybridization will amost certainly be less favorable and therefore selected against in the original environments' (Stebbins 1950, p. 256). The extreme version of this argument was formulated by Anderson (1948), who suggested that each recombinant type would have its own narrow habitat requirements and that 'only by a hybridization of the habitat can the hybrid recombinations be preserved in nature'. Moore (1977) proposed that hybrid zones will be associated with ecotones, which he considered to be marginal or transitional habitats in which parental phenotypes are relatively unsuccessful.

The best evidence to support the bounded hybrid superiority model would be the direct demonstration that, within a hybrid zone, individuals of mixed ancestry survive longer and/or reproduce more successfully than either parental type. I know of no case where this has been clearly shown. The alternative approach, inferring process from pattern (e.g. from hybrid zone shape and position with respect to ecotones or unstable/disturbed habitats), has not been successful in providing a clear test of the bounded hybrid superiority model. Moore (1977, p. 275) acknowledges that the model 'can account for a stable hybrid zone of almost any size or shape' and suggests that hybrid zones are often narrow because they are associated with narrow ecotones. Barton and Hewitt (1985, 1989) contend that it is unlikely that habitats in which hybrids are most fit would be distributed as long, narrow strips (which is the case for many of the zones discussed by Moore (1977), e.g. grackles in eastern North America). They conclude that a balance between dispersal and selection against hybrids is the more likely basis for such zones (see below). On the other hand, some examples cited by Moore (1977), e.g. hybrid populations of towhees in Mexico (Sibley 1954; Sibley and West 1958), do exhibit the 'broken, patchy distribution' that Barton and Hewitt (1985) predict for hybrid zones maintained by hybrid superiority (although such distributions are not proof of hybrid superiority). Unfortunately, in most of these cases, there is inadequate ecological information. Although many zones are said to coincide with ecotones or to parallel gradients in moisture or temperature, few hybrid zone studies have actually involved detailed mapping of environmental gradients or mosaics. Even where hybridization is obviously associated with disturbance (as is the case in many groups of plants and in the towhees), the explanation may simply be that only in these circumstances are individuals of mixed ancestry actually produced.

Stable hybrid zones may be maintained in dynamic equilibrium as a result of a balance between dispersal and selection. Selection may be a reflection of environmental gradients, with the two parental types having higher fitness at opposite ends of the gradient (or on opposite sides of an environmental discontinuity). Both gradient models and ecotone models can produce 'stepped clines' (hybrid zones), with properties that depend on the magnitude of the selection gradient and the amount of dispersal (Slatkin 1973; Endler 1977). Although these models differ from the hybrid superiority model in how they define spatial variation in selection coefficients, they share with it the assumption that the fitness of genotypes is dependent on the environment.

Models will be relevant to natural situations only in so far as the pattern of environmental heterogeneity they assume corresponds to patterns seen in nature. The usual practice is to describe selection coefficients that change gradually (gradient models) or abruptly (step models) along a one-dimensional transect (Slatkin 1973; Endler 1977). The justification for such an approach is that 'environmental factors are most often found in gradients' (Endler 1977, p. 53) and that individual organisms (genotypes) will have ecological tolerances that give them a relative advantage within some restricted part of the gradient. In fact, patterns of environmental heterogeneity are often more complex than implied by gradient models. The environment is perhaps more realistically modelled as a mosaic of patches (Fig. 2b) that represent different resources, habitats or physical environmental factors (Harrison and Rand 1989). Transitions from one environment to another may not involve a series of intermediate environments, but instead a changing proportion of patches of the two types. If organisms differ in their preference for or fitness in the two (or more) patch types, hybrid zone structure may reflect the mosaic nature of the environment. In fact, the relevance of simple gradient models or mosaic models will depend on how far individuals are likely to disperse compared with the average size of a patch, i.e. whether the environment is 'fine-grained' or 'coarse-grained' (Levins 1968). In a one-dimensional environment, Slatkin (1973) has shown that variation will reflect the presence of a patch only if the size of the patch is greater than a characteristic length ($l_c = l/s$, where l is average dispersal distance and s is the magnitude of selection). Similar scaling arguments apply in two-dimensional patches (N. Barton, personal communication).

The importance of spatial variation in selection coefficients as a determinant of hybrid zone structure in animals has been clearly demonstrated in relatively few cases. There are many hybrid zones that appear to occur along environmental gradients or to coincide with ecotones – e.g. house mice (Hunt and Selander 1973), grackles (Yang and Selander 1968), chickadees (Robbins *et al.* 1986), lizards (Jackson 1973) and frogs (Sage and Selander 1979) – but in none of these cases has a causal relationship been established. A clear habitat association is evident in hybrid zones between hybridizing species of threespine sticklebacks (*Gasterosteus*) in coastal streams in British Columbia (Hagen 1967). The two species appear both to prefer and to be well-adapted to their characteristic habitat. Evidence for a mosaic structure comes from work on a hybrid zone in crickets, in which there is a clear association of different genotypes with loam and sand soils (Harrison 1986; Rand and Harrison 1989). In this case, environmental heterogeneity is almost certainly responsible for the complex structure of the zone (either because crickets select different habitats or because there is differential survival and reproductive success).

Examples of environmental control of patterns of hybridization in plants are more numerous (e.g. Anderson and Hubricht 1938, Muller 1952; Briggs 1962; Alston and Turner 1963; Levin 1967, 1975; Heywood 1986). In each of these cases, species with narrow ecological tolerances hybridize in patterns that reflect an environment that is a patchwork of habitat types.

Stable hybrid zones can persist in the absence of environmental heterogeneity, as a result of a balance between dispersal and either selection against hybrids, selection due to epistatic interactions, or frequency-dependent selection against rare alleles. The reduced fitness of individuals of mixed ancestry has been documented (or inferred) in many examples of natural hybridization (see Table 2). Consequently, many hybrid zones may be maintained in this way (Barton and Hewitt 1981*b*, 1985, 1989). In contrast to models of hybrid zone maintenance in which the relative fitness of genotypes depends on patterns of environmental variation (i.e. the hybrid superiority model and the gradient and mosaic models), models based on hybrid unfitness assume that fitness values are independent of the environment (Barton and Hewitt 1981*b*). Therefore, 'tension zones' can move (Key 1968; Bazykin 1969; Barton 1979*b*). They will tend to come to rest in areas where natural barriers occur and/or where population density is low (Hewitt and Barton 1980). For this reason, tension zones are likely to become associated with ecotones or ecological discontinuities and to remain trapped by such density troughs for long periods of time (Hewitt 1988). This means that such associations cannot be taken as evidence for a direct effect of the environment.

Single-locus models suggest that a cline set up by selection against hybrids will often be indistinguishable from a cline due to an underlying environmental gradient (Barton and Hewitt 1985, 1989). However, Barton and Hewitt (1985, 1989) give three reasons why they believe a majority of hybrid zones are tension zones maintained by hybrid unfitness. First, hybrids are often less fit than parental types. Secondly, even if initial divergence is a result of adaptation to different environments, differential co-adaptation is likely to evolve (Clarke 1966; Endler 1977). Thirdly, the observation of concordant clines for many independent characters is consistent with expectations for tension zones but is not easily explained by models that invoke a direct response to the environment. In the latter case, concordant clines would imply that each of the independent characters exhibits an identical response to the environment. For example, in one of the best documented cases, the *Bombina bombina*/*B. variegata* hybrid zone (Fig. 1c; Szymura and Barton 1986; Szymura *et al.* 1986), natural selection would have to produce parallel changes in morphology, alleles at 5 enzyme loci and mtDNA genotype. Independently maintained clines would tend to attract (Slatkin 1975), but this is a relatively weak force.

Frequency-dependent selection against the rare form can also maintain a stable hybrid zone (Mallet 1986; Mallet and Barton 1989*a*,*b*). At the level of individual alleles, results from models of frequency-dependent selection are identical to those for the case of hybrid unfitness (Mallet 1986). Frequency-dependent selection is certainly important in hybrid zones between color-pattern races of *Heliconius* butterflies (Fig. 1e), a premiere example of a Mullerian mimicry system in which the color pattern serves as a 'warning' to avian predators (Mallet 1986; Mallet and Barton 1989*a*). Selection against rare forms may also operate in hybrid zones involving populations that differ in sexually selected characters (Moore 1981, 1987).

4.2 Are hybrid zones stable?

In theory, hybrid zones can be stable over long periods of time. In fact, the evidence for long-term stability of hybrid zones is meager, with the same few examples cited repeatedly. The absence of convincing data may simply reflect the difficulty of deciphering the history of hybrid zones. However, it is clear that many zones could not have occupied their present positions during the last major glaciation. These zones may have formed since the last ice age (the traditional secondary contact scenario), or they may have 'moved' as the climate ameliorated.

In a few cases, evidence for stability comes from the current positions of hybrid zones that are now discontinuous. Chromosome races of *Warramaba viatica* form narrow hybrid zones on Kangaroo Island and on the adjacent Australian mainland (White 1978). The configuration of the zones implies that they must have been in the same position during the last ice age, when Kangaroo Island was connected to the mainland (it has subsequently been cut off by the rise in sea level). Similar evidence exists for the *Podisma pedestris* hybrid zone in the French Alps (Hewitt 1989).

For animals that were collected by early naturalists/explorers (e.g. birds and butterflies), collections and records from the eighteenth and nineteenth centuries provide evidence for stability over the past few hundred years. The Northern flicker hybrid zone (Fig. 1d) has apparently not moved since the late nineteenth century (Moore and Buchanan 1985). *Heliconius* butterflies have been hybridizing in South America for at least 200 years (Turner 1971).

Other claims for hybrid zone stability are not compelling. One frequently-cited example of a stable hybrid zone is that between *Sceloporus woodi* and *S. undulatus* in Florida (Jackson 1973). These lizards occur in different habitats with hybrid zones found along ecotones. The argument for stability is simply that the two vegetation types have probably been present in Florida for at least 100 000 years. But the availability of suitable habitat provides no information about the presence of the lizards, their past distributions, or the existence of hybrids. Many other claims of 'ancient' hybrid zones are simply plausible scenarios based on suspected late Pleistocene and post-Pleistocene events. In arguing for the antiquity of the flicker hybrid zone in the Great Plains, Rising (1983, p. 151) writes that 'since the climate there has not changed greatly in the past 3000 years or so. . . there is no reason to doubt that. . . [the flickers] have been in contact there for at least that long'. Unfortunately, the paucity of real evidence will not be easy to remedy.

Some hybrid zones are known to have moved. Direct evidence comes from comparisons of specimens collected along the same transects at different times, e.g. orioles and buntings in the Great Plains (Rising 1983) and grackles in Louisiana (Yang and Selander 1968). Displacement of these zones is apparently not accompanied by any change in width. Changes in the avian hybrid zones appear to represent replacement of one form by the other, suggesting that there are selective differences between the parental types (Rising 1983).

4.3 Evolution within hybrid zones

Stable hybrid zones can be maintained by natural selection or by a balance between selection and dispersal. Of the models considered above, only tension zones are in any way independent of the environment. The hybrid superiority model and the gradient/mosaic models result in stability only if the environment remains constant. Moreover, 'pure' tension zones, in which selection coefficients are completely invariant, are useful theoretical constructs, but probably do not occur in the real world. The temporal heterogeneity of the environment is certain to have a major impact on hybrid zone structure and dynamics. The persistence of hybrid zones over long periods of evolutionary time (millions of years) is highly improbable (and in some cases clearly impossible: see Hewitt 1989). The real issue is whether hybrid zones are likely to change over hundreds or thousands of generations (a time-scale appropriate for population genetics models).

Hybrid zones may be modified in several ways:

1. Reinforcement results in the evolution of pre-mating barriers leading to elimination of hybridization and the formation of 'good' species.
2. Introgressive hybridization results in fusion, the formation of a single homogeneous (perhaps polymorphic) population (this includes the case of neutral diffusion). Selection may, in principle, facilitate the process, through an increase in frequency of recombinant genotypes that serve as a bridge between the two parental types or through the spread of modifiers that reduce hybrid unfitness.
3. Competition leads to extinction of one of the two parental types.

Hybrid zone interactions have frequently been seen as a 'race' between fusion and speciation (Sibley 1957; Wilson 1965; Remington 1968; Crosby 1970; Paterson 1978; Templeton 1981), with the outcome depending on the fitness of hybrids and the initial level of positive assortative mating (see Wilson 1965). The arguments for and against reinforcement are discussed in detail in a later section.

Within a tension zone, selection can lead to an increase in the frequency of those variants that show the least reduction in viability and fertility. This has apparently occurred in hybrid zones between chromosomally differentiated forms in which hybrids have reduced fertility. *Clarkia nitens* and *C. speciosa* differ by 6–7 reciprocal translocations and hybrids have very low fertility because of non-disjunction in meiosis (Bloom 1974). Where the two species hybridize in California, new reciprocal translocations have arisen that, when back-crossed to either parent, produce progeny with fertilities much higher than that of the F1 (Bloom 1976; Hauber and Bloom 1983). The hybrid zone involves clinal distribution of these new chromosome arrangements, acting as a bridge between the two species and apparently leading to extensive introgression. In the shrew

Sorex araneus, chromosomal races are characterized by different Robertsonian fusions (Searle 1984). Near Oxford, two races abut, producing some hybrids that are double Robertsonian heterozygotes with monobrachial homology (expected to have reduced fertility) (Searle 1986). However, the hybrid zone contains a high frequency of acrocentric karyotypes, which when crossed with either of the parental metacentric karyotypes produce offspring with less fertility loss. Because the acrocentric karyotypes are uncommon outside of the hybrid zone, the pattern is thought to be a consequence of selection (Searle 1986, 1988). As with the *Clarkia* example, evolution within the hybrid zone has apparently led to a weakening of barriers to gene exchange.

Fusion and extinction are not necessarily mutually exclusive outcomes to hybrid zone interactions. To understand the relationship requires merging ecological and genetical perspectives. When 'good' species interact (in the absence of any hybridization), ecological tolerances and competitive relationships determine patterns of co-existence and extinction. In hybrid zones, there is the additional complication that the 'species' are not distinct entities, that recombinant types are produced. Extinction, therefore, must be viewed not in terms of species or multi-locus genotypes, but in terms of individual alleles (or chromosomal segments). Introgressive hybridization can lead to the fusion of the two gene pools, but such fusion will almost certainly involve selective allelic extinction as well. As will become evident, differential introgression is a common characteristic of hybrid zones. Relatively few hybrid zones appear to represent simple neutral diffusion following secondary contact.

5. INTROGRESSION: PATTERN AND PROCESS

The term 'introgressive hybridization' was first used by Anderson and Hubricht (1938) to describe the consequences of repeated back-crossing of hybrids to the parental types. It has come to be defined as the incorporation of alleles of one species (subspecies, race) into the gene pool of another and is not always strictly limited to cases of back-crossing (see Heiser 1973). To demonstrate introgression requires at least two independent markers, one of which must be used to define the parental types. When multiple markers distinguish allopatric populations of two taxa, one or more of these must be chosen to serve as a point of reference for comparison of patterns of variation. If two hybridizing taxa differ in morphology, allozymes and mtDNA, then comparisons of the extent of introgression of allozymes and mtDNA must be made with respect to morphological differences. Alleles determining the morphological characters are assumed to provide an appropriate frame of reference, i.e. not to be introgressing themselves. Alternatively, mtDNA genotype could be held constant and the extent to which morphological characters and allozyme variants become associated with different maternal lineages could be assessed. As in the theory of relativity, the frame of reference is all important.

Patterns of variation that appear to reflect introgression may have other explanations (Heiser 1973). (The problem of inferring process from pattern again rears its ugly head!) The most obvious difficulty is that the presence of allele **A**, characteristic of species 1, in populations of species 2 (nearly fixed for allele **B**), could be a consequence of introgression but could also reflect either the retention of a polymorphism found in the common ancestor or an independent mutation event. Analyzing patterns of spatial variation provides the only simple approach for resolving the issue. If the frequency of allele **A** in species 2 is highest near the hybrid zone and declines away from the zone, the pattern is most likely due to introgression. Neither of the other explanations predict this pattern. If the allelic markers are DNA sequence variants, then analysis of neighboring (closely linked) sequences can provide a method for ruling out the parallel mutation hypothesis.

The extent of introgression varies considerably, among hybrid zones and among markers within hybrid zones (Table 1). These patterns will be discussed in greater detail in the context of understanding the nature of 'species boundaries'. Introgression has been documented in many different hybrid zones and is not restricted to certain taxonomic groups.

Of great interest to many of the early students of hybridization was whether introgression was an important source of genetic variation for natural populations (relative to mutation). The matter was of particular interest to botanists, as hybridization appears to be such a common phenomenon in plants. Anderson (1953) was the strongest proponent of the importance of introgression, arguing that it was far more important than mutation. Stebbins (1950) was more circumspect, and indeed seems to have taken both sides. At one point, he suggests that introgressive hybridization 'represents the crossing of genes from one "adaptive peak" to another and makes possible the formation of gene combinations capable of climbing new peaks' (p. 278), but on the next page comments that introgression is 'not a way of producing new morphological or physiological characteristics and therefore of progressive evolution'.

Clearly, recombination is an important force in evolution. Hybridization and introgression increase the 'field of recombination' by bringing together differentiated genomes. If these genomes are 'co-adapted', recombination is unlikely to produce successful new variants except perhaps in a disturbed or changing environment. That 'hybrid swarms' or populations of stable introgressants often appear to dominate in certain habitats (e.g. Epling 1947; Levin 1963) suggests that introgression may have a significant role in the evolution of some groups of plants. Many hybrid zones, however,

Table 1
Patterns of introgression in hybrid zones

Species[a]	Markers[b]	Differential introgression[c]	Clines displaced	Asymmetry	References
Plants					
Clarkia nitens × *C. speciosa*	m, c	m > c	Yes	—	Bloom (1976), Hauber and Bloom (1983)
Impatiens capensis × *I. ecalcarata*	m, e	yes	—	—	Ornduff (1967)
Phlox drummondi × *P. cuspidata*	a	a^*	—	Yes	Levin (1975)
Insects					
Warramaba viatica	m, c	m > c	—	—	Key (1974)
Caledia captiva	c, a, mt, r	$mt = r > a^* >> c$	Yes	Yes	Moran (1979), Moran *et al.* (1980), Marchant *et al.* (1988)
Podisma pedestris	c, d	c = d	No	No	Barton (1980), Barton and Hewitt (1981*a*)
Chorthippus parallelus	m, a	a > m	Yes	—	Butlin and Hewitt (1985a)
Gryllus firmus × *G. pennsylvanicus*	m, a, mt	$mt > a^* > m$	—	Yes	Harrison (1986), Harrison *et al.* (1987)
Crustaceans					
Menippe mercenaria × *M. adina*	m, a	$a^* > m$	No	Yes	Bert and Harrison (1988)

Fish					
Notropis cornutus × *N. chrysocephalus*	m, a, mt	Yes	—	Yes	Dowling *et al.* (1989)
Cyprinodon pecoensis × *C. variegatus*	a	No	No	Yes	Echelle and Connor (1989)
Amphibians					
Geocrinia laevis × *G. victoriana*	s, a	a > s	Yes	Yes	Littlejohn and Watson (1985)
Littoria ewingi × *L. verreauxi*	m, s, a	a > m = s	No	No	Littlejohn and Watson (1985)
Littoria ewingi × *L. paraewingi*	s, d, a	a > d = s	No	No	Gartside (1972), Littlejohn and Watson (1985)
Pseudophryne bibroni × *P. marmorata*	m, d, a	a = d > m	Yes	Yes	McDonnell *et al.* (1978)
Rana berlandieri × *R. sphenocephala*	a	No	No	No	Sage and Selander (1979), Kocher and Sage (1986)
Bombina bombina × *B. variegata*	m, a, mt	m = a = mt	No	No	Szymura and Barton (1986), Szymura *et al.* (1986)
Dicamptodon ensatus × *D. tenebrosus*	a	No	No	No	Good (1989)
Ranidella insignifera × *R. pseudoinsignifera*	s, m, a	s = m = a	No	No	Blackwell and Bull (1978)

(*continued*)

Table 1 *continued*

Species[a]	Markers[b]	Differential introgression[c]	Clines displaced	Asymmetry	References
Birds					
Quiscalus quiscula	m	No	No	Yes	Yang and Selander (1968)
Molothrus ater	m	No	No	—	Fleischer and Rothstein (1988)
Mammals					
Geomys bursarius	c, a, mt, r	—	No	—	Baker *et al.* (1989)
Geomys bursarius × *G. lutescens*	m, c, a	—	No	No	Heaney and Timm (1985)
Uroderma bilobatum	c, a	c = a	No	Yes	Baker (1981), Greenbaum (1981)
Thomomys bottae	m, c, a	a > m	—	No	Patton *et al.* (1979)
Thomomys bottae	m, c	—	No	No	Haffner *et al.* (1983)
Peromyscus leucopus	c, a, mt	c = a = mt?	No	Yes	Stangl (1986), Nelson *et al.* (1987)
Mus musculus × *M. domesticus* (Denmark)	m, a, mt	mt > a*	No	Yes	Hunt and Selander (1973), Ferris *et al.* (1983)
Mus musculus × *M. domesticus* (Bulgaria)	a, mt, Y	mt = a* > Y	No	Yes	Vanlerberghe *et al.* (1986)

[a] Where a single species name is given, the hybrid zone involves an interaction between forms that are considered to be subspecies or races.
[b] The following abbreviations are used: m, morphology; s, song; d, development; e, ecological characteristics; c, chromosomes; a, allozymes; mt, mitochondrial DNA; r, ribosomal DNA; Y, Y-chromosome markers.
[c] An asterisk indicates that there is heterogeneity in the extent of introgression within a particular class of markers (e.g. allozymes). A dash means that no information is available or that the question cannot be answered.

appear to represent substantial barriers to gene exchange, with limited introgression occurring in the face of selection against hybrids or recombinants. Patterns of introgression will reflect the genetic architecture of these barriers. Although hybrid zones are often significant barriers for neutral alleles, alleles with even a small selective advantage (independent of genetic and external environments) can penetrate relatively quickly (Barton 1979*a*; Barton and Hewitt 1985).

6. HYBRID ZONES AND THE NATURE OF SPECIES

6.1 Species concepts

Species are considered fundamental units in the evolutionary process, but there is no consensus about how they should be defined. Indeed, the debate about species concepts seems to have intensified in recent years. The biological species concept (BSC) has long held the high ground in battles over species concepts, but has been under constant attack from many sides. The essence of the BSC is that species are groups of populations, isolated from other such groups by one or more barriers to gene exchange (Mayr 1963, Dobzhansky 1970). Because the definition focuses on genetic or reproductive isolation as the criterion for establishing species boundaries (and because other definitions are equally 'biological'), Paterson (1980, 1982*a*,*b*) has termed this the isolation concept (IC). He strenuously objects to this view of species, arguing that 'isolation' is not a property of species and that the term 'isolating mechanism' (often used by some advocates of the IC) implies a process (reinforcement) that does not occur. In its place, he offers the recognition concept (RC), in which species are defined as groups of individuals that share a common fertilization system or a common specific-mate recognition system (SMRS) (Paterson 1982*a*,*b*, 1985). Paterson and his disciples (e.g. Lambert *et al.* 1987; Masters *et al.* 1987) have touted the RC as a great advance over the IC, but the advantages of the RC seem greatly exaggerated and recent critiques (Coyne *et al.* 1988; Chandler and Gromko 1989) seem very much to the point.

Several issues are of particular interest for this discussion. The IC is found wanting because species are defined in relation to other species (from which they are isolated). But, how does one decide which individuals actually share a 'common' fertilization system or SMRS and which individuals fall outside this grouping? In practice, the boundaries will be defined in much the same way whether one applies the IC or RC, i.e. the two concepts are 'equally relational' (Coyne *et al.* 1988). Proponents of the RC desire to make species 'self-defining' (Lambert *et al.* 1987), with the unfortunate consequence that post-mating barriers to gene exchange (hybrid sterility and inviability) are not relevant for defining species. Yet, such barriers are clearly as effective in maintaining independent gene pools as are differences in SMRSs.

Hybridization and hybrid zones represent cases of partial (incomplete) reproductive or genetic isolation, and therefore are not easily accommodated within the IC. Hybrid zones also pose serious problems for the RC. In some zones, groups of individuals apparently share a common SMRS (i.e. there is no evidence for positive assortative mating) and yet remain distinct. In other hybrid zones, there is positive assortative mating but it is not complete. In these cases, it is difficult to judge whether or not the hybridizing forms share a common SMRS.

Tempo and mode of speciation will influence the probability that partial barriers to gene exchange will be encountered in studies of natural populations. During allopatric divergence, disturbance or climatic change may bring diverging populations into secondary contact before reproductive isolation is complete. For parapatric divergence, hybrid zones will always be a stage in the speciation process (see Fig. 1.2 in Endler 1977). Because hybrid zones can be stable for long periods (at least within the context of a stable environment), the expectation that all populations can be assigned to one species or to two (on the basis of reproductive isolation or continuity) is naive.

Hewitt (1989) suggests that many species are subdivided by hybrid zones into 'compartments' between which gene exchange is severely restricted. Although not stated so explicitly, similar views were expressed by earlier evolutionary biologists. The 'rassenkreis' (Rensch 1959) of the zoologist and the 'syngameon' (Grant 1981) of the botanist are essentially units above the species level within which semi-permeable barriers occur between component subunits.

Clearly, one remedy for these problems is to abandon reproductive continuity (discontinuity) as the principal or sole criterion for defining species. The cohesion concept (Templeton 1989) and the phylogenetic species concept (Cracraft 1983, 1989), although they differ dramatically in their approach, both claim to avoid the shortcomings of the IC and RC in dealing with cases of natural hybridization.

It is not my intent to evaluate alternative species concepts. However, an analysis and understanding of hybrid zone pattern and process can help to clarify the nature of species and thereby aid in the development of appropriate species concepts. I will review a number of important lessons that emerge from a careful reading of the hybrid zone literature.

6.2 Reproductive isolation and genetic isolation

Hybrid zones provide unequivocal evidence that genetic isolation (absence of genetic exchange) is not necessarily equivalent to reproductive isolation (absence of interbreeding) (Bigelow 1965; Hunt and Selander 1973; Hall and Selander 1973). The clearest example (and perhaps a trivial one) is when two 'species' hybridize freely (no positive assortative mating), but produce F1 that are completely sterile. In this case, no gene exchange will occur. In many hybrid zones, effective pre-mating barriers are apparently absent and individuals of mixed ancestry are common. Yet introgression is often very limited, i.e. outside of a narrow zone the hybridizing taxa remain distinct (e.g. Hall and Selander 1973; Patton *et al.* 1979). In such situations, the designated status of the hybridizing populations (species, subspecies, races) is frequently a function of the extent of differentiation and whether the taxa were recognized as distinct prior to the discovery of the hybrid zone (in which case they are more likely to be called 'species'). There is little consistency among taxonomists (and even considerable disagreement among those working on the same group, e.g. witness the debate about whether *Mus musculus* and *M. domesticus* should be considered species or subspecies). Proponents of the RC would certainly argue that the hybridizing taxa are conspecific when patterns of mating reveal that they share a common SMRS. Many would disagree. For example, Hall and Selander (1973) propose that 'the important criterion of species status is not the absence of interbreeding *per se* but, rather, of genetic exchange between populations'. In a similar vein, Carson *et al.* (1989) conclude from studies of hybridization in Hawaiian *Drosophila* that 'the coherence of each gene pool is a much better basis for the characterization of a species than reproductive isolation, especially during periods of initial divergence of the populations'.

In addition to the debate about the status of hybridizing populations, there is also a debate about the evolutionary forces acting to restrict genetic exchange. One very popular view has been that co-adaptation (in response to selection pressures imposed both by the internal and external environments) prevents the flow of alleles from one stable equilibrium to the other.

> In. . . [hybrid zones] it appears that the gene-complexes which come together are so well balanced within themselves that combinations with alien genes lead to combinations of inferior viability and are eliminated by selection. This counter-selection reduces introgressive gene-flow drastically. (Mayr 1954, p. 166).

> . . . it has several times been shown that the intermediates produced by crossing are restricted to a narrow belt. . .. For almost by definition, the mixture between two gene-complexes, each harmoniously but differently adapted by selection, will be at a disadvantage compared with either one of them. (Ford 1954, p. 105).

> Selection seems to have very effectively weeded out those migrant genes that left their own well-integrated, harmonious gene constellations to threaten the well-integrated (but different) harmony of the other. (Bigelow 1965, p. 450).

Unfortunately, there is as yet little direct evidence for co-adaptation or for its role in limiting genetic exchange. Analysis of hybrid inviability in a grasshopper (*Podisma pedestris*) hybrid zone suggests that most (87 per cent) of the observed reduction in fitness is due to the break-up of co-adapted gene complexes and not simply to heterozygous disadvantage (Barton and Hewitt 1981*a*). Detailed studies of introgression (reviewed below) reveal consistent patterns in the extent to which alleles can penetrate hybrid zones. This phenomenon of differential introgression certainly implicates selection as an important force, but does not directly support the notion of co-adaptation. More elaborate studies of the distribution of multi-locus genotypes within and adjacent to hybrid zones may begin to provide the data needed to resolve this issue.

6.3 Semi-permeable barriers to gene exchange

Consider a hybrid zone between two taxa that exhibit fixed allelic differences at some arbitrary number of gene loci. For each of these loci, selective constraints and linkage relationships will determine the magnitude of gene flow across the hybrid zone (Bazykin 1969; Barton 1979*a*, 1983; Barton and Hewitt 1985). Selection against heterozygotes or selection that favors different alleles in different environments will limit introgression at the loci in question. The flow of neutral alleles will also be retarded to the extent that they are linked to loci under selection. To cross a barrier successfully, 'a neutral allele must recombine into the new genetic background before it is eliminated by selection against the alleles with which it is initially associated' (Barton and Hewitt 1985, p. 120). Introgression of neutral alleles embedded in 'co-adapted gene complexes' or linked to genes that cause positive assortative mating will likewise be impeded. Linkage relationships may not significantly reduce the flux of neutral alleles if gene conversion between allelic forms is an important evolutionary force. Gene conversion events involving only short DNA segments can introduce an allele into a new genetic background without introducing closely linked markers. Therefore, under gene conversion models, linkage to loci under selection will not retard the flow of a neutral allele across a hybrid zone (as it will in models that invoke more traditional recombination events).

The consequences of hybridization and recombination are that genomes are fragmented and the contents shuffled. The probability of producing new allelic combinations is a function of patterns of mating and recombination. The subsequent success of these combinations depends on the nature of natural selection (e.g. the distribution of underdominant loci, the extent of co-adaptation, etc.). If loci under selection are not uniformly distributed or if selection coefficients vary, the effectiveness of a hybrid zone as a barrier to genetic exchange will not be a constant for all loci (even for all neutral markers). Boundaries are, therefore, semi-permeable, the permeability depending on the genetic marker (Key 1968; Harrison

1986). Genetic isolation must be considered as a property of individual genes (or chromosome segments), not as a characteristic of entire genomes (Bazykin 1969; Barton and Hewitt 1981*b*). Carson (1975) suggested that diploid, sexually reproducing species may have two 'systems of genetic variability', open and closed. The former 'could accept genetic components due to gene flow. . . even through introgressive hybridization', whereas the latter would 'remain rigidly resistant to disintegration during the introgressive process'. I see no reason to partition the genome into two components, but Carson's open and closed systems represent ends of the continuum defined above.

Patterns of variation provide evidence that hybrid zones are semi-permeable barriers. Although coincident clines of equal width are observed in some cases (e.g. *Bombina bombina/B. variegata*, Fig. 1c), there are many examples of differential introgression, characterized by variation in the extent to which alleles penetrate the hybrid zone (Table 1, Fig. 1b). Where hybrid zones are not simple clines, evidence for semi-permeability comes from the non-random distribution of recombinant types within and immediately adjacent to the zone (e.g. *Phlox drummondi/P. cuspidata, Gryllus pennsylvanicus/G. firmus, Notropis cornutus/N. chrysocephalus*; see Table 1). When differential introgression is observed, alleles at loci coding for soluble enzymes tend to introgress more than morphological characters, suggesting that selection on the enzyme loci is weak. Clines for enzyme markers are usually broader than those for morphological traits and are sometimes displaced. However, there is often marked variation among allozymes in the extent of introgression. Within this class of markers there must be substantial differences in the constraints imposed by selection. This does not imply selection acting on the enzyme loci themselves, as the pattern can be produced by variation in selection at closely linked loci.

Mitochondrial DNA (mtDNA) segregates independently of all nuclear genes and therefore may be able to flow across a hybrid zone more rapidly than most nuclear gene markers (Barton and Jones 1983). This argument assumes that many genes contribute to the hybrid zone barrier, so that neutral markers encoded in the nucleus (allozymes?) are likely to be closely linked to one or more of these genes. Studies of the distributions of mtDNA genotypes in examples of natural hybridization do not reveal a consistent pattern. In some cases, there appears to be differential introgression of mtDNA; in others, patterns of mtDNA variation are perfectly concordant with those for nuclear gene markers (Table 1; Harrison 1989). In the *Mus musculus/M. domesticus* hybrid zone, fixation of *domesticus* mtDNA in *musculus* populations has occurred in Scandinavia (Ferris *et al*. 1983; Gyllensten and Wilson 1987), but in Bulgaria the extent of mtDNA introgression is no greater than that observed for several enzyme markers (Vanlerberghe *et al*. 1986). The most likely explanation is that the Scandinavian situation reflects a founder event during the initial colonization of Sweden by house mice. Clearly, nuclear gene markers vary in the extent of introgression; mtDNA is likely to fall at one end of the continuum (Harrison *et al.* 1987).

For any single genetic marker, asymmetric introgression might represent the wave of advance of an advantageous allele. However, this mechanism cannot easily explain situations in which many markers show the same asymmetric pattern, because this requires that advantageous alleles always appear first on one side of the hybrid zone. Consistently asymmetric introgression is seen in a number of well-studied hybrid zones (Table 1). In *Caledia, Quiscalus, Geocrinia* and *Pseudophryne*, the asymmetry may have been produced by the recent movement of the boundary, leaving in its wake a 'trail' of neutral or weakly selected markers. Recent movement would explain why all markers show similar asymmetric patterns. Differences in dispersal tendencies between the two hybridizing taxa can also lead to a fundamental asymmetry in the structure of a hybrid zone.

7. HYBRID ZONES AND THE NATURE OF BARRIERS TO GENE EXCHANGE

Hybrid zones provide evolutionary biologists with unique opportunities to study barriers to genetic exchange. Through a combination of field studies and laboratory crosses, it is possible to assess the relative importance of pre-mating and post-mating barriers, to determine how specific ecological and behavioral differences contribute to reproductive isolation (or serve as components of the mate recognition system), and to investigate the genetic basis of isolation. Although the differences between hybridizing taxa may in some cases be different from those that characterize 'good' species, the insights gained from the analysis of hybrid zones are certainly relevant to understanding the speciation process.

It is convenient, for purposes of discussion, to follow the traditional classification of barriers to gene exchange established by the strong adherents of the BSC (e.g. Mayr 1963). Post-mating (or post-zygotic) barriers are those that result in hybrid unfitness (or, more generally, the reduced fitness of individuals or mixed ancestry). This may be manifest as a decrease in either fertility, viability or reproductive success of F1, F2 or back-cross individuals. Pre-mating barriers are those that reduce the probability that two individuals will meet and mate, i.e. barriers resulting from differences in mate recognition systems, phenology or habitat/resource associations.

7.1 Fitness of individuals of mixed ancestry

The reduced fitness of individuals of mixed ancestry may be independent of the environment, the result of either (1) irregularities in meiosis or (2) problems in development (often in embryogenesis) in heterozygous and recombinant genotypes. Individuals of mixed ancestry may also be at a disadvantage only in the context of specific environments, e.g. they may be more susceptible to predators or pathogens. Finally, hybrids and backcrosses may be less successful in attracting mates.

The claim that most hybrid zones are tension zones depends on indirect inference from patterns of variation as well as on direct demonstrations of the reduced fitness of hybrids (Barton and Hewitt 1985). Evidence for hybrid unfitness comes from:

1. Deficiencies of F1 hybrids or recombinant genotypes within hybrid zones (this assumes no assortative mating, no assortative fertilization, and little or no migration).
2. Higher frequencies within hybrid zones of phenotypes with reduced survivorship or reproductive success (e.g. developmental anomalies, high frequencies of meiotic non-disjunction, increased susceptibility to predation or parasitism).
3. Reduced viability and/or fertility of hybrids from laboratory crosses.
4. Changes in genotype frequencies within a single generation in hybrid zone populations (cohort analysis).

Table 2 summarizes data from selected examples.

Chromosomal hybrid zones, those initially identified on the basis of karyotypic differences, are prominently represented (e.g. *Clarkia, Sceloporus, Caledia, Podisma*). However, only in *Clarkia* is hybrid unfitness due to meiotic abnormalities attributable to karyotypic heterozygosity and in this case there appears to have been selection within the hybrid zone for new rearrangements that reduce the strength of the barrier. It should not be assumed that chromosomal differences contribute directly to the post-mating barrier, although they are frequently good markers for the differentiated taxa. Shaw (1981) reviewed five cases of chromosomal hybrid zones in orthopteroid insects and concluded that 'the chromosomal rearrangement differences. . . offer minimal isolation via mechanical impairment of meiosis'.

Studies comparing the fitness of parentals and hybrids sometimes reveal no differences. For example, Bull (1979) found no reduction in fertility or viability in laboratory crosses between two hybridizing species of frogs (*Ranidella*). Moore and Koenig (1986) saw no evidence of reduced reproductive success of hybrids at four sites across the northern flicker hybrid zone. It is not clear what to make of such data. It can be argued that laboratory studies may not reveal differences that are important in the field

Table 2
Hybrid zones for which there is evidence that individuals of mixed ancestry have reduced fitness

Species[a]	Evidence	References
Plants		
Clarkia speciosa × *C. nitens*	Meiotic non-disjunction and reduced pollen stainability in translocation heterozygotes	Bloom (1974)
Liatris spp.	Reduced pollen fertility	Levin (1967)
Gaillardia pulchella	Reduced pollen fertility in artificial hybrids	Heywood (1986)
Populus fremontii × *P. angustifolia*	Hybrids more susceptible to aphid pests	Whitham (1989)
Populus × *parryi*	More pests and pathogens on hybrid than on either of the parental species	Eckenwalder (1984)
Insects		
Warramaba viatica	Meiotic abnormalities	Mrongovius (1979)
Podisma pedestris	Reduced viability of F1, reduced viability of insects from center of hybrid zone	Barton (1980), Barton and Hewitt (1981a)
Caledia captiva	Inviability of F2, reduced viability of back cross	Shaw and Wilkinson (1980), Shaw *et al.* (1982), Coates and Shaw (1982)
Chorthippus parallelus	Hybrid male sterility, reduced fertility of back cross males	Hewitt *et al.* (1987)
Limnoporus dissortis × *L. notabilis*	Reduction in egg hatch in F1 and back cross	Sperling and Spence (1991)

Fish		
Notropis cornutus × *N. chrysocephalus*	Decrease in proportion of hybrids in successive age classes	Dowling and Moore (1985)
Amphibians		
Geocrinia laevis × *G. victoriana*	Higher percentage of abnormal embryos in the hybrid zone	Littlejohn *et al.* (1971)
Littoria ewingi × *L. paraewingi*	Poor egg hatching, high proportion of abnormal embryos (only in *L. p.* male × *L. e.* female) in lab hybrids	Watson (1972), Watson and Littlejohn (1978)
Pseudophyrne bibroni × *P. semimarmorata*	Higher percentage of egg mortality in egg masses from the hybrid zone	Woodruff (1979), McDonnell *et al.* (1978)
Rana berlandieri × *R. sphenocephala*	Decrease in proportion of hybrids within a single cohort	Kocher and Sage (1986)
Bombina bombina × *B. variegata*	Higher embryonic mortality in hybrid zone	Szymura and Barton (1986)
Reptiles		
Sceloporus grammicus	No F2 or B2 genotypes in hybrid zone	Hall and Selander (1973)
Birds		
Sturnella magna × *S. neglecta*	Infertility of lab hybrids	Lanyon (1979)
Mammals		
Mus musculus × *M. domesticus*	Higher parasite loads in hybrid zone	Sage *et al.* (1986)
Geomys bursarius	Deficiency of F2 and back cross in hybrid zone	Tucker and Schmidly (1981)

[a]Where a single species name is given, the hybrid zone interaction involves forms that are considered subspecies or races.

and that field studies focus (as they often must) on only one component of fitness. Estimating fitness in the field is notoriously difficult and, consequently, subtle differences between hybrids and parental types are unlikely to show up unless sample sizes are large. Therefore, it is exceedingly difficult to prove that individuals of mixed ancestry are *not* at a disadvantage. Because negative data are less likely to get published, a survey of the literature will not be a reliable indicator.

Although cohort analysis provides a direct way of assessing the relative viabilities of different genotypes in natural populations, the approach has seen only limited use. In the two reported cases, shifts in the frequencies of phenotypic classes are consistent with relatively strong selection against individuals of mixed ancestry (Dowling and Moore 1985; Kocher and Sage 1986). It is surprising that this technique has not beem employed more frequently. It certainly will be productive in studies of plant hybrid zones and in animal species for which cohorts can be clearly identified.

7.2 Positive assortative mating in hybrid zones

Examples of positive assortative mating within hybrid zones are remarkably rare. Hybridizing taxa often differ in identifiable components of communication systems – advertisement calls in frogs and toads (Blair 1941; Bull 1978; Littlejohn and Watson 1985; Hillis 1988), calling songs in (Howard and Furth 1986; Butlin and Hewitt 1985*b*) – but in these taxa there is no data on the extent of assortative mating where the two parental types occur together. Hybridizing subspecies of northern flicker differ in plumage characteristics and hybridizing chickadees differ in song, yet in both cases mating within the hybrid zone appears to be at random (Moore 1987; Robbins *et al.*, 1986). In soldier beetles *Chauliognathus pennsylvanicus*, there is clear evidence for positive assortment with respect to elytral spot pattern at some localities but apparently random mating at other sites (McLain 1985, 1988; Kochmer, unpublished). Positive assortative mating has been documented in a field cricket (*Gryllus*) hybrid zone, based on comparisons of genotypes of field-inseminated females and their offspring (Harrison 1986). Again, this pattern of mating is only evident in some populations. In both beetles and crickets, the results of laboratory mate preference experiments are consistent with observations from the field. In the grasshopper *Chorthippus parallelus*, laboratory tests result in assortative mating, but the absence of natural contact between parental types prevents evaluation of assortative mating in the field (Butlin 1989). Laboratory mate selection experiments involving chromosomal races of the mole rat *Spalax ehrenbergi* show that females prefer males of their own chromosomal form (Nevo and Heth 1976). This may account for significant deficiencies of heterozygotes in the hybrid zone (Nevo and

Bar-El 1976), although other explanations (e.g. hybrid unfitness, Wahlund effect) cannot be excluded.

The paucity of examples of behavioral isolation is puzzling. Coyne and Orr (1989) have shown that pre-zygotic isolation (mating discrimination) and post-zygotic isolation evolve at about the same rate in allopatric species pairs of *Drosophila*. The data from hybrid zones appear to suggest that post-zygotic barriers evolve more rapidly. *Drosophila* may be unusual, or insects as a group may have more labile mate recognition systems than other animals. There are alternative explanations. First, data on assortative mating are difficult to obtain and relatively few hybrid zones have been examined closely. The observed pattern may be a consequence of the paucity of reliable data. Secondly, post-zygotic barriers may result in more stable hybrid zones (i.e. within hybrid zones post-zygotic barriers may decay more slowly than pre-zygotic barriers); those that persist will be a biased sample of those that are formed initially. Patterns of assortative mating within hybrid zones deserve more attention.

Arguments based on the economics of mate choice (Wilson and Hedrick 1982) can be invoked to explain a breakdown of reproductive isolation between 'good' species or a reduction in the level of positive assortative mating in hybrid zones. The expectation is that individuals of the rarer species will be more likely to hybridize. This appears to be a general pattern in fish hybridization (Hubbs 1955; Avise and Saunders 1984; Dowling *et al.* 1989). A similar argument has been used to explain hybridization between *Drosophila silvestris* and *D. heteroneura* on the island of Hawaii (Carson *et al.* 1989).

7.3 Temporal isolation

Life-cycle variation in animals and variation in flowering time in plants can lead to the temporal displacement of periods of reproductive activity. Seasonal isolation has been particularly well documented for frogs and toads, although it is rarely the only pre-mating barrier to gene exchange between a pair of species (Blair 1941; Frost and Bagnara 1977; Hillis 1981, 1988). At sites where habitat disburbance has led to the disruption of ecological barriers, differences in breeding season clearly restrict the amount of genetic exchange. In crickets (*Gryllus*), temporal isolation appears to be an important factor limiting genetic exchange in one segment of an extensive hybrid zone (Harrison 1985). Differences in these cricket life-cycles are genetically based and are probably adaptations to climatic differences between the coastal plain and upland sites occupied by 'pure' populations of the two species. If local environment molds the life-cycle, sympatric populations will tend to converge, resulting in a reduction in the extent of temporal isolation (i.e. a weakening of the pre-mating barriers).

In closely related plant species, the temporal displacement of flowering times is well documented, although there is often significant overlap so that hybridization occurs when other barriers break down, e.g. *Phlox* (Levin 1963) and *Liatris* (Levin 1967). In grass species growing across boundaries between pasture soils and soils contaminated with heavy metals, steep clinal variation in metal tolerance is paralleled by a variation in flowering time (McNeilly and Antonovics 1968). In this situation, differences in flowering time would appear to have arisen *in situ*, either as an adaptation to the local environment, an 'isolating mechanism', or both.

7.4 Ecological isolation and mosaic hybrid zones

In many hybrid zones, the interacting taxa exhibit clear habitat associations (Table 3). Examples have already been encountered in the context of discussion of how disturbance promotes hybridization. Disturbance often leads to a change in the distribution of habitat patches (usually resulting in many smaller patches and thus increasing the extent of the boundary across which two taxa can interact). It is evident that habitat segregation is not restricted taxonomically, as Table 3 includes studies of mammals, birds, reptiles, amphibians, fish, insects, crustacea and plants.

Ecological isolation is often a direct consequence of habitat association. However, habitat associations can arise as a result of differences in habitat preferences or as a result of selection within each habitat patch. In a habitat preference model, the association is a consequence of intrinsic differences in the behavior of the two genotypes, and clines much narrower than those maintained by dispersal/selection balance can be established. The alternative is that genotypes distribute themselves at random, and habitat association arises as a result of differential survival and/or reproductive success within patches. In this case, the extent of ecological isolation will be a function of dispersal, selection, and patch size and distribution (i.e. the graininess of the environment).

For only one of the examples in Table 3 is it clear that the observed habitat association is due to differences in habitat preference. Hagen (1967) used laboratory experiments to show that hybridizing species of sticklebacks have different preferences for substrate and vegetation. The choices exhibited by the two species in aquaria corresponded to the natural habitats in which they occur.

Mosaic hybrid zones occur when closely related species (races) that differ in habitat utilization patterns interact in a patchy environment (Rand and Harrison 1989; Harrison and Rand 1989). Many of the hybrid zones listed in Table 3 have this structure. As discussed above (and reviewed in detail in Harrison and Rand 1989), the dynamics of such zones may be significantly different from the dynamics of simple multi-locus clines. Hybridization will occur at the boundaries between the two patch types. Because many mosaic hybrid zones have a complex internal structure (reflecting the patchwork of habitats distributed across the landscape), local contacts may be effectively independent, with separate evolutionary trajectories. The fate of any interaction will depend on local selection pressures, random drift, and the extent of genetic exchange between patches. Because these vary from site to site within the hybrid zone, global extinction of either species (allele) is unlikely. In other words, patchy environments will tend to maintain diversity.

Mosaic hybrid zones have probably received less attention than they deserve, because, as patterns of variation, they require mapping of both genotype and habitat distributions and because mosaic structure is often not apparent from single linear transects. They are clearly useful natural laboratories for investigating the importance of ecological differentiation as a barrier to gene exchange.

7.5 The significance of pre-mating barriers to gene exchange

Two general conclusions emerge from a survey of factors determining patterns of interbreeding in hybrid zones. First, many zones are characterized by the existence of multiple barriers, which (in undisturbed situations) may lead to nearly complete reproductive isolation. However, these barriers are fragile and can easily be disrupted by environmental change. The consequences may be the direct elimination of the barrier (as in habitat disturbance) or a weakening of the barrier through the action of natural selection. Behavioral isolation, due to changes in the mate recognition system, is probably the most robust, but relatively few examples of positive assortative mating in hybrid zones are available for study.

The second conclusion, perhaps not obvious from the data summarized above, is that patterns of gene exchange often vary geographically. Such variation may be due to intrinsic factors (genetic variation within one or both of the interacting taxa) or to extrinsic factors (variation in the environmental context in which the taxa interact). Thus, in the field cricket, *Gryllus*, hybrid zone in eastern North America, temporal isolation is evident in Virginia but not in Connecticut, whereas behavioral isolation is important in Connecticut but not in Virginia (Harrison 1985, 1986; Harrison and Rand 1989). In the Mexican towhees studied by Sibley (1954; Sibley and West 1958), the two species hybridize extensively at some sites and remain completely distinct at others. The explanation given is that the extent of hybridization is a function of habitat disturbance. Geographic variation in the nature of hybrid zone interactions has also been documented for (1) the frogs *Geocrinia* (Littlejohn and Watson 1985), *Littoria* (Watson and Littlejohn 1978; Littlejohn and Watson 1985)

Table 3
Hybrid zones involving taxa that are ecologically differentiated.

Species[a]	Ecological differences	References
Plants		
Phlox drummondi × *P. cuspidata*	Moisture requirements	Levin (1975)
Liatris spicata × *L. aspera*	Soils with poor drainage and aeration/dry soils	Levin (1967)
Tradescantia subaspera × *T. canaliculata*	Full sun at top of cliffs/deep shade in rich soil at bottom	Anderson and Hubricht (1938)
Gaillardia pulchella	Carbonate-rich and carbonate-poor soils	Heywood (1986)
Ranunculus lappaceus group	Soil drainage, moisture regime, slope	Briggs (1962)
Quercus spp.	Soil type and moisture	Muller (1952)
Populus fremontii × *P. trichocarpa*	Separated along elevational gradient	Eckenwalder (1984)
Populus fremontii × *P. angustifolia*	Separated along elevational gradient	Whitham (1989)
Insects		
Allonemobius socius × *A. fasciatus*	Altitudinal separation, ridge-tops/valleys	Howard (1986)
Gryllus firmus × *G. pennsylvanicus*	Sandy soils/loam soils	Rand and Harrison (1989)
Papilio machaon	Host plant differences	Sperling (1987)
Crustaceans		
Menippe mercenaria × *M. adina*	Mixed seagrass, rocky outcrops/oyster bars, mud, sand	Bert and Harrison (1988)

Fish		
Gasterosteus leiurus × *G. trachurus*	No current, marshy, mud bottom/mild current, sandy bottom	Hagen (1967)
Amphibians		
Rana pipiens complex	Lentic habitats/lotic habitats	Hillis (1981, 1988)
Scaphiopus holbrooki × *S. couchi*	Sandy soils/loam or clay soils	Wasserman (1957)
Reptiles		
Sceloporus woodi × *S. undulatus*	Pine scrub/sandhill vegetation	Jackson (1973)
Birds		
Pipilo ocai × *P. erythrophthalmus*	Coniferous woodland/oak woodland and brushy undergrowth	Sibley (1954)
Quiscalus quiscula	Pine, mixed pine-hardwoods/cypress-tupelogum swamp forest and coastal marshes	Yang and Selander (1968)
Mammals		
Geomys bursarius × *G. lutescens*	Loam soils/sandy soils	Heaney and Timm (1985)
Thomomys bottae	Deep, humic soils in mountain meadows, coniferous forest/rockier soils in pinon-juniper woodland	Patton *et al.* (1979)
Thomomys bottae × *T. umbrinus*	Light-colored, fine soils/darker, coarser soils	Patton (1973)

[a]Where a single species name is given, the hybrid zone interaction involves forms that are considered subspecies or races.

and *Pseudacris* (Gartside 1980), (2) the salamander *Plethodon* (Highton and Henry 1970), (3) the minnow *Notropis* (Dowling *et al.* 1989), (4) the cricket *Allonemobius* (Howard 1986) and (5) the snail *Partula* (Murray and Clarke 1968).

7.6 Genetic architecture of reproductive isolation

Genetic architecture describes the number, effect and chromosomal distribution (linkage relationships) of genes responsible for observed phenotypic differences. Because hybrid zones are the product of many generations of hybridization and introgression, they can be exploited to gain information about the genetic architecture of barriers to gene exchange.

Cline theory provides methods for estimating the number of genes that are responsible for maintaining a cline (e.g. genes that contribute to hybrid unfitness) (Barton and Hewitt 1981*b*, 1985, 1989). These methods depend on knowing cline width, dispersal rate and patterns of linkage disequilibrium. They have been applied to data from two extremely well-characterized hybrid zones, one between *Bombina bombina* and *B. variegata* (Szymura and Barton 1986) and the second between chromosomal races of *Podisma pedestris* (Barton and Hewitt 1981*a*,*b*). Estimates of the number of genes responsible for isolation are 150 in *Podisma* and 50–300 in *Bombina*. These results are consistent with the view that speciation involves many small genetic changes and are not compatible with models that postulate one or a few major gene changes. As data accumulate from other hybrid zones, the generality of these conclusions can be tested. Of course, it can be argued that the finding that many genes contribute to hybrid unfitness in these relatively stable hybrid zones is not relevant to the origin of 'good' species.

That differential introgression is a common attribute of hybrid zones (discussed at length above) suggests that a detailed analysis of patterns of introgression can be used to dissect the genetic architecture of hybrid zone barriers (or the genetic architecture of SMRSs). If only a few major genes are responsible for isolation, introgression of alleles that happen to be closely linked to these genes will be restricted, but most markers will easily cross the hybrid zone. At the other extreme, if isolation is due to many genes, each of small effect and distributed randomly throughout the genome, all neutral markers will be retarded to the same extent. With techniques now available to generate a large array of genetic markers (using restriction fragment length polymorphisms, RFLPs), it is theoretically possible to actually map the genes responsible for selected components of the SMRS or for post-zygotic barriers. The basic idea is to search for RFLP markers that always co-occur with the trait(s) of interest. Hybrid zones that include a diverse array of recombinant types (i.e. those with little disequilibrium between unlinked genes) provide an excellent sample for such a search.

8. HYBRID ZONES AND THE PROCESS OF SPECIATION

Attempts to develop a coherent 'theory' of speciation, a set of generalizations that might apply at least to all sexually reproducing organisms, have consistently reached an impasse when confronted with questions about the geographic context in which populations diverge and barriers to gene exchange arise. The debate about geographic context has had two major themes, traditionally seen as distinct, but in fact having much in common. The first concerns the likelihood of non-allopatric speciation events (Futuyma and Mayer 1980; Bush and Howard 1986), i.e. how frequently (if at all) do species arise in the absence of geographic isolation, through divergence in sympatry or parapatry? The second issue is that of reinforcement (Paterson 1982*a*; Butlin 1987, 1989): Do pre-zygotic barriers to gene exchange arise in allopatry, simply as a byproduct of the divergence of isolated populations, or are they 'isolating mechanisms' molded in response to selection against hybrids (or F2 and back-cross progeny) when previously isolated populations come together in secondary contact?

Information about hybrid zone pattern and process is obviously important for resolving both of these issues. I have already discussed the question of hybrid zone origins and its relevance for evaluating models of parapatric divergence and speciation. Theoretical arguments clearly lead to the conclusion that parapatric divergence can occur (Endler 1977). Therefore, the initial bias in favor of secondary contact as an explanation for hybrid zones and clines must be abandoned in favor of an approach that acknowledges that either scenario is possible. However, in very few cases is there direct evidence to suggest that divergence has occurred *in situ*. The hybrid zones in *Cerion* appear to be good examples (Gould *et al*. 1975; Galler and Gould 1979), as does clinal variation in the milkweed *Asclepias tuberosa* (Wyatt and Antonovics 1981), but in these cases there is no evidence of incipient genetic isolation.

8.1 Hybrid zones and reinforcement

The idea of reinforcement was first clearly formulated and widely promoted by Dobzhansky (1940, 1941). He argued that when hybridization and recombination between differentiated genomes results in reduced fitness, mutations 'which make their carrier less likely to mate with representatives of the other species' will spread (Dobzhansky 1941, p. 286). He went on to suggest that 'to test this hypothesis one would like to know whether the isolating genes are encountered most frequently in those parts of the distribution areas of the respective species where a danger exists of hybridization with related species' (p. 287). The clear prediction, then, is that traits leading to a decreased probability of hybridization should be present at highest frequencies within and adjacent to hybrid zones, i.e. there should be reproductive character displacement. A model developed by Bossert (Wilson 1965) suggests that encounters will end up as a race between fusion and reinforcement, with the probability of reinforcement increasing with hybrid unfitness and with increased initial positive assortative mating.

Although speciation by reinforcement has had many adherents, it has also always attracted a loyal opposition. Paterson (1978, 1982*a*) has summarized many of the criticisms of reinforcement. His uncompromising rejection of this model of speciation and its vocabulary ('isolating mechanism') has catalyzed renewed interest in modeling the process and in carefully evaluating the evidence from laboratory experiments and natural populations. There are several theoretical objections to reinforcement in hybrid zones (see Butlin 1987, 1989; Spencer *et al.* 1986; Sanderson 1989 for a more complete discussion):

1. Mate recognition systems are likely to be under strong stabilizing selection (Paterson 1978, 1982*a*). This has two consequences. First, there will be little heritable variation in SMRSs. Secondly, alleles that lead to reproductive character displacement within a hybrid zone will be at a disadvantage outside the zone and will not tend to spread (Moore 1957).
2. Gene flow from outside of the hybrid zone (immigration into the zone of naive parental types that have never interacted with the 'other' form) will swamp any changes in allele frequency within the zone due to selection against hybridization (Bigelow 1965; Sanderson 1989).
3. In a closed population, simple population genetics theory predicts extinction of the rarer form (allele) when hybrids (heterozygotes) are at a disadvantage. Paterson (1978) extends this result to natural populations and suggests that extinction rather than reinforcement is the expected outcome.
4. If hybrid unfitness is due to many gene differences (each of small effect), then recombination of each generation will break up associations of these genes with each other and with the gene (or genes) causing positive assortative mating (e.g. see Felsenstein 1981).

I consider each of these objections in turn.

Evidence for stabilizing selection on mate recognition systems is limited, and the view espoused by Paterson seems to ignore a large literature (both theoretical and empirical) on sexual selection. Moreover, reproductive isolation does not need to involve changes in mate recognition. For example, temporal isolation can arise through the displacement of flowering time in plants. Clearly, more data are needed both on the amounts of heritable variation in traits (behavioral and ecological) that can lead to reproductive isolation and on the extent to which male and female components of communication systems are genetically coupled (Butlin and Ritchie 1989).

Models of reinforcement suggest that if alleles causing isolation are neutral outside the zone, they will ultimately spread (Caisse and Antonovics 1978), whereas if the alleles causing reinforcement are at a disadvantage outside the hybrid zone, gene flow will tend to swamp evolutionary changes within the zone (Sanderson 1989). Crosby (1970) emphasized that selection against hybrids is a 'second-order effect' and therefore easily overwhelmed by direct selection on the traits that lead to isolation. There seems little doubt that reinforcement will be more likely when the influx of genes from outside the zone is minimal (Howard 1986).

The argument that extinction is a more likely outcome than reinforcement (Paterson 1978) is based on a model of a closed population in which the two subpopulations are not regulated independently. Attempts to select for reproductive isolation in *Drosophila* populations have been successful (e.g. Koopman 1950; Knight *et al.* 1956), but in these systems populations were maintained at a constant size and/or were prevented from going extinct. Similarly, Crosby (1970) found that his simulations would result in extinction unless populations were occasionally supplemented by pure parental types. In the real world, stable polymorphisms can persist either as a result of a balance between dispersal and selection (i.e. tension zones a la Barton and Hewitt 1985) or if the two interacting populations are ecologically differentiated and therefore regulated independently. Maynard Smith (1966) has emphasized the similarity to sympatric speciation models (see also Diehl and Bush 1989). He argues that if two populations in secondary contact with hybrids of reduced fitness 'differ by at least one allelic pair. . . adapting them to different niches. . . this would lead to a stable polymorphism in the zone of overlap and hence to speciation'. In these situations, Paterson's objection no longer holds – extinction is not a likely outcome.

Recombination poses serious problems for reinforcement models, but under some circumstances the problems disappear or are reduced in magnitude. If hybrids are completely sterile or inviable, then recombination does not occur and the two taxa remain distinct. In this case, speciation is already complete, but selection can still lead to the evolution of pre-mating barriers. Recombination also does not interfere with reinforcement if positive assortative mating is due to fixation of the same allele in both populations, rather than to fixation of alternative alleles (see, e.g. Felsenstein 1981; Rice 1984). Finally, if hybrid unfitness is due to one or two genes with major effects, then 'pure' parental types (with respect to the post-mating barrier) will continue to segregate at high frequency in the population. That is, the post-mating barrier will not be destroyed by recombination, allowing more time for evolution of reproductive isolation.

Mosaic hybrid zones are often broad, with the center of the zone far from large regions of 'pure' populations (Howard 1986; Rand and Harrison 1989; Harrison and Rand 1989). They are most often encounters between species that occupy different habitats and/or utilize different resources and therefore are likely to have populations that are regulated independently. Global extinction is unlikely. Moreover, the multiple local contacts provide an opportunity for many independent 'trials' (Littlejohn 1981). These considerations suggest that mosaic hybrid zones are logical places to look for reinforcement.

8.2 The evidence for reinforcement

Despite an abundance of data consistent with the reinforcement hypothesis, there are few (if any) really convincing examples. In two hybrid zones, the extent of interbreeding appears to have declined over time (Jones 1973; Corbin and Sibley 1977), but in both cases other explanations seem equally likely (Butlin 1987). Patterns of reproductive character displacement have been documented for a variety of frog species (Blair 1955; Littlejohn 1965; Fouquette 1975; Littlejohn and Watson 1985), *Drosophila* (Wasserman and Koepfer 1977) and dragonflies (Waage 1979). In each case, a pair of species exhibits greater difference in some component of the mate recognition system (or greater pre-mating isolation) when individuals from within the hybrid zone are compared than when similar comparisons are made between allopatric populations. In none of the cases is there any direct evidence that selection against hybrids has been the driving force producing the pattern of variation. Displacement in flowering time (McNeilly and Antonovics 1968; Millar 1983) and in breeding season of amphibians (Hillis 1981) have also been offered as possible evidence of reinforcement. A survey of barriers to gene exchange in *Drosophila* (Coyne and Orr 1989) reveals patterns consistent with reinforcement. Comparisons of pre-zygotic and post-zygotic isolation indicate that in sympatric species pairs, but not in allopatric pairs, pre-zygotic barriers evolve more rapidly.

Negative results have also been published. Walker (1974) looked for examples of reproductive character displacement in acoustic insects and found none. In a number of hybrid zones where reinforcement might be expected, patterns consistent with reinforcement have not been seen (e.g. Heth and Nevo 1981; Dowling and Moore 1985).

Bush and Howard (1986) point out several reasons why the phenomenon of reproductive character displacement may appear to be rare. To demonstrate reinforcement not only requires very careful documentation of patterns of variation and relative fitnesses, but also that the two can be related (i.e. that a relationship between pattern and process can be clearly established). Furthermore, it is not obvious what components of the mate recognition system to examine. Thus, it may seem logical to look for song differences in an acoustic insect, but isolation may in fact be a result of different chemical cues or habitat preferences. Butlin (1987) also suggests that the process would appear to be infrequent if it occurs rapidly. It is clearly premature to reject the reinforcement model, but certainly there is little available evidence to suggest that it is a common evolutionary pathway. In most cases of secondary contact, pre-zygotic barriers to gene exchange are probably a simple byproduct of divergence in allopatry.

8.3 Hybridization and the origin of species

Although this is not the place for a detailed review, it is important to emphasize that hybridization events are central to several distinctive modes of speciation. In animals, hybridization events are clearly associated with the origin of parthenogenetic forms that are both genetically isolated and ecologically distinct (White 1978). In plants, hybridization has apparently played a fundamental role in the origin of diversity. Grant (1981) reviews a variety of different mechanisms of 'hybrid speciation', which he classifies into two major groups – hybrid speciation with sexual reproduction and hybrid speciation without sexual reproduction. The former includes (1) the origin of new recombinant types that remain distinct from the parental species due to 'external isolating mechanisms' (ecological, temporal and behavioral isolation), (2) recombinational speciation, the origin from chromosomally semi-sterile hybrids of homozygous recombinants that are fertile themselves but sterile in crosses with the parental types or with other recombinants, and (3) allopolyploidy, the origin of polyploids by doubling of the chromosome number in sterile hybrids. Speciation with asexual reproduction includes apomictic speciation, vegetative propagation of interspecific hybrids and permanent translocation heterozygosity. Some of these pathways to speciation are clearly quite common (e.g. allopolyploidy), whereas others are presumably rare (e.g. recombinational speciation) (Grant 1981).

Far more speculative is another possible route to the origin of species that might be catalyzed by hybridization events. A number of observations suggest that hybrid zones are potentially the source of new variants: (1) hybrid zone populations often contain alleles not found in the allopatric populations (data from protein electrophoretic studies in a wide variety of organisms (reviewed by Woodruff 1989)); (2) hybridization between chromosomal races of *Caledia captiva* results in the production of novel chromosome rearrangements (Shaw *et al.* 1983); (3) hybridization between P and M (or I and R) strains of *Drosophila* result in an increased mutation rate due to increased rates of transposition (part of a syndrome known as hybrid dysgenesis) (Kidwell *et al.* 1977; Engels 1983). Hybridization can increase diversity not only through recombination but also as a consequence of higher mutation rates. Hybrid zones may not simply be proving grounds or way-stations but are potential sites of species origination, sources not just sinks.

Imagine a hybrid zone in which distinct 'species' interact and produce hybrid (or recombinant) offspring of reduced fitness (or zero fitness). Selection leads to reproductive character displacement – components of the SMRS of one of the species change significantly such that the extent of hybridization is reduced (perhaps hybridization is eliminated). One of the traditional arguments against this scenario is that individuals with the new SMRS will be at a disadvantage outside the hybrid zone. An alternative view is to see reproductive character displacement as leading to the origin of an entirely new species, distinct from either of the two parental types. Clearly, such a scenario demands a particular population structure (to minimize gene flow from outside the zone) and ecological setting (a disturbed or changing environment?). The process will also be facilitated if the male and female components of the mate recognition system are genetically coupled (Butlin and Ritchie 1989). The basic idea is that a locally isolated population of one of the hybridizing pair evolves in response to its interaction with the other member such that it no longer shares a SMRS with the parent species from which it was derived. It is interesting to note that in several of the best examples of reproductive character displacement (reinforcement), only one of the two interacting species shows significant change within the hybrid zone (Littlejohn 1965; Wasserman and Koepfer 1977).

9. CONCLUSIONS

It is obvious that the hybrid zone literature is large and diverse and impinges upon many of the most important issues in evolutionary biology. Much of the literature represents simple descriptions of patterns of variation, often starting with morphological characters or with chromosomes and then graduating to 'more sophisticated' approaches, allozymes and ultimately DNA sequences. Many of the patterns are remarkable, difficult to understand intuitively. Why should hybrid zones be so narrow (tens or hundreds of meters) and yet not obviously related to any environmental discontinuity? Must these patterns represent recent chance encounters between forms that have diverged in allopatry? In response to these observations and the questions they have provoked, hybrid zone theory has matured quickly.

Hybrid zones are clearly not a single phenomenon – they have various origins, distinctive dynamics and presumably different fates. Attempts to force all examples into a single evolutionary scenario are bound to fail. Hybrid zones are windows on evolutionary process, but each window opens on a different landscape. Evolutionary biologists have peered through many windows, but only in a few cases have they done more than crudely sketch the scene outside. Although careful documentation of patterns of variation has been the hallmark of many recent studies, this must be supplemented by:

1. Data on the frequencies of multi-locus genotypes (linkage disequilibrium) within and adjacent to hybrid zones.
2. Estimates of fitness (both of hybrids and recombinant types relative to parentals and of all genotypes across environments).
3. Analysis of patterns of interbreeding (components of the mate recognition system, influence of temporal and ecological isolation).
4. The use of experimental approaches (e.g. transplants).

Hybrid zones in organisms for which genetic resources are well developed (e.g. *Drosophila, Mus*) can be especially valuable, but these are not always the best systems for ecological or behavioral analysis. A number of animal hybrid zones have been or are currently being studied in great detail, using a variety of approaches, but there are very few long-term, multi-faceted studies of hybrid zones in plants. Plant hybrid zones offer many advantages over their animal counterparts: (1) plant hybrid zones are very common; (2) data on spatial distribution, demography and relative fitness are easier to obtain (because individuals are sedentary); and (3) plants are more amenable to experimental manipulation.

For hybrid zones, the confident answers of the 1960s have been abandoned, to be replaced by considerable uncertainty about both origin and dynamics. This uncertainty has, in fact, resulted in enormous creativity. A robust theoretical framework has emerged (especially for hybrid zones that can be modeled as clines) and with it the opportunity to test alternative hypotheses about the role of selection in speciation and about the genetic architecture of species boundaries. Modern molecular genetic techniques will certainly play an important role in unraveling the secrets of hybrid zones. These techniques open up new dimensions of genetic analysis, and in so doing bring hybrid zone research to a new threshold.

ACKNOWLEDGMENTS

I am indebted to many students and colleagues for patiently discussing data and ideas about hybrid zones and speciation. I am especially grateful for the excellent graduate students who have worked with me, both at Yale and

Cornell. Of these, Dan Howard, David Rand, Terri Bert, John Kochmer, Felix Sperling and Ned Young have been intimately involved in hybrid zone research and each has influenced my thinking in diverse ways. Janis Antonovics, Nick Barton, Godfrey Hewitt, Jim Mallet and Jeremy Searle provided careful critiques of an earlier draft of the manuscript. Their comments and queries have been extremely helpful. The Systematic Biology Program at NSF has generously supported my own work on cricket hybrid zones for more than a decade. This review was written while I was a Visiting Research Fellow at Merton College, Oxford and an Academic Visitor in the Department of Zoology. My special thanks to Dick Southwood for arranging my visit and to Jeremy Searle for helping me find my way about and for stimulating conversations about hybrid zones and evolutionary genetics. Special thanks to Margie Nelson for drawing the figures.

REFERENCES

Alston, R. E.and Turner, B. L. (1963). Natural hybridization among four species of *Baptisia* (Leguminosae). *Amer. J. Bot.* **50,** 159–73.

Anderson, E. (1948). Hybridization of the habitat. *Evolution* **2,** 1–9.

—— (1953). Introgressive hybridization. *Biol. Rev.* **28,** 280–307.

—— and Hubricht, L. (1938). Hybridization in Tradescantia. III. The evidence for introgressive hybridization. *Amer. J. Bot.* **25,** 396–402.

Avise, J. C. and Saunders, N. C. (1984). Hybridization and introgression among species of sunfish (genus *Lepomis*): Analysis by mitochondrial DNA and allozyme markers. *Genetics* **108**, 237–55.

Baker, R. J. (1981). Chromosome flow between chromosomally characterized taxa of a volant mammal, *Uroderma bilobatum* (Chiroptera: Phyllostomatidae). *Evolution* **35,** 296–305.

——, Davis, S. K., Bradley, R. D., Hamilton, M. J. and Van Den Bussche, R. A. (1989). Ribosomal-DNA, mitochondrial DNA, chromosomal and allozymic studies of a contact zone in the pocket gopher. *Geomys. Evolution* **43,** 63–75.

Barrowclough, G. F. (1980). Genetic and phenetic differentiation in a wood warbler (genus *Dendroica*) hybrid zone. *The Auk* **97,** 655–68.

Barton, N. H. (1979*a*). Gene flow past a cline. *Heredity* **43,** 333–9.

—— (1979*b*). The dynamics of hybrid zones. *Heredity* **43,** 341–59.

—— (1980). The fitness of hybrids between two chromosomal races of the grasshopper *Podisma pedestris. Heredity* **45,** 47–59.

—— (1983). Multilocus clines. *Evolution* **37,** 454–71.

—— and Hewitt, G. M. (1981*a*). The genetic basis of hybrid inviability in the grasshopper *Podisma pedestris. Heredity* **47,** 367–83.

—— and Hewitt, G. M. (1981*b*). Hybrid zones and speciation. In *Evolution and speciation* (ed. W. R. Atchley and D. S. Woodruff), pp. 109–145. Cambridge University Press, Cambridge.

—— and Hewitt, G. M. (1981*c*). A chromosomal cline in the grasshopper *Podisma pedestris. Evolution* **35,** 1008–1018.

—— and Hewitt, G. M. (1985). Analysis of hybrid zones. *Ann. Rev. Ecol. Syst.* **16,** 113–48.

—— and Hewitt, G. M. (1989). Adaptation, speciation and hybrid zones. *Nature* **341,** 497–503.

—— and Jones, J. S. (1983). Mitochondrial DNA: New clues about evolution. *Nature* **306,** 317–18.

Bazykin, A. D. (1969). Hypothetical mechanism of speciation. *Evolution* **23,** 685–7.

Bert, T. M. (1986). Speciation in western Atlantic stone crabs (genus *Menippe*): The role of geological patterns and climatic events in the formation and distribution of species. *Marine Biol.* **93,** 157–70.

—— and Harrison, R. G. (1988). Hybridization in western Atlantic stone crabs (genus *Menippe*): Evolutionary history and ecological context influence species interactions. *Evolution* **42,** 528–44.

Bigelow, R. S. (1965). Hybrid zones and reproductive isolation. *Evolution* **19,** 449–58.

Blackwell, J. M. and Bull, C. M. (1978). A narrow hybrid zone between two western Australian frog species *Ranidella insignifera* and *R. pseudinsignifera*: The extent of introgression. *Heredity* **40,** 13–25.

Blair, A. P. (1941). Variation, isolating mechanisms, and hybridization in certain toads. *Genetics* **26,** 398–417.

Blair, W. F. (1955). Mating call and stage of speciation in the *Microhyla olivaceae - M. carolinensis* complex. *Evolution* **9,** 469–80.

Bloom, W. L. (1974). Origin of reciprocal translocations and their effects in *Clarkia speciosa. Chromosoma* **49,** 61–76.

—— (1976). Multivariate analysis of the introgressive replacement of *Clarkia nitens* by *Clarkia speciosa polyantha* (Onagraceae). *Evolution* **30,** 412–24.

Briggs, B. G. (1962). Interspecific hybridization in the *Ranunculus lappaceus* group. *Evolution* **16,** 372–90.

Brown, K. S., Turner, J. R. G. and Sheppard, P. M. (1974). Quarternary forest refugia in tropical America: Evidence from race formation in *Heliconius* butterflies. *Proc. Roy. Soc. Lond.* **B187,** 369–78.

Bull, C. M. (1978). The position and stability of a hybrid zone between the western Australian frogs *Ranidella insignifera* and *R. pseudinsignifera. Aust. J. Zool.* **26,** 305–322.

—— (1979). A narrow hybrid zone between two western Australian frog species *Ranidella insignifera* and *R. pseudinsignifera*: The fitness of hybrids. *Heredity* **42,** 381–9.

Bush, G. L. and Howard, D. J. (1986). Allopatric and non-allopatric speciation: Assumptions and evidence. In *Evolutionary processes and theory* (ed S. Karlin and E. Nevo), pp. 411–38. Academic Press, New York.

Butlin, R. K. (1987). Speciation by reinforcement. *Trends Ecol. Evol.* **2,** 8–13.

—— (1989). Reinforcement of premating isolation. In *Speciation and its consequences* (ed. D. Otte and J. Endler), pp. 158–79. Sinauer Associates, Sunderland, Mass.

—— and Hewitt, G. M. (1985*a*). A hybrid zone between *Chorthippus parallelus parallelus* and *Chorthippus parallelus erythropus* (Orthoptera: Acrididae): Morphological and electrophoretic characters. *Biol. J. Linn. Soc.* **26,** 269–85.

—— and Hewitt, G. M. (1985*b*). A hybrid zone between *Chorthippus parallelus parallelus* and *Chorthippus parallelus erythropus* (Orthoptera: Acrididae): Behavioural characters. *Biol. J. Linn. Soc.* **26,** 287–99.

—— and Ritchie, M. G. (1989). Genetic coupling in mate recognition systems: What is the evidence? *Biol. J. Linn. Soc.* **37,** 237–46.

Caisse, M. and Antonovics, J. (1978). Evolution in closely adjacent plant populations. IX. Evolution of reproductive isolation in clinal populations. *Heredity* **40,** 371–84.

Carson, H. L. (1975). The genetics of speciation at the diploid level. *Amer. Nat.* **109,** 83–92.

——, Kaneshiro, K. Y. and Val, F. C. (1989). Natural hybridization between the sympatric Hawaiian species *Drosophila silvestris* and *Drosophila heteroneura. Evolution* **43,** 190–203.

Chandler, C. R. and Gromko, M. H. (1989). On the relationship between species concepts and speciation processes. *Syst. Zool.* **38,** 116–25.

Chapin, J. P. (1948). Variation and hybridization among the paradise flycatchers of Africa. *Evolution* **2,** 111–26.

Clarke, B. (1966). The evolution of morph-ratio clines. *Amer. Nat.* **100,** 389–402.

Coates, D. J. and Shaw, D. D. (1982). The chromosomal component of reproductive isolation in the grasshopper *Caledia captiva. I.* Meiotic analysis of chiasma distribution patterns in two chromosomal taxa and their F1 hybrids. *Chromosoma* **86,** 509–531.

Corbin, K. W. and Sibley, C. G. (1977). Rapid evolution in Orioles of the genus *Icterus. Condor* **79,** 335–42.

Coyne, J. A. and Orr, H. A. (1989). Patterns of speciation in *Drosophila. Evolution* **43,** 362–81.

——, Orr, H. A. and Futuyma, D. J. (1988). Do we need a new species concept? *Syst. Zool.* **37,** 190–200.

Cracraft, J. (1983). Species concepts and speciation analysis. *Current Ornithol.* **1,** 159–87.

—— (1989). Speciation and its ontology: The emprical consequences of alternative species concepts for understanding patterns and processes of differentiation. In *Speciation and its consequences* (ed. D. Otte and J. Endler), pp. 28–59. Sinauer Associates, Sunderland, Mass.

Crosby, J. L. (1970). The evolution of genetic discontinuity: Computer models of the selection of barriers to interbreeding between subspecies. *Heredity* **25,** 253–97.

Diehl, S. R. and Bush, G. L. (1989). The role of habitat preference in adaptation and speciation. In *Speciation and its consequences* (ed. D. Otte and J. Endler), pp. 345–65. Sinauer Associates, Sunderland, Mass.

Dobzhansky, T. (1940). Speciation as a stage in evolutionary divergence. *Amer. Nat.* **74,** 312–21.

—— (1941). *Genetics and the origin of species*. Columbia University Press, New York.

—— (1970). *Genetics of the evolutionary process*. Columbia University Press, New York.

Dowling, T. E. and Moore, W. S. (1985). Evidence for selection against hybrids in the family Cyprinidae (genus *Notropis*). *Evolution* **39,** 152–8.

——, Smith, G. R. and Brown, W. M. (1989). Reproductive isolation and introgression between *Notropis cornutus* and *Notropis chrysocephalus* (family Cyprinidae): Comparison of morphology, allozymes, and mitochondrial DNA. *Evolution* **43,** 620–34.

Echelle, A. A. and Connor, P. J. (1989). Rapid, geographically extensive genetic introgression after secondary contact between two pupfish species (*Cyprinodon*, Cyprinodontidae). *Evolution* **43,** 717–27.

Eckenwalder, J. E. (1984). Natural intersectional hybridization between North American species of *Populus* (Salicaceae) in sections *Aigeiros* and *Tacamahaca*. I. Population studies of *P.* x. *parryi. Can. J. Bot.* **62,** 317–24.

Endler, J. A. (1973). Gene flow and population differentiation. *Science* **179,** 243–50.

—— (1977). *Geographic variation, speciation, and clines*. Princeton University Press, Princeton.

—— (1982). Problems in distinguishing historical from ecological factors in biogeography. *Amer. Zool.* **22,** 441–52.

Engels, W. R. (1983). The P family of transposable elements in *Drosophila. Ann. Rev. Genet.* **17,** 315–44.

Epling, C. (1947). Natural hybridization between *Salvia apiana* and *S. mellifeara. Evolution* **1,** 69–78.

Felsenstein, J. (1981). Skepticism towards Santa Rosalia or why are there so few kinds of animals. *Evolution* **35,** 124–38.

Ferris, S. D., Sage, R. D., Huang, C.-M., Nielsen, J. T., Ritte, U. and Wilson, A. C. (1983). Flow of mitochondrial DNA across a species boundary. *Proc. Natl. Acad. Sci. USA* **80**, 2290–4.

Fisher, R. A. (1950). Gene frequencies in a cline determined by selection and diffusion. *Biometrics* **6,** 353–61.

Fleischer, R. C. and Rothstein, S. I. (1988). Known secondary contact and rapid gene flow among subspecies and dialects in the brown-headed cowbird. *Evolution* **42,** 1146–58.

Ford, E. B. (1954). Problems in the evolution of geographical races. In *Evolution as a process* (ed. J. Huxley, A. C. Hardy and E. B. Ford), pp. 99–108. George Allen and Unwin, London.

Fouquette, M. J., Jr. (1975). Speciation in chorus frogs. I. Reproductive character displacement in the *Pseudacris nigrita* complex. *Syst. Zool.* **24,** 16–22.

Frost, J. S. and Bagnara, J. T. (1977). An analysis of reproductive isolation between *Rana magnaocularis* and *Rana berlandieri forreri* (*Rana pipiens* complex). *J. Exp. Zool.* **202,** 291–306.

Futuyma, D. J. and Mayer, G. C. (1980). Non-allopatric speciation in animals. *Syst. Zool.* **29,** 254–71.

Galler, L. and Gould, S. J. (1979). The morphology of a 'hybrid zone' in *Cerion*: Variation, clines, and an ontogenetic relationship between two 'species' in Cuba. *Evolution* **33,** 714–27.

Gartside, D. F. (1972). The *Litoria ewingi* complex (Anura: Hylidae) in south-eastern Australia. III. Blood protein variation across a narrow hybrid zone between *L. ewingi* and *L. paraewingi. Aust. J. Zool.* **20,** 435–43.

—— (1980). Analysis of a hybrid zone between chorus frogs of the *Pseudacris nigrita* complex in the southern United States. *Copeia* **1980,** 56–66.

Good, D. A. (1989). Hybridization and cryptic species in *Dicamptodon* (Caudata: Dicamptodontidae). *Evolution* **43,** 728–44.

Gould, S. J., Woodruff, D. S. and Martin, J. P. (1975). Genetics and morphometrics of *Cerion* at Pongo Carpet: A new systematic approach to this enigmatic land snail. *Syst. Zool.* **23,** 518–35.

Grant, V. (1981). *Plant speciation.* Columbia University Press, New York.

Greenbaum, I. F. (1981). Genetic interactions between hybridizing cytotypes of the tent-making bat (*Uroderma bilobatum*). *Evolution* **35,** 306–321.

Grudzien, T. A., Moore, W. S., Cook, J. R. and Tagle, D. (1987). Genic population structure and gene flow in the Northern Flicker (*Colaptes auratus*) hybrid zone. *The Auk,* **104,** 654–64.

Gyllensten, U. and Wilson, A. C. (1987). Interspecific mitochondrial DNA transfer and the colonization of Scandinavia by mice. *Genet. Res. Camb.* **49,** 25–29.

Haffer, J. (1969). Speciation in Amazonian forest birds. *Science* **165,** 131–7.

Haffner, J. C., Haffner, D. J. Patton, J. L. and Smith, M. F. (1983). Contact zones and the genetics of differentiation in the pocket gopher *Thomomys bottae* (Rodentia: Geomyidae). *Syst. Zool.* **32,** 1–20.

Hagen, D. W. (1967). Isolating mechanisms in threespine sticklebacks (*Gasterosteus*). *J. Fish. Res. Bd Canada* **24,** 1637–92.

Haldane, J. B. S. (1948). The theory of a cline. *J. Genetics* **48,** 277–84.

Hall, W. P. and Selander, R. K. (1973). Hybridization of karyotypically differentiated populations in the *Sceloporus grammicus* complex (lguanidae). *Evolution* **27,** 226–42.

Harrison, R. G. (1985). Barriers to gene exchange between closely related cricket species. II. Life cycle variation and temporal isolation. *Evolution* **39,** 244–59.

—— (1986). Pattern and process in a narrow hybrid zone. *Heredity,* **56,** 337–49.

—— (1989). Animal mitochondrial DNA as a genetic marker in a population and evolutionary biology. *Trends Ecol. Evol.* **4,** 6–11.

—— and Arnold, J. (1982). A narrow hybrid zone between closely related cricket species. *Evolution* **36,** 535–52.

——and Rand, D. M. (1989). Mosaic hybrid zones and the nature of species boundaries. In *Speciation and its consequences* (ed. D. Otte and J. Endler), pp. 111–33. Sinauer Associates, Sunderland, Mass.

——, Rand, D. M. and Wheeler, W. C. (1987). Mitochondrial DNA variation in field crickets across a narrow hybrid zone. *Mol. Biol. Evol.* **4,** 144–58.

Hauber, D. P. and Bloom, W. L. (1983). Stability of a chromosomal hybrid zone in the *Clarkia nitens* and *Clarkia speciosa* ssp. *polyantha* complex (Onagraceae). *Amer. J. Bot.* **70,** 1454–9.

Heaney, L. R. and Timm, R. M. (1985). Morphology, genetics, and ecology of pocket gophers (genus *Geomys*) in a narrow hybrid zone. *Biol. J. Linn. Soc.* **25,** 301–317.

Heiser, C. B., Jr (1973). Introgression reexamined. *Bot. Rev.* **39,** 347–66.

Heth, G. and Nevo, E. (1981). Origin and evolution of ethological isolation in subterranean mole rats. *Evolution* **35,** 259–74.

Hewitt, G. M. (1988). Hybrid zones – natural laboratories for evolutionary studies. *Trends Ecol. Evol.* **3,** 158–67.

—— (1989). The subdivision of species by hybrid zones. In *Speciation and its consequences* (ed. D. Otte and J. Endler), pp. 85–110. Sinauer Associates, Sunderland, Mass.

—— and Barton, N. H. (1980). The structure and maintenance of hybrid zones as exemplified by *Podisma pedestris*. In *Insect cytogenetics* (ed. R. L. Blackman, G. M. Hewitt and M. Ashburner), pp. 149–69. Blackwell Scientific, Oxford.

——, Butlin, R. K. and East, T. M. (1987). Testicular dysfunction in hybrids between parapatric species of the grasshopper *Chorthippus parallelus*. *Biol. J. Linn. Soc.* **31,** 25–34.

Heywood, J. S. (1986). Clinal variation associated with edaphic ecotones in hybrid populations of *Gaillardia pulchella*. *Evolution* **40,** 1132–40.

Highton, R. (1977). Comparison of microgeographic variation in morphological and electrophoretic traits. *Evol. Biol.* **10,** 397–436.

—— and Henry, S. A. (1970). Evolutionary interactions between species of North American salamanders of the genus *Plethodon*. *Evol. Biol.* **4,** 211–56.

Hillis, D. M. (1981). Premating isolating mechanisms among three species of the *Rana pipiens* complex in Texas and southern Oklahoma. *Copeia* **1981,** 312–19.

—— (1988). Systematics of the *Rana pipiens* complex: Puzzle and paradigm. *Ann. Rev. Ecol. Syst.* **19,** 39–63.

Howard, D. J. (1986). A zone of overlap and hybridization between two ground cricket species. *Evolution* **40,** 34–43.

—— and Furth, D. G. (1986). Review of the *Allonemobius fasciatus* (Orthoptera: Gryllidae) complex with description of two new species separated by electrophoresis, songs, and morphometrics. *Ann. Entomol. Soc. Amer.* **79,** 472–81.

Hubbs, C. L. (1955). Hybridization between fish species in nature. *Syst. Zool.* **4,** 1–20.

Hunt, W. G. and Selander, R. K. (1973). Biochemical genetics of hybridization in European house mice. *Heredity* **31,** 11–33.

Jackson, J. F. (1973). The phenetics and ecology of a narrow hybrid zone. *Evolution* **27,** 58–68.

Jones, J. M. (1973). Effects of thirty years hybridization on the toads *Bufo americanus* and *Bufo woodhousii fowleri* at Bloomington, Indiana. *Evolution* **27,** 435–48.

Key, K. H. L. (1968). The concept of stasipatric speciation. *Syst. Zool.* **17,** 14–22.

—— (1974). Speciation in the Australian Morabine grasshoppers – taxonomy and ecology. In *Genetic mechanisms of speciation in insects* (ed M. J. D. White), pp. 43–56. Australia and New Zealand Book Co., Sydney.

Kidwell, M. G., Kidwell, J. F. and Sved, J. A. (1977). Hybrid dysgenesis in *Drosophila melanogaster*: A syndrome of aberrant traits including mutation, sterility, and male recombination. *Genetics* **86,** 813–33.

Knight, G. R., Robertson, A. and Waddington, C. H. (1956). Selection for sexual isolation within a species. *Evolution* **10,** 14–22.

Kocher, T. D. and Sage, R. D. (1986). Further genetic analysis of a hybrid zone between leopard frogs (*Rana pipiens* complex) in central Texas. *Evolution* **40,** 21–33.

Koopman, K. F. (1950). Natural selection for reproductive isolation between *Drosophila pseudoobscura* and *Drosophila persimilis*. *Evolution* **4,** 135–48.

Lambert, D. M., Michaux, B. and White, C. S. (1987). Are species self-defining? *Syst. Zool.* **36,** 196–205.

Lanyon, W. E. (1979). Hybrid sterility in meadowlarks. *Nature* **279,** 557–8.
Levin, D. A. (1963). Natural hybridization between *Phlox maculata* and *Phlox glaberrima* and its evolutionary significance. *Amer. J. Bot.* **50,** 714–20.
—— (1967). An analysis of hybridization in *Liatris*. *Brittonia* **19,** 248–60.
—— (1975). Interspecific hybridization, heterozygosity and gene exchange in *Phlox*. *Evolution* **29,** 37–51.
Levins, R. (1968). *Evolution in changing environments*. Princeton University Press, Princeton.
Littlejohn, M. J. (1965). Premating isolation in the *Hyla ewingi* complex (Anura:Hylidae). *Evolution* **19,** 234–43.
—— (1981). Reproductive isolation: A critical review. In *Evolution and speciation* (ed. W. R. Atchley and D. S. Woodruff), pp. 298–334. Cambridge University Press, Cambridge.
—— and Watson, G. F. (1985). Hybrid zones and homogamy in Australian frogs. *Ann. Rev. Ecol. Syst.* **16,** 85–112.
——, Watson, G. F. and Loftus-Hills, J. J. (1971). Contact hybridization in the *Crinia laevis* complex (Anura: Leptodactylidae). *Aust. J. Zool.* **19,** 85–100.
Mallet, J. (1986). Hybrid zones of *Heliconius* bufferflies in Panama and the stability and movement of warning colour clines. *Heredity* **56,** 191–202.
—— and Barton, N. H. (1989*a*). Strong natural selection in a warning-color hybrid zone. *Evolution* **43,** 421–31.
—— and Barton, N. H. (1989*b*). Inference from clines stabilized by frequency-dependent selection. *Genetics* **122,** 967–76.
Marchant, A. D., Arnold, M. L. and Wilkinson, P. (1988). Gene flow across a chromosomal tension zone. I. Relicts of ancient hybridization. *Heredity* **61,** 321–8.
Masters, J. C., Rayner, R. J., McKay, I. J., Potts, A. D., Nails, D., Ferguson, J. W., Weissenbacher, B. K., Allsopp, M. and Anderson, M. L. (1987). The concept of species: Recognition versus isolation. *S. Afr. J. Sci.* **83,** 534–7.
Maynard Smith, J. (1966). Sympatric speciation. *Amer. Nat.* **100,** 637–50.
Mayr, E. (1942). *Systematics and the origin of species*. Columbia University Press, New York.
—— (1954). Change of genetic environment and evolution. In *Evolution as a process* (ed. J. Huxley, A. C. Hardy and E. B. Ford), pp. 157–80. George Allen and Unwin, London.
—— (1963). *Animal species and evolution*. Belknap Press, Cambridge.
McDonnell, L. J., Gartside, D. F. and Littlejohn, M. J. (1978). Analysis of a narrow hybrid zone between two species of *Pseudophryne* (Anura: Leptodactylidae) in south-eastern Australia. *Evolution* **32,** 602–612.
McLain, D. K. (1985). Clinal variation in morphology and assortative mating in the solder bettle, *Chauliognathus pennsylvanicus* (Coleoptera: Cantharidae). *Biol. J. Linn. Soc.* **25,** 105–117.
—— (1988) Male mating preferences and assortative mating in the soldier beetle. *Evolution* **42**, 729–35.
McNeilly, T. and Antonovics, J. (1968). Evolution in closely adjacent plant populations. IV. Barriers to gene flow. *Heredity* **23,** 205–18.
MIllar, C. I. (1983). A steep cline in *Pinus muricata. Evolution* **37,** 311–19.

Moran, C. (1979). The structure of the hybrid zone in *Caledia captiva. Heredity* **42,** 13–32.

—, Wilkinson, P. and Shaw, D. D. (1980). Allozyme variation across a narrow hybrid zone in the grasshopper, *Caledia captiva. Heredity* **44,** 69–81.

Moore, J. A. (1957). An embryologist's view of the species concept. In *The species problem* (ed. E. Mayr). American Association for the Advancement of Science.

Moore, W. S. (1977). An evaluation of narrow hybrid zones in vertebrates. *Quart. Rev. Biol.* **52,** 263–77.

—— (1981). Assortative mating genes selected along a gradient. *Heredity* **46,** 191–5.

—— (1987). Random mating in the Northern Flicker hybrid zone: Implications for the evolution of bright and contrasting plumage patterns in birds. *Evolution* **41,** 539–46.

—— and Buchanan, D. B. (1985). Stability of the Northern Flicker hybrid zone in historical times: Implications for adaptive speciation theory. *Evolution* **39,** 135–51.

—— and Koenig, W. D. (1986). Comparative reproductive success of yellow-shafted, red-shafted and hybrid flickers across a hybrid zone. *The Auk,* **103,** 42–51.

Mrongovius, M. J. (1979). Cytogenetics of the hybrids of three members of the grasshopper genus *Vandiemenella* (Orthoptera: Eumastacidae: Morabinae). *Chromosoma* **71,** 81–107.

Muller, C. H. (1952). Ecological control of hybridization in *Quercus*: A factor in the mechanism of evolution. *Evolution* **6,** 147–61.

Murray, J. and Clarke, B. (1968). Partial reproductive isolation in the genus *Partula* (Gastropoda) on Moorea. *Evolution* **22,** 684–98.

Nei, M. (1987). *Molecular evolutionary genetics*. Columbia University Press, New York.

Neigel, J. E. and Avise, J. C. (1985). Phylogenetic relationships of mitochondrial DNA under various demographic models of speciation. In *Evolutionary processes and theory* (ed. S. Karlin and E. Nevo), pp. 515–34. Academic Press, New York.

Nelson, K., Baker, R. J. and Honeycutt, R. L. (1987). Mitochondrial DNA and protein differentiation between hybridizing cytotypes of the white-footed mouse, *Peromyscus leucopus. Evolution* **41,** 864–72.

Nevo, E. and Bar-El, H. (1976). Hybridization and speciation in fossorial mole rats. *Evolution* **30,** 831–40.

—— and Heth, G. (1976). Assortative mating between chromosome forms of the mole rat, *Spalax ehrenbergi. Experientia* **32,** 1509–1511.

Ornduff, R. (1967). Hybridization and regional variation in Pacific northwestern *Impatiens* (Balsaminacea). *Brittonia* **19,** 122–8.

Palma-Otal, M., Moore, W. S., Adams, R. P. and Joswiak, G. R. (1983). Morphological, chemical and biogeographical analyses of a hybrid zone involving *Juniperus virginiana* and *J. horizontalis* in Wisconsin. *Can. J. Bot.* **61,** 2733–46.

Paterson, H. E. H. (1978). More evidence against speciation by reinforcement. *S. Afr. J. Sci.* **74,** 369–71.

—— (1980). A comment on 'mate recognition systems'. *Evolution* **34,** 330–31.

—— (1982*a*). Perspective on speciation by reinforcement. *S. Afr. J. Sci.* **78,** 53–7.

—— (1982*b*). Darwin and the origin of species. *S. Afr. J. Sci.* **78,** 272–5.

—— (1985). The recognition concept of species. In *Species and speciation* (ed. E. S. Vrba), pp. 21–9. Transvaal Museum, Pretoria.

Patton, J. L. (1973). An analysis of natural hybridization between the pocket gophers *Thomomys bottae* and *Thomomys umbrinus* in Arizona. *J. Mammal.* **54,** 561–84.

——, Hafner, J. C., Hafner, M. S. and Smith, M. F. (1979). Hybrid zones in *Thomomys bottae* pocket gophers: Genetic, phenetic and ecologic concordance patterns. *Evolution* **33,** 860–76.

Rand, D. M. and Harrison, R. G. (1989). Ecological genetics of a mosaic hybrid zone: Mitochondrial, nuclear and reproductive differentiation of crickets by soil type. *Evolution* **43,** 432–49.

Remington, C. L. (1968). Suture zones of hybrid interaction between recently joined biotas. *Evol. Biol.* **2,** 321–28.

Rensch, B. (1959). *Evolution above the species level.* Methuen, London.

Rice, W. R. (1984). Disruptive selection on habitat preference and the evolution of reproductive isolation: A simulation study. *Evolution* **38,** 1251–60.

Rising, J. D. (1983). The Great Plains hybrid zones. *Current Ornithol.* **1,** 131–57.

Robbins, M. B., Braun, M. J. and Tobey, E. A. (1986). Morphological and vocal variation across a contact zone between the chickadees *Parus atricapillus* and *P. carolinensis. The Auk* **103,** 655–66.

Rosen, D. E. (1979). Fishes from the uplands and the intermontane basins of Guatemala: Revisionary studies and comparative geography. *Bull. Amer. Mus. Nat. Hist.* **162,** 267–376.

Sage, R. D. and Selander, R. K. (1979). Hybridization between species of the *Rana pipiens* complex in central Texas. *Evolution* **33,** 1069–88.

——, Heyneman, D., Lim, K.-C. and Wilson, A. C. (1986). Wormy mice in a hybrid zone. *Nature* **324,** 60–63.

Sanderson, N. (1989). Can gene flow prevent reinforcement? *Evolution* **43,** 1223–35.

Searle, J. B. (1984). Three new karyotypic races of the common shrew *Sorex araneus* (Mammalia: Insectivora) and a phylogeny. *Syst. Zool.* **33,** 184–94.

—— (1986). Factors responsible for a karyotypic polymorphism in the common shrew, *Sorex araneus. Proc. Roy. Soc. Lond.* **B229,** 277–98.

—— (1988). Selection and Robertsonian variation in nature: The case of the common shrew. In *The cytogenetics of mammalian autosomal rearrangements* (ed. A. Daniel), pp. 507–531. Alan R. Liss, New York.

Shaw, D. D. (1981). Chromosomal hybrid zones in orthopteroid insects. In *Evolution and speciation* (ed. W. R. Atchley and D. S. Woodruff), pp. 146–70. Cambridge University Press, Cambridge.

—— and Wilkinson, P. (1980). Chromosome differentiation, hybrid breakdown and the maintenance of a narrow hybrid zone in *Caledia. Chromosoma* **80,** 1–31.

——, Wilkinson, P. and Coates, D. J. (1982). The chromosomal component of reproductive isolation in the grasshopper *Caledia captiva*. II. The relative viabilities of recombinant and non-recombinant chromosomes during embryogenesis. *Chromosoma* **86,** 533–49.

——, Wilkinson, P. and Coates, D. J. (1983). Increased chromosomal mutation rate after hybridization between two subspecies of grasshoppers. *Science* **220,** 1165–7.

Short, L. L. (1969). Taxonomic aspects of avian hybridization. *The Auk* **86,** 84–105.

Sibley, C. G. (1954). Hybridization in the red-eyed towhees of Mexico. *Evolution* **8,** 252–90.

—— (1957). The evolutionary and taxonomic significance of sexual dimorphism and hybridization in birds. *Condor* **59,** 166–91.

—— and West, D. A. (1958). Hybridization in the red-eye towhees of Mexico: The Eastern plateau populations. *Condor* **60,** 85–104.

Slatkin, M. (1973). Gene flow and selection in a cline. *Genetics* **75,** 733–56.

——(1975). Gene flow and selection in a two-locus system. *Genetics* **31,** 787–802.

Spencer, H. G., McArdle, B. H. and Lambert, D. J. (1986). A theoretical investigation of speciation by reinforcement. *Amer. Nat.* **128,** 245–62.

Sperling, F. A. H. (1987). Evolution of the *Papilio machaon* species group in western Canada (Lepidoptera: Papilionidae). *Quaest. Entomol.* **23,** 198–315.

—— and Spence, J. R. (1991). Structure of an asymmetric hybrid zone between two water strider species (Hemiptera: Gerridae: *Limnoporus*). *Evolution*, submitted.

Stangl, F. B., Jr (1986). Aspects of a contact zone between two chromosomal races of *Peromyscus leucopus* (Rodentia: Cricetidae). *J. Mammal.* **67,** 465–73.

Stebbins, G. L. (1950). *Variation and evolution in plants*. Columbia University Press, New York.

Szymura, J. M. and Barton, N. H. (1986). Genetic analysis of a hybrid zone between the fire-bellied toads, *Bombina bombina* and *Bombina variegata*, near Cracow in southern Poland. *Evolution* **40,** 1141–59.

——, Spolsky, C. and Uzzell, T. (1986). Concordant changes in mitochondrial and nuclear genes in a hybrid zone between two frog species (genus *Bombina*). *Experientia* **41,** 1469–70.

Templeton, A. R. (1981). Mechanisms of speciation – a population genetic approach. *Ann. Rev. Ecol. Sys.* **12,** 23–48.

—— (1989). The meaning of species and speciation: A genetic perspective. In *Speciation and its consequences* (ed. D. Otte and J. Endler), pp. 3–27. Sinauer Associates, Sunderland, Mass.

Thorpe, R. S. (1984). Primary and secondary transition zones in speciation and population differentiation: A phylogenetic analysis of range expansion. *Evolution* **38,** 233–43.

Tucker, P. K. and Schmidly, D. J. (1981). Studies of a contact zone among three chromosomal races of *Geomys bursarius* in east Texas. *J. Mammal.* **62,** 258–72.

Turner, J. R. G. (1971). Two thousand generations of hybridization in a *Heliconius* butterfly. *Evolution* **25,** 471–82.

Vanlerberghe, F., Dod, B., Boursot, P., Bellis, M. and Bonhomme, F. (1986). Absence of Y-chromosome introgression across the hybrid zone between *Mus musculus domesticus* and *Mus musculus musculus*. *Genet. Res.* **48,** 191–7.

Waage, J. K. (1979). Reproductive character displacement in *Calopteryx* (Odonata: Calopterygidae). *Evolution* **33,** 104–116.

Wake, D. B. and Yanev, K. P. (1986). Geographic variation in allozymes in a 'ring species', the plethodontid salamander, *Ensatina eschscholtzii* of western North America. *Evolution* **40,** 702–715.

——, Yanev, K. P. and Frelow, M. M. (1989). Sympatry and hybridization in a 'ring species': The plethodontid salamander *Ensatina eschscholtzii*. In *Speciation and its consequences* (ed. D. Otte and J. Endler), pp. 134–57. Sinauer Associates, Sunderland, Mass.

Walker, T. J. (1974). Character displacement and acoustic insects. *Amer. Zool.* **14,** 1137–50.

Wasserman, A. O. (1957). Factors affecting interbreeding in sympatric species of spadefoots (genus *Scaphiopus*). *Evolution* **11,** 320–38.

Wasserman, M. and Koepfer, H. R. (1977). Character displacement for sexual isolation between *Drosophila mojavensis* and *Drosophila arizonenesis*. *Evolution* **31,** 812–23.

Watson, G. F. (1972). The *Litoria ewingi* complex (Anura: Hylidae) in southeastern Australia. II. Genetic incompatibility and delimitation of a narrow hybrid zone between *L. ewingi* and adjacent taxa. *Aust. J. Zool.* **20,** 423–33.

—— and Littlejohn, M. J. (1978). The *Litoria ewingi* complex (Anura: Hylidae) in southeastern Australia. V. Interactions between northern *L. ewingi* and adjacent taxa. *Aust. J. Zool.* **26,** 175–95.

White, M. J. D. (1978). *Modes of speciation*. W. H. Freeman, San Francisco.

Whitham, T. G. (1989). Plant hybrid zones as sinks for pests. *Science* **244,** 1490–93.

Wilson, D. S. and Hedrick, A. (1982). Speciation and the economics of mate choice. *Evol. Theory* **6,** 15–24.

Wilson, E. O. (1965). The challenge from related species. In *The genetics of colonizing species* (ed. H. G. Baker and G. L. Stebbins), pp. 7–27. Academic Press, New York.

Woodruff, D. S. (1973). Natural hybridization and hybrid zones. *Syst. Zool.* **22,** 213–18.

—— (1979). Postmating reproductive isolation in *Pseudophryne* and the evolutionary significance of hybrid zones. *Science* **203,** 561–3.

—— (1981). Toward a genodynamics of hybrid zones: Studies of Australian frogs and West Indian land snails. In *Evolution and speciation* (ed. W. R. Atchley and D. S. Woodruff), pp. 171–97. Cambridge University Press, Cambridge.

—— (1989). Genetic anomalies associated with *Cerion* hybrid zones: The origin and maintenance of new electromorphic variants called hybrizymes. *Biol. J. Linn. Soc.* **36,** 281–94.

Wyatt, R. and Antonovics, J. (1981). Butterflyweed revisited – spatial and temporal patterns of leaf shape variation in *Asclepias tuberosa*. *Evolution* **35,** 525–42.

Wynn, A. H. (1986). Linkage disequilibrium and a contact zone in *Plethodon cinereus* on the Del-Mar-Va peninsula. *Evolution* **40,** 44–54.

Yang, S. Y. and Selander, R. K. (1968). Hybridization in the grackle *Quiscalus quiscula* in Louisiana. *Syst. Zool.* **17,** 107–143.

Evolutionary process as studied in population genetics: clues from phylogeny

HAMPTON L. CARSON

1. HOW HAS PHYLOGENY INFLUENCED POPULATION GENETICS?

The population geneticist centers his evolutionary research on immediate genetic processes operating within the populations of a selected species. The work deals with genetic change in populations from one generation to the next. The complexities of the genetic system are so far removed from the phylogenetic and ecological milieu of the species, that the influence of biological factors external to the genetic system of the species tends to be overlooked by the population geneticist. When emphasis turns to the use of population genetics for evolutionary study, such ignorance can lead to unfortunate choices of species and characters for experimental study. This chapter recounts some of the difficulties and challenges experienced by those who have worked on the genus *Drosophila*.

1.1 The choice of a species to study

A major problem is that the choice of a species to investigate genetically seems usually to have been made without regard to specific evolutionary problems. There are, of course, some very real and practical constraints on population genetics work. The chosen organism, for example, must display genetic variation and must be amenable to genetic analysis, that is, have rapidly succeeding sexual generations that can be monitored both in the laboratory and in natural populations. Sometimes, as was the case originally with *Drosophila melanogaster*, species thrust themselves forward for strictly analytical reasons: Morgan chose *Drosophila* as a tool to study genetics, not evolution. When the early *Drosophila* workers, including the writer, contended that descent with change could indeed be studied in such a species, cries of derision arose. Although apparently not recorded in the literature, the thrust of the criticism was as follows. How can evolution be studied in a species collected from garbage cans and then confined to glass bottles on a laboratory shelf? Can we afford to ignore ecological clues to the parameters of natural selection? It was further argued that to understand evolution, historical information must be drawn from phylogeny in such a way that will permit the identification of relevant evolutionary processes that can be scrutinized experimentally. Indeed, in *Drosophila* work, the necessary influence of phylogeny and ecology were a long time in coming.

With nearly a century of genetic experience now behind us, the time has come to make sophisticated decisions as to what species and what characters to study and just how to approach the dynamics of the evolutionary process. This can be done only by going through a long preliminary process in order to gather data on the general biological characteristics of the organism that is being chosen for study.

In this chapter, 'phylogeny' is used as a catchword referring to very broad information on the comparative biology of the organism. It includes not only the characters that are necessary for the technical aspects of observational and experimental systematics, but also demographic, ecological, geographic and historical facts about the species and its close relatives. Such comparative information must be obtained and then sifted, looking for clues that might help to focus the population genetics efforts.

For virtually every organism that is workable genetically – even *Drosophila* – there now exists a modest backlog of phyletic and ecological information about many species and their relatives that can be drawn upon. This was not true 50 years ago. Phylogenetic information can be useful to the population geneticist in two quite different ways. First, inferences may be made about populational modes pertinent to the origin of species. The second aspect has to do with the study of the genetic origin in populations of specific novel character states.

1.2 Stable versus unstable populations

Considerable evidence exists that many species have large populations that appear to be relatively ancient, stable and in equilibrium with the environment. The populations of certain other species, however, may be basically unstable in that they are small or frequently pass through bottlenecks of population size. Among these can be identified what have been called 'neospecies' (see Carson 1976), as various kinds of evidence indicate that they have been relatively newly formed.

Diagnosis as to whether a species is ancient or new, however, can only be made after data are available on the phylogenetic context of the species. This should include its genetic distance from close relatives and data on shared polymorphisms and other historical, distributional and ecological information. In some cases, this information may make it possible to infer that certain demes or intraspecific populations are actually in the process of directional genetic change, or may be species *in statu nascendi*, to use Dobzhansky's term (Dobzhansky and Spassky, 1959).

Secondly, the population geneticist may wish to concentrate on the origin of novel characters or character states. Data should be sought from phylogenetic information that gives clues as to which characters are relatively newly formed in time. Can recently evolved apomorphic characters as opposed to relatively older ones (plesiomorphic) be identified? Is there a difference in their genetic structure? The population geneticist often appears to be studying characters that may have been genetically stable for millions of generations. Systematic study can help identify recently evolved characters.

2. *DROSOPHILA* POPULATION GENETICS

For various reasons that need not be recounted here, flies of the genus *Drosophila* were especially useful for the study of the formal laws of inheritance during the early decades of this century. As many new species of the genus *Drosophila* were discovered and shown to be amenable to laboratory culture, it seemed quite logical to begin to use such species for the analysis of evolutionary problems, employing the techniques of population genetics. Some of the difficulties that arose have already been hinted at; the genus *Drosophila* will now be used as a paradigm for exposing these problems in detail and suggesting ways in which they may be overcome in future work.

2.1 Early developments (1930–70)

2.1.1 *The genetics of populations of* Drosophila *developed in the virtual absence of phylogenetic or ecological data*

Using *Drosophila melanogaster*, A. H. Sturtevant made many pioneering contributions to formal genetics. It is less well known that, almost single-handedly, he began systematic and ecological studies of the genus (1921), including the first published key to the North American species. His goal, and that of those who followed his lead, such as the group at the University of Texas (e.g. Patterson and Stone 1952), was to increase the usefulness of these species for the study of evolutionary processes through genetic analyses. As data accumulated, two evolutionary goals emerged as exciting possibilities. To the bolder investigators, it appeared to be possible to decipher the genetic bases and modes of origin of both adaptations and species. This was to be accomplished through the application of simple Mendelian population genetics to the solution of these problems (Dobzhansky 1950). Following early studies of the geographical distribution of recessive mutants in natural populations, there came a succession of novel techniques for detecting natural genetic variability. Chromosomal inversions, translocations, heterochromatic and polygenetic variability, viability-affecting genes and allozymes were discovered and catalogued for the wild populations of many different species. Just prior to the introduction of molecular analyses of DNA in the early 1980s, a series of partial reviews of this literature required five substantial volumes (Ashburner *et al.* 1981–6). The complexities of the genetic system did much to dispel simplistic notions that evolution commonly proceeded by the simple fixation by selection of single mutant genes having large effects.

In the early years, the work concentrated on and emphasized widespread, near-cosmopolitan or continentally distributed species or groups of species. All those selected were easy to raise in the laboratory. Almost without exception, these studies were conceived and carried out in the absence of any relevant basic phylogenetic, geographic, ecological and systematic information. Until the first of the Ashburner volumes devoted to these matters (1981), this material was widely scattered in the periodical literature and was without much ecological base; indeed it is still far from complete. Almost none of the early work on the genetics of natural populations sought any guidance from what little phylogenetic or ecological data were available.

2.1.2 *Data on the population genetics of* Drosophila *prior to 1970*

The early studies of natural populations of *Drosophila* emphasized the use of species that are endemic to large continental areas and appear to have natural distributions not much affected by human activities. These include *D. melanica, pseudoobscura, robusta, subobscura* and *willistoni* to name only a few. 'Domestic' or near-cosmopolitan forms such as *D. melanogaster* and *virilis* were rarely used, on the assumption, now judged to be erroneous, that their populations would not show regional genetic differentiation. On the other hand, the species confined to single hemispheres or continents consistently show extensive balanced polymorphism for substantial sections of the genome in the form of chromosomal inversions. Their local populations form inversion clines that are sensitive to altitudinal and latitudinal differences. When allozyme techniques were developed in the 1960s, all of the species studied showed exuberant variation. Their pattern of distribution, however, was generally rather uniform in the species and tended to be mostly independent with regard to the inversions. A few of these allelic polymorphisms, however, were shown to be 'hitch-hiking', that is, they were evidently linked to inverted segments that were being balanced in the population by major selective forces. Evidence now strongly supports the hypothesis that most allozyme loci are neutral or quasi-neutral to selection (Kimura 1983). The paradox of the persistence of the majority of these genes in natural populations, along with lethal and semi-lethal gene loci, may also be mostly resolved by invoking hitch-hiking (see Hedrick 1980). If not demonstrably associated with inversions, they may reside as neutral elements within major polygenic blocks underlying important quantitative traits (Parsons 1979; Lande 1980*a*; Takano *et al.* 1987). The matter is unresolved; in any event, the data are consistent in showing that great stores of genetic variability are tenaciously retained in natural populations.

2.1.3 *Character change and speciation*

As mentioned above, genetic adjustments to latitude and altitude show great regularity in, for example, *D. robusta* and *pseudoobscura* of North America and *subobscura* of Europe. When the latter species was accidentally introduced into the Western Hemisphere in the late 1970s (Prevosti *et al.* 1988), these clinal adjustments were reconstituted in some detail as the species colonized the North and South American continents. It is tempting to refer to this kind of adjustment as an adaptive response to local selective forces, even though the genes involved cannot be recognized individually and their physiological effects are unknown. But we have learned not to oversimplify the problem of regional genetic adjustment. As indicated earlier, single, mendelizing gene loci are definitely not involved, and polygenes are often invoked.

Genetic work in a separate area may be pertinent here. A large literature has developed that deals with metric characters such as the numbers of scutellar, abdominal or sternopleural bristles and wing or body size. The information that comes from their study has revealed details about the genetics of quantitative variation in populations (Thompson and Thoday 1979). However, in the absence of ecological data and field studies of fitness, such data and the characters dealt with are hard to relate directly to character differences either within or between species. Therefore, we do not know whether or not most bristle numbers in *Drosophila* relate to differential adaptational or speciational trends within the species, the species group or the genus. All these findings have greatly altered the early but still tenacious simplistic notion that character evolution, either within a species or in speciation, proceeds by a simple series of fixed genes of additive effect. Complexities in the genetic basis of characters must be dealt with as a problem in quantitative inheritance (see Lande 1980*b*).

Just as in the case of character evolution, inference of modes of speciation by intense scrutiny of the natural populations of widespread *Drosophila* species have been only partly successful. Partial gene pool isolation has occurred, for example, in the South American subspecies of *pseudoobscura*, but the mode of origin of such discontinuities is open to question. Part of the problem may be attributable to the fact that most of the species selected for study are apparently relatively ancient, suggesting that the details of the events responsible for divergence and speciation are lost in antiquity. For example, *D. robusta* is a species widespread in temperate North America and has been the subject of much population genetic study since the early 1940s (see review in Levitan 1982). Recent phylogenetic information indicates that it is the only North American member of the *robusta* group, the rest being Palearctic. The other North American forms, including *colorata*, appear to belong to the *melanica* group (Beppu 1988). Chromosomal and systematic similarities indicate that *robusta* is derived from the Far East, probably reaching the North American continent via the Bering Strait (Narayanan 1973). However, whether the species *robusta* was formed before, during or after its arrival in North America is, at this point in time, difficult to infer. The same appears to be true of many of the other widespread species of the genus.

The above examples and perfunctory reviews have been presented in order to emphasize the fact, perhaps obvious to many people, that the evolutionary population geneticist needs a clear phylogenetic and ecological perspective before planning evolutionary genetics work. It appears that the study of speciation in anciently diverged populations and species will continue to go unrewarded because of the difficulty of reconstructing processes involved in character change and speciation.

2.1.4 *Sibling species not a critical emphasis*

The sibling species that were identified in many of the species groups mentioned above seemed at first to be critical indicators of the mode of species origin, because such species were at one time thought to be either incipient species or neospecies. More recently, these interpretations of sibling species seem less certain; most now appear to be full species of rather long standing, being separated by substantial genetic distances (for a discussion, see Carson 1976).

2.2 New approaches since 1970 in view of phylogenetic and ecological data on *Drosophila*

In the 50 years that have ensued since work began on natural populations of *Drosophila*, considerable phylogenetic data have accumulated on the genus. The *Drosophila* population geneticist is now in a position to use these data to plan a more sophisticated experimental attack on the processes of evolution as they occur in populations. Two groups of species in particular have emerged as especially promising areas for population genetics research; these are the *melanogaster* subgroup and the picture-winged species of Hawaii.

2.2.1 *The* melanogaster *subgroup*

Extensive recent work on the natural populations of *D. melanogaster* and the seven most closely related species of its subgroup have opened up new opportunities for approaching evolutionary processes. In the first place, *melanogaster* itself is distributed virtually worldwide and has very extensive chromosome variability in the form of inversions. Whereas most of these appear to be new mutants that do not persist or spread widely in natural populations, there is a basic array of four specific inverted sections that are virtually ubiquitous in populations of the species. These show striking clines of distribution that reflect latitude: these clines are mirrored both north and south of the equator with all the relatively inverted sections increasing toward the equator (for a review, see Knibb 1982). In addition, there are clear latitudinal clines in the frequency of at least four allozyme polymorphisms (Oakeshott *et al.* 1983). These include *Adh, G6pd, alphaGpdh* and *Est-6*. The changes are precise: the first two show an increase of the F allele with latitude; the latter two show an increase of the S allele with latitude. Although some of these effects may be partly due to hitch-hiking of these alleles on inversions, a thorough study of the situation (Voelker *et al.* 1978) concluded that the inversions failed to account for the allozyme clines.

D. melanogaster has a high tolerance for ethanol, which also shows clinal change. The most tolerant populations are present in the temperate zones and show high frequencies of *Adh*-F. Vouidibio *et al.* (1989) have reported a striking local situation in the tropical African city of Brazzaville. Whereas countryside populations are characteristic of the tropics (about 3 per cent *Adh*-F), highly localized brewery populations in the city were profoundly different (i.e. about 90 per cent *Adh*-F).

Such changes suggest that natural selection is responsible for this situation but clearly more than just the *Adh* locus alone is involved. As mentioned previously, adaptation to local conditions does not involve the production of fixed sets of genes in the homozygous state but rather the basis of this type of local adjustment appears to be shifts in polymorphic states. Indeed, a morphometric cline between the African and French populations has also been demonstrated for *melanogaster* (David and Bocquet 1975) and further experiments to integrate the morphological and biochemical data are awaited with interest.

New information has become available on the phylogenetic and geographical background of *D. melanogaster* and the seven most closely related species of its subgroup. These comprise a cluster of very close species (Lemeunier *et al.* 1986; Lachaise *et al.* 1988). Several *melanogaster* relatives are endemic to continental Africa and the fact that some are endemic to certain islands in the Indian Ocean adds additional interest. Only two of the species (*melanogaster* and *simulans*) are found worldwide; the recognition of locally endemic populations of certain species has opened up the field for the study of the dynamics of both character change and speciation in this group. The extensive behavioral, developmental, ecological and genetic information now becoming available on the species of this subgroup, and the ease of their culture in the laboratory, should stimulate a thorough study of the population dynamics of these species as related to their evolutionary origin. A partially missing element is thorough knowledge of the ecology of some of these species, especially natural oviposition sites in different parts of their ranges.

2.2.2 *The Hawaiian picture-winged* Drosophila *species*

Continuing to use *Drosophila* as a paradigm, let us turn to a consideration of a microcosm, the Hawaiian *Drosophila*. Before publication of the taxonomic monograph of Hardy (1965) on the Drosophilidae of Hawaii, little was known about the fauna either of Hawaii or of Pacific Oceania in general. Hardy's work provided names, descriptions and island origins of about 400 morphologically determined species endemic to the six high islands of Hawaii. This information stimulated the organization of a major project for the study of the fauna of this highly isolated archipelago.

It is noteworthy that except for inversion surveys in populations of a few selected species (e.g. Carson 1966; Carson and Sato 1969), the work of the first 10 years of the project (see review in Carson and Kaneshiro 1976) was largely phylogenetic, systematic and ecological in orientation. These studies were multidisciplinary, involving many aspects of the biology of the flies. Studies included internal and external anatomical features, descriptions of new species, sexual and agonistic behavior, laboratory culture, physiology and nutrition, ecology, oviposition sites, altitudinal and geographical distribution of species, comparative metaphase and polytene chromosomal studies, hybridization of species, allozymes, mitochondrial and nuclear DNA and immunological probes. This material has been presented in the various reviews cited earlier. Where is it all leading?

During the course of the study of the Hawaiian *Drosophila*, a crucial theory relating to the origin of the islands was independently developed (Wilson 1963; McDougall and Swanson 1972; Clague and Dalrymple (1987). It proposes that the islands of the archipelago have been sequentially formed as the Pacific tectonic plate has moved northwestward over an upwelling 'hot spot'. Potassium–argon dating of lava flows can be used to establish the age of each of the volcanoes. The highly detailed polytene chromosomal phylogenies of the *Drosophila* species, most of which show single-island endemism, gave several powerful new dimensions to the phylogenetic inferences. Stalker (1972) made the remarkable discovery that a number of intact Hawaiian polytene sequences were shared by several continental species, i.e. *moriwakii* of Japan and *colorata* of North America, species that are now considered to be primitive members of the *melanica* group (Beppu 1988). This suggests an ultimate continental source of the Hawaiian fauna, derived from either North America or Asia. Secondly, a broad picture of the sequential colonization of the islands from northeast to southwest in the last 5 million years can be inferred from the inversion data (for a recent short review, see Carson 1987*a*).

The extensive data on the biology and inter-island patterns of distribution of the Hawaiian species posed a serious problem of choice for the population geneticist, who could at first find little guide to the choice of a species to study. Partly for this reason, population genetics took a back seat to other studies that were involved in clarifying the shapes of the evolutionary problems at hand. Phylogenetic and ecological studies, however, came to be concentrated on the 'picture-wing' species and it is only recently that population genetics has come to be re-emphasized. Particular emphasis has been placed on the species endemic to the newest island, Hawaii, because this would seem to be the best place to recognize and select a neospecies for study (Carson 1984). Particular use came to be made of *D. silvestris* on the island of Hawaii. It was selected as a major object of study (Sene and Carson 1977; Craddock and Johnson 1979; Carson 1982*a*; Craddock and Carson 1989). Some idea of the wealth of material and the difficulty of the choosing process is set out in Fig. 1.

3. RESEARCH IN POPULATION GENETICS OF HAWAIIAN *DROSOPHILA* AS INFLUENCED BY PRIOR DATA ON GEOGRAPHY, ECOLOGY AND PHYLOGENY

A review of the phylogenetic, geographical and ecological information on the Hawaiian *Drosophila* has suggested the development of three lines of investigation in population genetics. Two of these will test, in experimental populations, certain hypotheses suggested by the data; (1) the founder hypothesis, a proposal that ascribes a crucial role to a founder event in the formation of new species and (2) the sexual selection hypothesis, that proposes a powerful role for sexual selection in the speciation process. A third line of investigation will be to carry out a genetic analysis of apomorphic character sets as a means of understanding the origin of recent evolutionary novelties.

3.1 Founder events and founder effects

3.1.1 *Definitions*

As developed by Mayr (1954), Carson (1971), Templeton (1980) and others, a *founder event* occurs when a geographically isolated area, not at that time occupied by a particular species, is successfully colonized by that species, following the rare dispersion of one or a very few sexual propagules. The term *founder effect* refers to the genetic changes that transpire in the new daughter population and that are set in motion by chance processes associated with the founder event. The severity of the

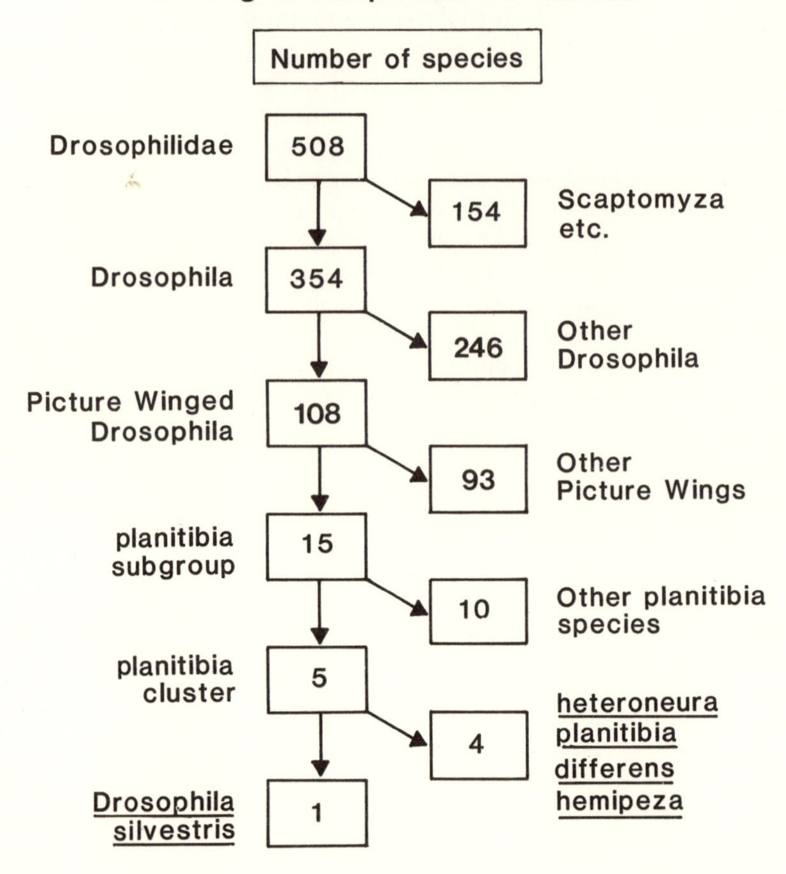

Fig. 1. The process of elimination whereby *Drosophila silvestris* was chosen as a species for laboratory study.

founder effects is expected to be positively correlated with the degree of isolation of the new population and inversely correlated with the number of propagules that either originally accomplish or subsequently reinforce the colonization event. Founder effects are probably negligible if the propagules are asexual or stem from sexual populations that are strongly inbred, self-fertilized or initially genetically homogeneous. In sexually reproducing populations, it has been argued that the founder effect will be strongly affected by the genetic state of the ancestral population (Templeton 1980; but see Barton 1989).

3.1.2 *Geographical and geological factors conducive to founder events in Hawaii*

Most prominent in this regard are the data, mentioned previously, on the sequential ages of these volcanic oceanic islands. Kauai is the oldest of the 'high' Hawaiian islands and its lava flows have been dated at 5.1 million years (my). The islands to the northwest are older but are all of very low altitude and seem unlikely to harbor endemic drosophilids at the present time. These flies are characteristic of high altitudes on the newer islands. No real attempts to collect drosophilids there, however, have been made. There are data that suggest that some of these northwestern islands existed as high islands in the remote past, but the founder event hypothesis is here applied not to possible ancient colonizations of the archipelago as a whole but to events that appear to have occurred on the much more recent major islands or island groups at the southeastern end of the island chain.

The geographic and geologic situation at the newer end of the archipelago is somewhat simplified by the fact that only three deepwater channels exist separating these newer islands. The existence of these channels stresses the fact that certain of the islands have never been connected by land bridges since their emergence above sea level. Figure 2 illustrates the situation that would occur were the sea level lowered or raised by about 300 m. At low-water levels, Maui, Molokai and Lanai would, because of the shallow channels, be joined into a single island. Indeed, there is strong evidence for the existence of such a land mass several times in the past (Macdonald and Abbot 1970). This anicent island has been designated as Maui Nui (Greater Maui).

Accordingly, three channels are abysmally deep: (1) the Kauai Channel between Kauai and Oahu (3350 m), (2) the Kaiwi Channel between Oahu and Molokai (706 m) and (3) the Alenuihaha Channel between Maui and Hawaii (1865 m) (see Figs 2 and 3). In contrast, none of the channels separating the present islands of the Maui complex exceeds 67 m in depth. Accordingly, the three deep channels are geographically crucial in that they provide an opportunity for an inter-island, overwater founder event to occur. Moreover, the probability that such an event will involve the same ancestral lineage twice is extremely small, given the width of the channels.

3.2 Endemism of Hawaiian drosophilids

Approximately 110 picture-winged species are recognized on 11 separate volcanoes of the six major high islands (Fig. 3). Almost all of these species are highly distinctive morphologically, at least the males are. In the past 25 years, extensive collections of these species have been made.

Although such data are always subject to change with further collecting,

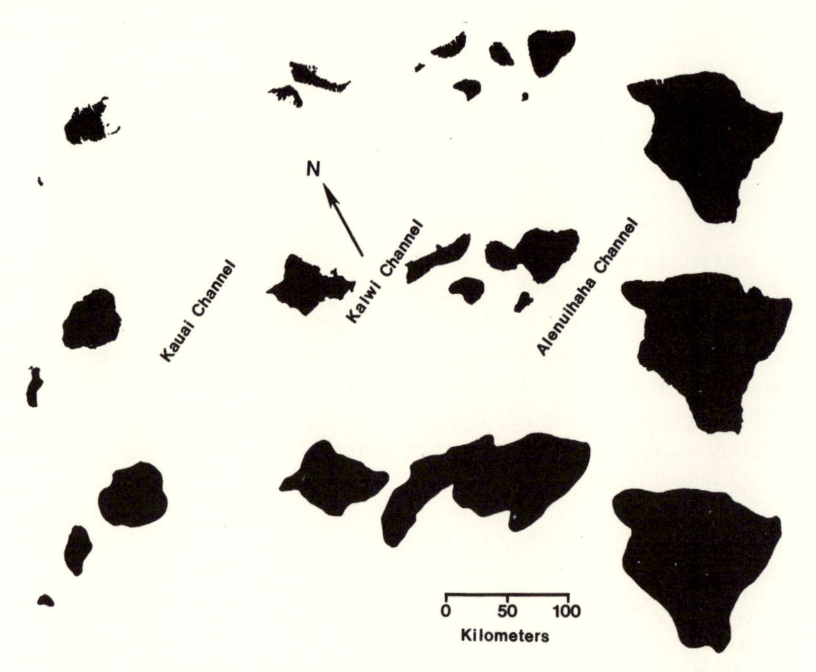

Fig. 2. The middle figure shows the main Hawaiian Islands at the present stage of sea level. The three major deep channels are shown. The top figure illustrates the islands at about 300 m above present sea level. The lower figure indicates their appearance at 300 m below present sea level. After Zimmerman (1948).

the distribution of species by island and volcano is now fairly well known. Like many other endemic Hawaiian insects, the degree of single-island and single-volcano endemism is high.

Colonization across any of the three overwater channels by a taxonomically single species has been recognized in only four instances. Two such crossings involved *D. grimshawi*, recognized under this name on 6 of the 11 volcanoes (see bottom of Fig. 3). Nevertheless, some of these island populations are strongly differentiated genetically (chorion sculpture, electrophoretic distance, behavior, yolk proteins, oviposition site).

Therefore, while the name '*grimshawi*' has been utilized to refer to classically determined, morphologically similar populations, it can be argued that the Kauai, Oahu and Maui-complex populations should be considered as three distinct species (Piano *et al.* 1988). This would remove two of the deep-channel crossings indicated in Fig. 3. A third case, that of *D. orthofascia*, is somewhat comparable to that of *grimshawi*. Thus, although the three Maui-complex populations appear to be quite similar in all respects so far studied, the morphologically similar Hawaii population displays unique cytological and behavioral differences from the popu-

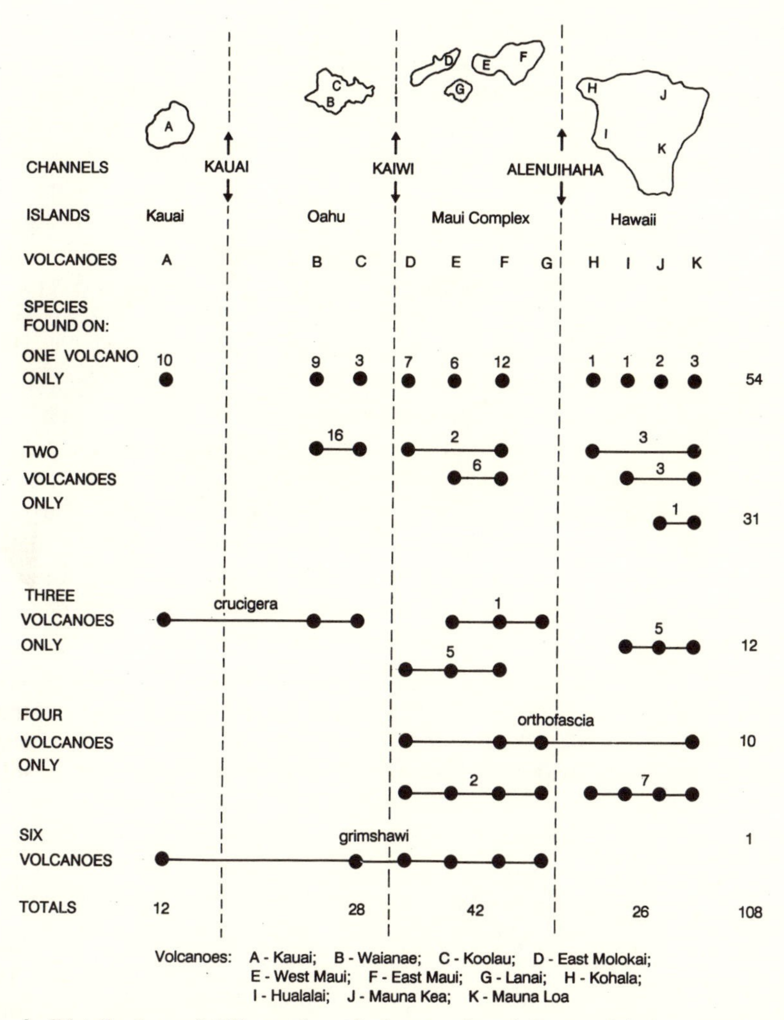

Fig. 3. Distribution of 108 species of picture-winged *Drosophila* by island and volcano. The dashed vertical lines mark the positions of the three abysmally deep channels.

lations on the Maui complex. Such characters could conceivably be used to justify specific rank. It should be pointed out, however, that in the diagnosis of specific rank among Hawaiian *Drosophila*, one cannot safely apply hybridization, pre-mating or post-mating isolation data. This is because many of the populations are strictly allopatric: reproductive isolation that has been reinforced by selection appears to be rare, even between genetically very distant species. Inter-island hybrids between widely different morphological forms are frequently surprisingly easy to obtain in the laboratory and are often fertile (Yang and Wheeler 1969).

The above considerations indicate that of all the 108 species, *crucigera* is the only single species that has apparently crossed an abysmal channel without undergoing species-level genetic changes (Fig. 3). On this basis, the endemism for islands or island groups that are separated by deep channels is about 99 per cent. Such facts, coupled with the clear affinities of the different subgroups on the two sides of the channels, suggest that founder events are often accompanied by speciation, and that experimental population studies of such possible effects might be profitable.

DeSalle and Templeton (1988) found a difference in evolutionary rates between species that have apparently been subject to strong founder events and those, differentiating within the Maui complex, which were apparently not so subjected. Although the rate differences are substantial, it should be pointed out that founder events can by no means be excluded in the history of species developing on neighboring volcanoes. This may even be true within the confines of a single volcano.

Certain of the islands currently consist of two or more adjacent volcanoes. There is evidence that they have grown at different times (Macdonald and Abbott 1970) so that they provide an opportunity for the occurrence of a founder event between an older one and a younger one. Wide water gaps do not appear to be necessary. Accordingly, inter-volcano founder events may have originally occurred between the two volcanoes of Oahu, Waianae (3.7 my) and the younger Koolau (2.6 my) or between the separate volcanoes of the Maui complex or Hawaii Island (details of the numbers of species involved is shown in Fig. 3). Colonization without speciation is frequently observed between such volcanoes. This is manifested in the data on the sharing of identical chromosomal polymorphisms. Nevertheless, each volcano, except Lanai, has at least one endemic species. These species are frequently more closely related to allopatric forms on an adjacent volcano than they are to species that are sympatric with them (Carson and Yoon 1982), an observation that argues against a sympatric mode of speciation.

Remoteness of colonization, therefore, forms a series of hierarchies. These may be carried one step further if consideration is given to the geological evidence indicating the manner in which an individual shield volcano grows. A study of the relatively new volcanoes of Maui and Hawaii suggest that barriers to colonization and the setting up of founder event possibilities exist even within the confines of a single volcano. The growth of shield volcanoes involves primarily the eruption of lava from fissures on the flanks. As the volcano grows, lava destroys large areas of existing forest an an irregular but relentless rate. Each new flow forms a cooled lava bed that is usually slightly elevated relative to the surrounding areas on the flank of the growing mountain. Each cycle of replacement is completed as the somewhat lower forested slopes eventually come to be inundated with lava.

3.3 Surface replacement in shield volcanoes

Surface replacement cycles may be observed in the Holocene eruptive history of currently active volcanoes. At Mauna Loa on the island of Hawaii, for example, the evidence is strong that the pattern of surface turnover extends back into the early stages of the history of the mountain. Using radiocarbon methods to record the ages of lava flows on the southwest rift, Lockwood and Lipman (1987) conclude that approximately 90 per cent of Mauna Loa's surface has been buried in the past 4000 years.

Where new lava has cut a swath through an established forest, the latter serves as a source of propagules for colonization of the bare cooled lava. The adjacent forest may contain all of the complex elements of the ecosystem characteristic of the altitudinal and rainfall profiles of that part of that particular region of the mountain. If this is the case, the process of colonization may proceed simply, much in the manner of old field succession, that is, without major changes in the gene pools of the organisms involved (Carson 1987*c*).

During very active growth of the shield, however, adjacent ecosystems tend to remain young and incomplete with regard to many biotic elements (Mueller-Dombois *et al.* 1981). A young, recently cooled flow, therefore, will be likely to find itself adjacent to a sparsely vegetated young flow, containing only elements of the wind-dispersed pioneer species. This situation appears to create an opportunity for long-distance dispersal to the new site, possibly even the arrival of successful propagules from an adjacent volcano.

Biotic colonization cycles, as described above, appear to occur as a result of the continual surface turnover. If complete surface turnover is estimated conservatively as requiring 8000 years, a shield volcano 100 000 years old will have undergone a large number of cycles of complete biotic replacement. During this time, it seems likely that the gene pools of many species, especially those that are secondary and slow to colonize, would be likely to be subjected to occasional founder events. As pointed out earlier, founder events are not necessarily accompanied by founder effects such as major genetic shifts or genetic reorganizations that might lead to species formation. Nevertheless, the potential for genetic change to occur, especially during the early very active growth phase of the volcano, appears to be substantial.

3.4 *Drosophila silvestris* as a paradigm

Although more data are highly desirable, the extraordinary genetic diversity encountered between populations of the single species *D. silvestris* expressed in hierarchical *F* statistics (Craddock and Carson 1989) suggests rapid evolutionary divergence in local populations. This species is very closely related to allopatric species on the adjacent older volcanoes of Molokai and Maui on the west side of the Alenuihaha Channel. Behavioral and morphological data suggest that it arose on Hualalai (see review in Carson 1982*a*).

Once formed, *silvestris* appears to have colonized in two directions, to the south along the southwest slope of Mauna Loa, and to the northeast, colonizing the Kohala volcano. The first appears to have occurred in a stepwise manner, as the populations show local genetic differentiation that increases with distance. The area has been continually crossed by new molten lava flows (see fig. 2 in Carson and Templeton 1984). The Kohala colonization produced a more profound break in the gene pool. From there it appears to have spread stepwise along the north and east slopes of Mauna Kea and ultimately to the new lava flows on the northeastern side of Mauna Loa. The first step (Hualalai to Kohala) appears to have involved a founder event across the low-lying dry saddle between the volcanoes. Unlike the southwest colonization, this was apparently accompanied by an abrupt change in foreleg bristle number and distribution (Carson 1982*a*). This bristle change is retained and, indeed, embellished in the later steps, which are clinal rather than abrupt. These morphological changes are paralleled by cumulative changes in sexual behavior (Kaneshiro and Kurihara 1981; Kaneshiro and Giddings 1987) and in mtDNA (DeSalle *et al.* 1986; DeSalle and Templeton 1987). All these data are consistent with this mode of geographical spread from Kohala to the east.

Although the specific name '*silvestris*' has been retained for all these populations, the apparent founder event between Hualalai and Kohala has been accompanied by genetic changes that might form the basis of specific recognition (Carson and Lande 1984). Nevertheless, assortative mating in the laboratory between these two forms is minimal (Spiess and Carson 1981). Further consideration is given to this case in the forthcoming discussion of sexual selection, especially with regard to the origin of novel secondary sexual character states.

3.5 Possible founder effects

So far, this account has served merely to suggest that the appearance of substantial genetic change often seems to coincide with founder events. Considerable attention, however, has been given to the theory that a major recombinational genetic reorganization is involved as a basis for speciation following population bottlenecks (e.g. Carson 1982*b*; Carson and Templeton 1984). Nevertheless, no demonstration has been made directly that any specific founder effects of a genetic nature are involved. The thrust of this chapter is that a population genetic approach may be used to investigate possible founder effects in a causal–analytical manner. Because laboratory populations of these species of *Drosophila* are easily made and maintained, empirical founder events are open to experimental study.

As will be described in Section 3.7, the mating system, both behavior and the morphological embellishments of behavior, seem vulnerable to change in Hawaiian *Drosophila*; variability in these systems is thus especially open to experimental study both in the laboratory and in nature.

Few critics deny the existence of founder *events* in the geographical history and demographics of the biota of such an archipelago as Hawaii. The main challenge to the founder hypothesis has come from geneticists arguing from theoretical models that it is neither necessary nor sufficient to invoke specific genetic effects, accompanying, or immediately following, the founder event (see Barton and Charlesworth 1984; Charlesworth and Rouhani 1988; and Barton 1989). These models make simplifying assumptions and there are limited genetic data against which to test these assumptions.

Experimental data on the possible genetic effects correlated with population bottlenecks (e.g. Powell 1978; Ringo *et al*., 1985; Bryant and Meffert 1988; Galiana *et al*. 1989) have so far been few, but they have shown that much genetic variability can pass through even a very narrow bottleneck. Most losses of variability involve alleles that were rare in the initial population (Nei *et al*. 1985). Some theoretical formulations have suggested that genetic variance may actually *increase* at a bottleneck (Goodnight 1987, 1988; Tachida and Cockerham 1989*a*). Indeed, such increases have been actually observed in several instances (Lints and Bourgois 1984; Bryant *et al*. 1986; Carson and Wisotzkey 1989). The further interesting possibility that the mutagenic and/or regulatory effects of transposable elements might be activated in a population destabilized by the genetic stress accompanying a bottleneck has been offered (McDonald 1989). There is little direct evidence of such effects following bottlenecks, although stress due to temperature shocks has strong genetic effects that are apparently mediated by transposable elements (Ratner and Vasilyeva 1989).

Although interest is increasing, more theoretical and empirical studies are needed to evaluate the role of founder effects in character evolution and speciation. Most particularly, experimental work to explore the precise genetic effects of bottlenecks on quantitative traits is needed. An overview of the genetics of *Drosophila* populations continues to indicate the importance of alterations in quantitative traits as the building blocks from which new adaptive genetic systems are forged by selection. Models of population-induced shifts in quantitative traits are especially difficult to make (Tachida and Cockerham 1989*b*). Nevertheless, the exuberant recent evolution observed in such traits in Hawaiian *Drosophila* suggests that this material is especially suitable for empirical study of the population genetics of this problem.

3.6 Discordance between morphological evolution and the ambient environment

In addition to criticizing the founder effect hypothesis on the basis of theoretical genetics, Barton (1989) has raised certain ecological arguments with respect to the evolution of the Hawaiian drosophilids. These stimulate a few comments here. It is contended that a population founded by either a large or a small set of propagules may find itself in a novel ambient environment. According to this view, genetic response to this new environment results from standard genetic processes similar to those that occur in the origin of adaptations in large populations.

Observations indicate, however, that the morphological, physiological and behavioral changes that occur after founding do not correlate clearly with many readily observable features in the ambient environment. For example, the early colonization of the island of Hawaii seems to have reached the Hualalai and Kohala volcanoes. The ecosystems on these volcanoes are extraodinarily similar to those present in the forests of the older volcanoes of Maui and Molokai. This includes most of the same major species of forest trees, tree ferns and understory elements. Yet each species of picture-winged *Drosophila* found on these volcanoes represents a novel evolutionary development. In most cases, host relationships of derived and ancestral species of flies are similar, the only change being a tendency for some species on the newest island, Hawaii, to show a somewhat broadened generalism in oviposition site.

The species *heteroneura* and *silvestris* may be mentioned as an example. These two are partially sympatric, utilize the same host plants for oviposition and occasionally produce hybrids, fertile in both sexes, that nevertheless do not result in hybrid swarms (Carson *et al.* 1989). Their morphological differences appear to relate to their differing modes of sexual selection rather than to features of the ambient environment. Accordingly, it is contended that the genetic changes that occur relate to the sexual and not the ambient environment.

3.7 The sexual selection hypothesis

As stated above, the extraordinary proliferation of species and species differences in the Hawaiian drosophilids appears to be discordant with the characteristics of the ambient environment. This discordance requires explanation. Even a cursory review of the taxonomic descriptions of the flies reveals that the most spectacular evolutionary developments between species in the Hawaiian drosophilids relate to the reproductive milieu, especially the development of secondary sexual characters in males. That these characters may be the outcome of a pervasive sexual selection is therefore a strong possibility (Ringo 1977; Carson 1978). It is very unlikely that niche differentiation, for example, would specifically affect the observed secondary sexual characters.

Initial studies have shown that local populations harbor genetic variability in both male-specific traits (Carson 1985; Carson and Lande 1984) and mate choice by females (Kaneshiro and Kurihara 1981). The reproductive success of individual males in experimental populations is highly variable, that is, about one-third of the males accomplish two-thirds of the matings (Carson 1986). Most simple theoretical schemes call for the exhaustion of genetic variability among males due to runaway directional selection for sexual characters. Further experiments with *silvestris*, however, suggest that genetic polymorphism for these characters is maintained. In a population that carried multiple inversion polymorphisms, parallel studies of survival and mating success revealed that complex inversion heterozygotes are very significantly higher among breeding individuals (Carson 1987*d*; see also Carson and Wisotzkey 1989).

The sexual selection process in *silvestris* thus appears to be associated with a high variance in male mating success (Carson 1987*b*). The majority of the males in the population appear to be relegated to a relatively low Darwinian fitness. This situation would be expected to reduce the effective population size. A further extreme reduction in population size may result from a founder event. In turn, such a chance event might lead to a breakdown in the normal choice mechanism that was inherent in the sexual selection system of the population from which the founder or founders were drawn. Realignment into a new sexual selection system may open the way to the origin of novel characters by temporary directional sexual selection (see Section 3.8).

Such a hypothesis is open to experimental scrutiny and there have already been some suggestive results. Thus Ahearn (1980) identified behavioral changes in mate choice in a laboratory population that had gone through a bottleneck. In particular, there was a loss of discrimination on the part of the females in the bottlenecked population. Kaneshiro (1989) reported that a laboratory population showed response to artificial selection for either high or low discrimination on the part of females. These behavioral changes were accompanied by an alteration in the numbers of tibial cilia in males, a secondary sexual character. Carson and Teramoto (1984) showed that the number and distribution of male tibial cilia can also be altered significantly by selection for high or low numbers.

In no case so far, however, has fixation of any of this genetic variability been observed. This implies that selection is impinging on and causing shifts in an underlying balanced polymorphism.

3.8 The genetics of apomorphic characters

As the Hawaiian drosophilids have evolved over the last 5 million years on the newer Hawaiian islands, many novel characters have appeared (Hardy 1965). Most are unlike those seen in continental species but some show convergence within Hawaii. These characters are manifested most strikingly in the male sex, which suggests that they are modified to promote reproductive success by their carriers. This includes characters that relate to agonistic male–male interactions as well as courtship displays. Examples include head shape, maxillary palp ornamentation and modification of mouthparts, antennae or aristae. On the thorax, the forelegs of males of various species are strikingly modified in dozens of different ways. In some cases, the mesothoracic leg is modified and is used in either courtship or agonistic display.

Both wings and abdomen also participate in this exuberant interspecific differentiation of males and in some cases are involved in sound production during courtship (Hoy *et al.* 1988). When the behavior of the insect is observed, the manner of deployment of these characters suggests that the great majority of these relate to the courtship display, although some relate to male–male combat (Spieth 1982). When the many different lineages of Hawaiian drosophilids are compared, possible convergence of character state arises. This is of course a problem with which all systematists continually struggle. In the case of the Hawaiian drosophilids, especially the picture-winged species, the chromosomal phylogeny provides a way to distinguish convergent characters from those that are more ancient and indicate bona fide common descent.

3.9 Limited usefulness of behavioral characters in the intergroup phylogeny of Hawaiian *Drosophila*

The thrust of this chapter has been to discuss guides to evolutionary process that the population geneticist can glean from phylogenetic considerations. In this final discussion, however, the procedure is reversed and some thoughts are offered on certain attempts that have been made to utilize behavioral characters in a broad phylogenetic context.

There is no question that males of closely related species tend to share courtship-related, morphologically based characters that are sometimes greatly and sometimes slightly modified from species to species within a subgroup. Thus, for example, 11 species of Hawaiian drosophilids are highly distinctive in that they have the second tarsomere of the male foreleg modified into a short, flattened, concave segment (the 'spoon tarsus' subgroup). The 'spoon' is applied to the sides of the female's abdomen as the male follows the female from behind during courtship. Like many similar foreleg characters, this serves as an excellent character for grouping these species, although the females lack the character.

Spieth (1966, 1982) shows how the Hawaiian drosophilids can be split into many small groupings in the above manner. In cases in which these narrow subgroups can be independently checked for chromosomal polytene sequence similarity, the cytological similarity confirms the morphological similarity very well. The occasional exceptions, however, are instructive in that they show that behavioral novelties can arise (or drop out) dramatically between very closely related species. Usually, these are characters that are not paralleled by novel morphological changes. A striking case involves the 'assault' mating behavior displayed by males of *D. neopicta* (Spieth 1982). Most of the males of the *planitibia* subgroup to which this species belongs are large in size, with striking secondary sexual characters in males that are specially deployed in lek behavior and in elaborate courtship movements. On the other hand, *D. neopicta* is a small species that has no lek behavior or special male morphology. The males seek out the females and attempt to achieve copulation by assault. Yet the characteristics of the polytene chromosomes tie this species extremely closely to the lek-forming species of the subgroup. In fact, *D. neopicta* even shares two autosomal inversion polymorphisms with *D. neoperkinsi*, a characteristic lek-forming species. This is unusually strong evidence of a close relationship (Carson 1973). In allozymes, furthermore, the two show a coefficient of genetic similarity of 1.0 (Johnson *et al.* 1975).

In view of the possibility of such localized and apparently rapid major shifts, behavioral attributes such as lek formation have a limited usefulness for broad phylogenetic comparisons unless they are tied to specific and unique morphological structures for which the possibility of convergent evolution can be eliminated. This is of importance, because Spieth (1981), basing his conclusions largely on the lack of lek behavior in *D. primaeva*, has removed the latter from close relationship with the picture-winged species, disregarding the strong chromosomal sequence similarities. Although the chromosomal affinities of *primaeva* with the continental *robusta–melanica* group are strong (Stalker 1972), Spieth and Heed (1975) have further argued for the position of *D. pinicola* of western North America as occupying a position more closely related to the ancestor of the Hawaiian *Drosophila*. This is based on display movements that are not rooted in morphology. Under sexual selection, similar behavioral attributes repeatedly appear as extreme apomorphic characters within several Hawaiian phylads. These characters may be very useful for the study of dynamic process in an intraspecific context, as Kaneshiro (1989) has shown, but are so subject to convergence as to be virtually useless for broad phylogenetic inference.

4. CONCLUSIONS

Phylogenetic studies permit inferences about mechanisms of evolution. Nevertheless, it is only by analyzing intraspecific populations that the genetic processes whereby evolution operates (mutation, recombination, selection, random drift) are open to direct experimental attack. The latter approach is the domain of the population geneticist. Curiously, phylogenetic and populational approaches to evolutionary study have tended to proceed along disparate lines, yet it is obvious that each approach has much to learn from the other. This chapter explores what population genetics can learn from phylogeny. If the origin of adaptations and/or species are to be studied in a single human life-span, a poor choice of experimental material can be a serious handicap for the researcher. Many – perhaps most – species are unsuitable for observing the accumulation of genetic change over time. They are characterized by long generation times and large populations with complex gene pools. An investigation of processes resulting in species formation encounters even greater difficulties and is best served by ferreting out for study either newly formed species (neospecies) or incipient species with short life-cycles. All available geological, geographical, ecological, systematic and phylogenetic information must be brought to bear on the selection of a species for the analytical genetic study of evolutionary process. Examples are cited from past work on continental *Drosophila* in which the choice of species was often unsuitable for the study either of species formation or of the origin of novel characters. The reasons for the current research stress on populations of Hawaiian species such as *D. silvestris* (endemic to the island of Hawaii, age $< 4 \times 10^5$ years) are recounted. Such recent single-island endemic species are interpreted as neospecies, having been recently formed allopatrically from ancestral populations on the older islands. In the study of neospecies, three research emphases recommend themselves; all are amenable to an experimental approach. First, biogeographical data suggest an important role for a population size bottleneck (founder event) in the emergence of genetic changes that result in the formation of a new species of Hawaiian *Drosophila*. Such bottlenecks can be easily engineered in the laboratory. Secondly, phylogenetic data serve to focus attention on the origin of unique, sex-limited, secondary sexual character sets. Accordingly, attention both to the genetics of such characters and to the performance of experiments on the population genetics of mate choice and sexual selection are badly needed. Finally, interspecific comparisons of the clusters of new Hawaiian species suggest that newly evolved or apomorphic character sets can be identified. Such characters may be far more amenable to a genetic analysis of their origin than those that are more ancient in a lineage.

ACKNOWLEDGEMENTS

This research was supported by National Science Foundation grant BSR 8415633. I thank Dr K. Y. Kaneshiro for helping to assemble the data for Fig. 3.

REFERENCES

Ahearn, J. N. (1980). Evolution of behavioral isolation in a laboratory stock of *Drosophila silvestris*. *Experientia* **36,** 63–4.

Ashburner, M., Carson, H. L. and Thompson, J. N. Jr (ed). (1981–6). *The genetics and biology of Drosophila*, Vols 3a–3e. Academic Press, London.

Barton, N. H. (1989). Founder effect speciation. In *Speciation and its consequences* (ed. D. Otte and J. A. Endler), pp. 229–56. Sinauer, Sunderland, Mass.

—— and Charlesworth, B. (1984). Genetic revolutions, founder events and speciation. *Ann. Rev. Ecol. Syst.* **15,** 133–64.

Beppu, K. (1988). Systematic positions of three *Drosophila* species (*Diptera, Drosophilidae*) in the *virilis–repleta* radiation. *Proc. Japan Soc. Syst. Zool.* **37,** 55–8.

Bryant, E. H. and Meffert, L. M. (1988). Effect of an experimental bottleneck on morphological integration in the house-fly. *Evolution* **42,** 698–707.

——, McCommas, S. A. and Combs, L. M. (1986). The effect of an experimental bottleneck upon quantitative genetic variation in the housefly. *Genetics* **114,** 1191–1211.

Carson, H. L. (1966). Chromosomal races of *Drosophila crucigera* from the Islands of Oahu and Kauai, State of Hawaii. *University of Texas Publication* **6615,** 405–412.

—— (1971). Speciation and the founder principle. *University of Missouri Stadler Symposia* **3,** 51–70.

—— (1973). Ancient chromosomal polymorphism in Hawaiian *Drosophila. Nature* **241,** 200–202.

—— (1976). Genetic differences between newly formed species. *BioScience* **26,** 700–701.

—— (1978). Speciation and sexual selection in Hawaiian *Drosophila.* In *Ecological genetics, the interface* (ed. P. F. Brussard), pp. 93–107. Springer-Verlag, New York.

—— (1982*a*). Evolution of *Drosophila* on the newer Hawaiian volcanoes. *Heredity* **48,** 3–25.

—— (1982*b*). Speciation as a major reorganisation of polygenic balances. In *Mechanisms of speciation* (ed. C. Barigozzi), pp. 411–33. Alan J. Liss, New York.

—— (1984). Speciation and the founder effect on a new Oceanic island. In *Biogeography of the tropical Pacific* (ed. F. J. Radovsky, P. H. Raven and S. H. Sohmer), pp. 45–54. B.P. Bishop Museum, Honolulu.

—— (1985). Genetic variation in a courtship-related male character in *Drosophila silvestris* from a single Hawaiian locality. *Evolution* **39**, 678–86.

—— (1986). Sexual selection and speciation. In *Evolutionary processes and theory* (ed. S. Karlin and E. Nevo), pp. 391–409. Academic Press, London.

—— (1987*a*). Tracing ancestry with chromosomal sequences. *Trends Ecol. Evol.* **2,** 203–307.

—— (1987*b*). The contribution of sexual behavior to Darwinian fitness. *Behavior Genetics* **17,** 597–611.

—— (1987*c*). Colonization and speciation. In *Colonization, succession and stability* (ed. A. J. Gray, M. J. Crawley and P. J. Edwards), pp. 187–206. Blackwell, London.

—— (1987*d*). High fitness of heterokaryotypic individuals segregating naturally within a long-standing laboratory population of *Drosophila silvestris. Genetics* **116,** 415–22.

—— and Kaneshiro, K. Y. (1976). *Drosophila* of Hawaii: Systematics and ecological genetics. *Ann. Rev. Ecol. Syst.* **7,** 311–45.

—— and Lande, R. (1984). Inheritance of a secondary sexual character in *Drosophila silvestris. Proc. Natl Acad. Sci. USA* **81,** 6904–7.

—— and Sato, J. E. (1969). Microevolution within three species of Hawaiian *Drosophila. Evolution* **23,** 493–501.

—— and Templeton, A. R. (1984). Genetic revolutions in relation to speciation phenomena: The founding of new populations. *Ann. Rev. Ecol. Syst.* **15,** 97–131.

—— and Teramoto, L. T. (1984). Artificial selection for a secondary sexual character in males of *Drosophila silvestris* of Hawaii. *Proc. Natl Acad. Sci. USA* **81,** 3915–17.

—— and Wisotzkey, R. G. (1989). Increase in genetic variance following a population bottleneck. *Amer. Nat.* **134,** 668–73.

—— and Yoon, J. S. (1982). Genetics and evolution of Hawaiian *Drosophila.* In *The genetics and biology of Drosophila* (ed. M. Ashburner, H. L. Carson and J. N. Thompson, Jr), Vol. 3b, pp. 297–344. Academic Press, New York.

——, Kaneshiro, K. Y. and Val, F. C. (1989). Natural hybridization between the sympatric Hawaiian species *Drosophila silvestris* and *Drosophila heteroneura. Evolution* **43,** 190–203.

Charlesworth, B. and Rouhani, S. (1988). The probability of peak shifts in a founder population. II. An additive polygenic trait. *Evolution* **42,** 1129–45.

Clague, D. A. and Dalrymple, G. B. (1987). The Hawaiian-Emperor volcanic chain. Part 1. In *Volcanism in Hawaii* (ed. R. W. Decker, T. L. Wright and P. H. Stauffer), Vol. 1, pp. 5–54. U.S. Geological Survey Professional Paper No. 1350, U.S. Government Printing Office, Washington, D.C.

Craddock, E. M. and Carson, H. L. (1989). Chromosomal inversion patterning and population differentiation in a young insular species, *Drosophila silvestris. Proc. Natl Acad. Sci. USA* **86,** 4798–802.

—— and Johnson, W. E. (1979). Genetic variation in Hawaiian *Drosophila* V. Chromosomal and allozymic diversity in *Drosophila silvestris* and its homosequential species. *Evolution* **33,** 137–55.

David, J. R. and Bocquet, C. (1975). Similarities and differences in latitudinal adaptation of two *Drosophila* sibling species. *Nature* **257,** 588–90.

DeSalle, R. and Templeton, A. R. (1987). Comments on 'The Significance of Asymmetrical Sexual Isolation'. *Evol. Biol.* **21,** 21–7.

—— and Templeton, A. R. (1988). Founder effects and the rate of mitochondrial DNA evolution in Hawaiian *Drosophila. Evolution* **42,** 1076–84.

——, Giddings, L. V. and Kaneshiro K. Y. (1986). Mitochondrial DNA variability in natural populations of Hawaiian *Drosophila*. II. Genetic and phylogenetic relationships of natural populations of *D. silvestris* and *D. heteroneura. Heredity* **56,** 87–96.

Dobzhansky, Th. (1950). Mendelian populations and their evolution. *Amer. Nat.* **84,** 401–18.

—— and Spassky, B. (1959). *Drosophila paulistorum*, a cluster of species *in statu nascendi. Proc. Natl Acad. Sci. USA* **45,** 419–28.

Galiana, A., Ayala, F. J. and Moya, A. (1989). Flush-crash experiments in *Drosophila.* In *Evolutionary biology of transient unstable populations* (ed. A. Fontdevila), pp. 58–73. Springer-Verlag, New York.

Goodnight, C. J. (1987). On the effect of founder events on epistatic genetic variance. *Evolution* **41,** 80–91.

—— (1988). Epistasis and the effect of founder events on the additive genetic variance. *Evolution* **42,** 441–54.

Hardy, D. E. (1965). *Insects of Hawaii*, Vol. 12. University of Hawaii Press, Honolulu.

Hedrick, P. W. (1980). 'Hitchhiking': A comparison of linkage and partial selfing. *Genetics* **94,** 791-808.

Hoy, R. R., Hoikkala, A. and Kaneshiro, K. Y. (1988). Hawaiian courtship songs: Evolutionary innovation in communication signals of *Drosophila. Science* **240,** 217–19.

Johnson, W. E., Carson, H. L., Kaneshiro, K. Y., Steiner, W. W. and Cooper M. M. (1975). Allozyme differentiation in the *Drosophila planitibia* subgroup. In *Isozymes IV: Genetics and evolution* (ed. C. L. Markert), pp. 563–84. Academic Press, New York.

Kaneshiro, K. Y. (1989). Dynamics of sexual selection and founder effects in species formation. In *Genetics, speciation and the founder principle* (ed. L. V. Giddings, K. Y. Kaneshiro and W. W. Anderson), pp. 279–96. Oxford University Press, New York.

—— and Giddings, L. V. (1987). The significance of asymmetrical sexual isolation and the formation of new species. *Evol. Biol.* **21,** 29–43.

—— and Kurihara, J. S. (1981). Sequential differentiation of sexual behavior in populations of *Drosophila silvestris. Pacific Science* **35,** 177–83.

Kimura, M. (1983). *The neutral theory of molecular evolution*. Cambridge University Press, Cambridge.

Knibb, W. R. (1982). Chromosome inversion polymorphisms in *Drosophila melanogaster* II. Geographic clines and climatic associations in Australasia, North America and Asia. *Genetica (The Hague)* **58,** 213–21.

Lachaise, D., Cariou, M.-L., David, J. R., Lemeunier, F., Tsacas, L. and Ashburner, M. (1988). Historical biogeography of the *Drosophila melanogaster* species subgroup. *Evol. Biol.* **22,** 159–225.

Lande, R. (1980*a*). Genetic variation and phenotypic evolution during allopatric speciation. *Amer. Nat.* **116,** 463–9.

—— (1980*b*). The genetic covariance between characters maintained by pleiotropic mutations. *Genetics* **94,** 203–15.

Lemeunier, F., David, J. R. and Tsacas, L. (1986). The *melanogaster* species group. In *The genetics and biology of Drosophila* (ed. M. Ashburner, H. L. Carson and J. N. Thompson, Jr), Vol. 3e, pp. 147–256. Academic Press, New York.

Levitan, M. (1982). The *robusta* and *melanica* groups. In *The genetics and biology of Drosophila* (ed. M. Ashburner, H. L. Carson and J. N. Thompson, Jr), Vol. 3b, pp. 141–92. Academic Press, New York.

Lints, F. A. and Bourgois, M. (1984). Population crash, population flush and genetic variability in cage populations of *Drosophila melanogaster. Genet. Select. Evol.* **16,** 45–56.

Lockwood, J. P. and Lipman, P. W. (1987). Holocene eruptive history of Mauna Loa volcano. In *Volcanism in Hawaii* (ed. R. W. Decker, T. L. Wright and P. H. Stauffer), Vol. 1, pp. 509–35. U.S. Geological Survey Professional Paper 1350. U.S. Government Printing Office, Washington, D.C.

MacDonald, G. A. and Abbott, A.T. (1970). *Volcanoes in the sea*. University of Hawaii Press, Honolulu.

Mayr, E. (1954). Change of genetic environment and evolution. In *Evolution as a process* (ed. J. Huxley, A. C. Hardy and E. B. Ford), pp. 157–80. Allen and Unwin, London.

McDonald, J. F. (1989). The potential evolutionary significance of retroviral-like transposable elements in peripheral populations. In *Evolutionary biology of transient unstable populations* (ed. A. Fontdevila), pp. 190–205. Springer-Verlag, New York.

McDougall, I. and Swanson, D. A. (1972). Potassium–argon ages of lavas from Hawi and Pololu volcanic series, Kohala Volcano, Hawaii. *Geol. Soc. Amer. Bull.* **83,** 3731–8.

Mueller-Dombois, D., Bridges, K. W. and H. L. Carson (ed.) (1981). *Island ecosystems*. Hutchinson Ross, Stroudsburg, Penn.

Narayanan, Y. (1973). The phylogenetic relationships of the members of the *Drosophila robusta* group. *Genetics* **73,** 319–50.

Nei, M., Maruyama, T. and Chakraborty, R. (1975). The bottleneck effect and genetic variability in populations. *Evolution* **29,** 1–10.

Oakeshott, J. G., Chambers, G. K., Gibson, G. K., Eanes W. F. and Willcocks, D. A. (1983). Geographic variation in *G6pd* and *Pgd* allele frequencies in *Drosophila melanogaster. Heredity* **50,** 67–72.

Patterson, H. T. and Stone, W. S. (1952). *Evolution in the genus Drosophila*. Macmillan, New York.

Parsons, P. A. (1979). Polygenic variation in natural populations of *Drosophila*. In *Quantitative genetic variation* (ed. J. N. Thompson, Jr and J. M. Thoday), pp. 61–79. Academic Press, New York.

Piano, F., Seo, E. W., Craddock E. and Kambysellis, M. P. (1988). *Drosophila grimshawi*, a Hawaiian endemic: Island populations or incipient species? *Genome* **30,** 388 (suppl. 1).

Powell, J. R. (1978). The founder-flush theory: An experimental approach. *Evolution* **32,** 465–74.

Prevosti, A., Ribo, G., Serra, L., Aguade, M., Balana, J. Monclus, M. and Mestres, F. (1988). Colonization of America by *Drosophila subobscura*: Experiment in natural populations that supports the adaptive role of chromosomal-inversion polymorphisms. *Proc. Natl Acad. Sci. USA* **85,** 5597–600.

Ratner, V. A. and Vasilyeva, L. A. (1989). Mobile genetic elements and quantitative characters in *Drosophila*: Fast heritable changes under temperature treatment. In *Evolutionary biology of transient unstable populations* (ed. A. Fontdevila), pp. 165–89, Springer-Verlag, New York.

Ringo, J. M. (1977). Why 300 species of Hawaiian *Drosophila*? The sexual selection hypothesis. *Evolution* **31,** 695–6.

——, Wood, D., Rockwell, R. and Dowse, H. (1985). An experiment testing two hypotheses of speciation.*Amer. Nat.* **126,** 642–61.

Sene, F. M. and Carson, H. L. (1977). Genetic variation in Hawaiian *Drosophila* IV. Allozymic similarity between *D. silvestris* and *D. heteroneura* from the island of Hawaii. *Genetics* **86,** 187–98.

Spiess, E. B. and Carson, H. L. (1981). Sexual selection in *Drosophila silvestris* of Hawaii. *Proc. Natl Acad. Sci. USA* **78,** 3088–92.

Spieth, H. T. (1966). Courtship behavior of endemic Hawaiian *Drosophila. University of Texas Publication* **6615,** 245–313.

—— (1981). Courtship behavior and evolutionary status of the Hawaiian *Drosophila primaeva* Hardy and Kaneshiro. *Evolution* **35,** 815–17.

—— (1982). Behavioral biology and evolution of the Hawaiian picture-winged species group of *Drosophila. Evol. Biol.* **14,** 351–437.

—— and Heed, W.B. (1975). The *Drosophila pinicola* species group. *Pan-Pacific Entomol.* **51,** 287–95.

Stalker, H. D. (1972). Intergroup phylogenies in *Drosophila* species groups as determined by comparisons of salivary banding patterns. *Genetics* **70,** 457–74.

Sturtevant, A. H. (1921). *The North American species of Drosophila*. Carnegie Institution of Washington Publication 301, Washington, D.C.

Tachida, H. and Cockerham, C. C. (1989*a*). Effects of identity disequilibrium and linkage of quantitative variation in finite populations. *Genet. Res.* **53,** 63–70.

—— and Cockerham, C. C. (1989*b*). A building block model for quantitative genetics. *Genetics* **121,** 839–44.

Takano, T., Kusakabe, S. and Mukai, T. (1987). The genetic structure of natural populations of *Drosophila melanogaster*. XX. Comparison of genotype–environment interaction in viability between a northern and a southern population. *Genetics* **117,** 245–54.

Templeton, A. R. (1980). The theory of speciation via the founder principle. *Genetics* **94,** 1011–38.

Thompson, J. N., Jr and Thoday, J. M. (ed.) (1979). *Quantitative genetic variation*. Academic Press, London.

Voelker, R. A., Cockerham, C. C., Johnson, F. M., Schaffer, H. E., Mukai, T. and Mettler, E. (1978). Inversions fail to account for allozyme clines. *Genetics* **88,** 515–27.

Vouidibio, J., Capy, P., Defaye, D., Pla, E., Sandrin, J., Csink, A. and David, J. R. (1989). Short-range genetic structure of *Drosophila melanogaster* populations in an Afrotropical urban area and its significance. *Proc. Natl Acad. Sci. USA* **86,** 8442–6.

Wilson, J. T. (1963). A possible origin of the Hawaiian Islands. *Can. J. Physics* **41,** 135–8.

Yang, H. and Wheeler, M. R. (1969). Studies on interspecific hybridization within the picture-winged group of endemic Hawaiian *Drosophila. University of Texas Publication* **6918,** 133–70.

Zimmerman, E. C. (1948). *Insects of Hawaii*, Vol. 1. University of Hawaii Press, Honolulu.

Sexual selection, sensory systems and sensory exploitation

MICHAEL J. RYAN

1. INTRODUCTION

Sexual selection by female choice is a process involving communication. The male is the sender and his courtship display is the signal, whereas the female is the receiver, with both her sensory and endocrine systems responding to the signal. To the extent that variation in signals differentially influences receivers, there is the opportunity for sexual selection. Understanding the mechanisms of this communication process reveals how sexual selection by female choice operates. It can suggest how females might evolve preferences that result in adaptive mate choice, how sensory biases could determine the direction of the runaway process, and how males might evolve traits that exploit pre-existing sensory biases of the female (sensory exploitation). When combined with appropriate phylogenetic data, an understanding of the sensory basis of female choice can allow the test of hypotheses for the evolution of female choice. The purpose of this chapter is to consider sensory mechanisms in sexual selection.

2. FEMALE MATE CHOICE AS COMMUNICATION

In sexually reproducing species, mating rarely takes place without some form of communication between the two partners. In many mating systems, this communication takes place in the form of courtship, usually males courting females. Most studies of mate choice have dealt with species recognition. The early ethologists discovered that one of the most important functions of courtship was to communicate information about species identity; thus species-typical aspects of courtship behavior were documented extensively (e.g. Lorenz 1950; Tinbergen 1951; Morris 1956). However, in order to recognize conspecifics, not only must species differ in their displays, but females must be able to discriminate these differences; neuroethologists have been successful in demonstrating that sensory systems are biased towards species-specific information (e.g. Huber 1978; Capranica 1976; Hoy 1978; Kendrick and Baldwin 1987; Walkowiak 1988).

Studies of species recognition also have played an important role in studies of evolution. Speciation theory was the cornerstone of the Modern Synthesis in evolutionary biology (Dobzhansky 1937; Mayr 1942, 1963, 1982), and the role of courtship as an ethological isolating mechanism was the most important contribution of animal behavior to evolutionary theory (Blair 1964; Littlejohn 1965; Walker 1974; Alexander 1975). The value of considering interspecific mate choice as a problem in communication is apparent in the success of delineating this behavior at three levels of analysis: mechanism, current function and evolution.

Since the seminal work of Williams (1966) and the influential collection of contributed papers assembled by Campbell (1972), the choice of mates within a species (i.e. sexual selection by female choice) has become an important topic of research, attracting interest from behavioral ecologists and population geneticists. The parallels between species recognition and sexual selection by female choice are more obvious than are the distinctions. The two phenomena are similar in that both involve discrimination among potential mates (usually) by females, there is an obvious effect of the female's behavior on the mating success of males, and a possible influence of the female's choice on her own reproductive success. However, the process of male courtship and female choice as a problem in communication has been less appreciated in sexual selection studies than in studies of species recognition (but see Morris 1956; Barlow 1977; West-Eberhard 1979; Arak 1983; Burley 1985; Endler 1989; Rowland 1989*a*, 1989*b*). It is hoped that this chapter will convince readers of the value of considering sexual selection by female choice in the context of communication, especially regarding mechanisms of the receiver.

3. MECHANISMS OF FEMALE CHOICE

3.1 Mate choice based on male traits

Studies of female mate choice in the past two decades have helped resurrect an important aspect of sexual selection ignored or rejected for more than a century. The immediate reception of Darwin's theory of sexual selection (1859, 1871) was ambivalent. He suggested that sexual selection could operate in two modes: males could differ in their ability to gain access to females through direct competition, and males could differ in their attraction to females. Although willing to accept the role of male competition, many authors, including steadfast Darwinians such as Wallace (1905) and Huxley (1938), doubted the efficacy of female choice. Darwin himself seemed skeptical of the ability to actully demonstrate female choice: 'What the attractions may be which give an advantage to certain males in wooing . . . can rarely have been conjectured' (Darwin

1882). Sexual selection by female choice lay dormant for more than a century.

One of the first issues to receive rigorous attention in recent studies of sexual selection was that of female choice based on male traits. These studies of mechanisms of mate choice often required the combination of quantitative field observations and controlled experiments. Such studies have shown, for example, the importance of acoustic (crickets: Hedrick 1986; fish: Myrberg *et al.* 1986; frogs: Ryan 1980; Sullivan 1983; Gerhardt 1988; Morris and Yoon 1989; birds: Searcy and Marler 1981; Payne 1983; Searcy 1984; Gibson and Bradbury 1985), visual (crabs: Christy 1988; fish: Semler 1971; Kodric-Brown 1985; Houde 1987, 1988; Ryan and Wagner 1987; birds: Andersson 1982; Burley 1985) and chemical signals (moths: Conner *et al.* 1981; Boppré and Schneider 1985) of males that influence female mate choice.

Searcy and Andersson (1986) have argued that to demonstrate female mate choice based on male traits, studies must show a significant relationship between male traits and male mating success, and then demonstrate experimentally that the traits in question influence female mating decisions. Regarding the first criterion, there have been recent developments of statistical techniques to estimate the strength of direct and indirect selection on male traits, and these have greatly improved our resolution in estimating selection (reviewed in Endler 1986; Wade 1987). However, the current emphasis on measuring selection has sometimes resulted in ignoring the second criterion suggested by Searcy and Andersson (1986), that is, understanding the processes that generate selection (Endler 1986; Graffen 1987; Ryan 1988*b*). In fact, Endler (1986) states: 'A highly accurate measure of selection differentials or coefficients, combined with a lack of knowledge of the reasons for and the mechanisms of selection, is little more than refined alchemy.'

There are some cases in which experimental studies of female mate choice are informative even in the absence of data showing that mating preferences generate selection. For example, Forester and Czarnowsky (1985) showed that in a frog, *Hyla crucifer*, larger males produced calls with lower frequencies and females preferred lower frequency calls. However, in nature there was not a large-male mating advantage. The reason for this disparity, the authors suggested, is that smaller males adopt non-calling, satellite mating strategies and ambush females en route to calling males. Therefore, the lack of size-based variance in male mating success is due to opposition of two mechanisms of sexual selection, female choice and male interactions, rather than to the lack of selection. Andersson (1982) provides another motivation for probing female preferences without evidence of selection. Male widowbirds have long tails that might appear to have evolved under the influence of sexual selection by female mate choice. Andersson did not find a significant correlation in the field between male mating success and tail length, but when he artificially altered tail length such a relationship became apparent. This study makes two important points. First, examining a female preference in relation to only the extant variation in male traits informs us about current mechanisms generating (or not generating) selection, but does not tell us the potential for female choice to influence variation in male traits that might arise. Secondly, this study highlights the problem of addressing an historical question (why do males have long tails?) by studying only current effects (how does the current variation in the male trait affect female mate choice?). We will return to this issue later.

General conclusions from more than a decade of studies of sexual selection by female choice are that: in many systems there is significant variation among males in their courtship behavior; females are responsive to this variation; and, as a result, females generate sexual selection on male traits by their differential choice of mates. Although there are many systems in which sexual selection by female mate choice has little or no importance, there are now enough data to generally confirm Darwin's intuition regarding the efficacy of female mate choice.

3.2 Sensory systems and mate choice

Studies of mechanisms of female mate choice have established the importance of female choice, but such studies can offer much more. For example, understanding how signals used in male courtship influence a female's perception of potential mates can reveal much about the evolution of male traits under sexual selection. A trivial example is how the sensory modality that guides mate choice determines those aspects of the male's phenotype subject to selection. In nocturnally breeding animals such as many frogs and moths, visual signals are of limited value, and thus vocal or chemical cues are used in mate attraction, whereas in diurnally breeding animals, such as dart-poision frogs and butterflies, males are strikingly colored. Although the acoustic and chemical channels still play some role in mate choice in the diurnal species, visual signals have become elaborated under sexual selection. It is an obvious observation that only those aspects of the male's phenotype perceived by the female can be subject to sexual selection by female choice.

Although the sensory modality used in communication defines the type of signals used in sexual selection, the sensory system can be permissive about the precise form of the signal favored by selection. In bower birds, it has been suggested that there is a trade-off between elaboration of the bower and elaboration of male plumage; species with dull plumage have elaborate bowers and those with elaborate male plumage are characterized by less extravagant bowers (Gilliard 1956; Borgia 1986). Some studies have shown such permissiveness by presenting females with male phenotypes that do not exist in nature. For example, Burley (1986) has shown that female zebra finches favor males with orange beaks, and that the preference for orange can be transferred to the color of artificial leg bands. A similar phenomenon occurs in anuran mate recognition. Many examples upon which I will draw come from this literature, and thus I will quickly review some of the basic aspects of call recognition in anurans (for detailed discussions of mechanisms, function and evolution of anuran acoustic systems, see reviews in Fritzsch *et al.* 1988).

All frogs have two inner ear organs with distinct ranges of frequency sensitivity. The amphibian papilla is most sensitive to low-frequency sounds. Individual fibers emanating from this end organ are each most sensitive to one of a variety of frequencies below *c.* 1200 Hz, and as a population these fibers show one or two peaks of frequency sensitivity at threshold intensities. The basilar papilla is sensitive to higher frequencies, all the fibers from this end organ are tuned to the same frequency, and this best excitatory frequency is usually above 1500 Hz (Zakon and Wilczynski 1988). In the male's advertisement call, those frequencies that contain the most energy match the tuning of the peripheral auditory system. If the call has two frequency peaks of energy, they usually match the tuning of both the amphibian and basilar papillae. If the call has only one frequency peak, then it matches the tuning of either the amphibian or basilar papillae. At higher levels of the brain (e.g. the midbrain and thalamus) information from separate frequency channels is integrated to result in a synergistic neural response to stimuli from the conspecific mating call (Gerhardt 1988; Fuzessery 1988; Walkowiak 1988). Therefore, an important initial step in the processing of spectral information takes place in the auditory periphery, even before this information reaches the brain. Less is known about how temporal aspects of the mating call are processed, but it has been shown that neurons in the brainstem are similarly tuned to temporal properties of the call (Rose and Capranica 1983; Walkowiak 1988).

Most experimental studies of female mate choice that use synthetic stimuli restrict the form and the range of the stimulus to that observed in nature. However, such experiments can also provide an opportunity to probe the female's response properties for hypothetical male traits, and thus to determine how sexual selection might act if certain male variants were to arise. In studies of natural selection, such opportunities are rarely available and evaluation of the performance of hypothetical variants, such as changes in functional–morphological relationships, can only be evaluated by models (Hildebrand *et al.* 1985; Radinsky 1987). It seems that researchers have been too conservative in exploiting experimental methods available in studies of mate choice.

Sexual selection and communication have been investigated extensively in the túngara frog, *Physalaemus pustulosus* (e.g. Ryan 1985). Recently, we investigated how signal variants that are not part of the male's repertoire might influence female choice if these traits were to evolve.

Males of *P. pustulosus* can add a 'chuck' to the initial 'whine' portion of their call. This makes the call more attractive to females (Fig. 1). A natural chuck has a broad frequency range that includes the frequencies to which each of the inner ear organs is most sensitive. Ryan and Rand (1990) have shown that synthetic chucks with either only the lower frequency portion or only the higher frequency portion of the chuck, and thus matching the most sensitive frequencies of only one of the two inner ear organs, are as effective in eliciting preferential female phonotaxis as the natural chuck with the full frequency range, as long as all the stimuli contain the same total amount of energy. This is despite the fact that the

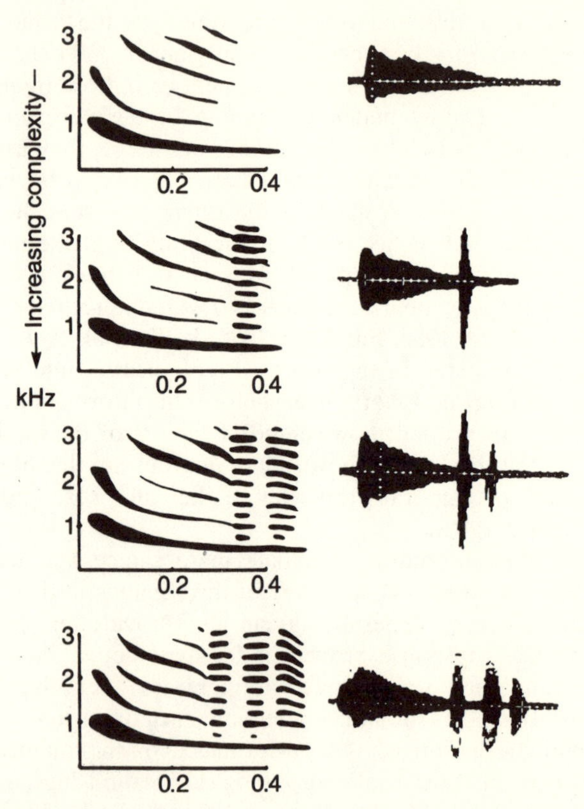

Fig. 1. Advertisement calls of the frog *Physalaemus pustulosus*. Calls increase in complexity from a whine with no chucks (top) to a whine with three chucks (bottom). Sonograms (left) exhibit how frequency (kHz) changes as a function of time (sec). Oscillograms (right) exhibit how amplitude of the same calls change with time.

low-frequency portion of the chuck has only 10 per cent of the total energy in natural calls.

The studies by both Burley (1986) and Ryan and Rand (1990) indicate that the sensory system might define a wide range of signals that could be favored by sexual selection. A female preference for a particular variant of a male trait might be due to a general preference for a certain kind of stimulation. Understanding preferences at this more general level would greatly enhance our understanding of the dynamics of sexual selection. In such cases, morphological constraints on signal evolution probably determine which of the favored forms of the signal evolve (Ryan and Drewes 1990).

I have discussed sexual selection acting in only a single modality, but many biologicaly relevant tasks facing an animal involve the potential use of more than one sensory modality, such as homing in pigeons (Keeton 1979) and foraging in bats (Ryan and Tuttle 1987). In both these examples there is a hierarchy of cues. Pigeons rely first on the sun and only switch to cues such as polarized light and magnetism if the sun is not visible, and the African bat, *Cardioderma cor*, ignores information from echolocation signals if the sound made by its prey is sufficient for localization.

The interaction of sensory modalities has hardly been explored in mate choice. In one example, cues in different modalites might be redundant. Conner (1987) showed that an arctiid moth uses both chemical and acoustic signals in courtship. Ablation experiments showed that the lack of one of the cues did not reduce the male's mating success, but he was unable to mate if he lost the ability to produce both signals.

Another example of multiple cues is conspecific mate choice in swordtails. Ryan and Wagner (1987) showed that when exposed only to visual cues, female *Xiphophorus pygmaeus* preferred larger, courting males of their allopatric, sister species, *X. nigrensis*, to their own smaller males which do not court. However, if exposed only to chemical cues, the females preferred the conspecific stimulus to that of the heterospecific. When female *X. pygmaeus* were simultaneously presented with both stimuli they showed neither a conspecific nor a heterospecific preference (Crapon de Caprona and Ryan 1990). In this case, the conflicting preferences from the two sensory modalities cancelled. It is not known if these cues are used simultaneously in nature or, for example, if females use chemical cues to locate areas with males and then base their final mate choice on visual cues. But this study shows that cues in different sensory modalities can result in different female preferences, and that the interaction of two modalities produces different results than when only one is considered. In general, these kinds of studies both define the signal modalities subject to selection, and allow investigations as to the fate of hypothetical traits that might be subject to sexual selection.

3.3 Sensory limits on signal divergence and the opportunity for speciation

The range of stimuli to which the sensory system is capable of responding will define the limits of signal evolution; this is more apparent in among-species comparisons. Because the divergence of courtship signals is often an important component of the speciation process (e.g. Mayr 1963), Ryan (1986) suggested that sensory constraints on signal divergence could also limit the opportunity for speciation. In many frogs, the amphibian papilla, which is sensitive to low-frequency sounds, is critically involved in the perception of the species-specific advertisement call. In primitive frogs, this organ consists of only a single, small patch of sensory epithelium that is sensitive to a fairly narrow range of frequencies. In advanced frogs, an embryonically distinct patch of sensory epithelium joins the homologous patch present in primitive frogs; the sensory epithelium is much longer, and the frequency range to which this inner ear organ is sensitive is much larger.

There are four character states for this inner ear organ, from small to greatly elongated. Because the length of the sensory epithelium is directly related to the frequency range of hearing, then it also determines the variation in call frequencies that can be perceived by females. Therefore, the state of the amphibian papilla imposes sensory limits on signal evolution. In primitive frogs, the sensory constraints are more severe and the number of species is smaller, whereas in advanced frogs the sensory constraints are most relaxed and the number of species is greatest. This relation holds for the two intermediate character states as well (Fig. 2). This is consistent with the notion that sensory constraints influence the opportunity for speciation.

A similar explanation involving morphological rather than sensory constraints on signal divergence may apply to the age-old question of why there are so many species of passerine birds. An obvious response is that there are many passerines because there are lots of oscines (song birds, a group within the passerines: Raikow 1986). This group has a complicated syrinx capable of producing internal duetting and thus an overwhelming diversity of song. It has been suggested that divergence of songs among populations can give rise to local mate preferences and thus genetic structuring of populations (Baker and Cunningham 1985; but see Baptista 1985; Brenowitz 1985; Chambers 1985). And it also has been suggested that the extreme ability of generating a variety of sounds in oscines might promote divergence of mate recognition signals and enhance the speciation process, in much the same way as was suggested for frogs (Fitzpatrick 1988; Vermeij 1988).

The assertion that constraints on signal divergence influence speciation rates in frogs and birds is difficult to evaluate because of the possibility

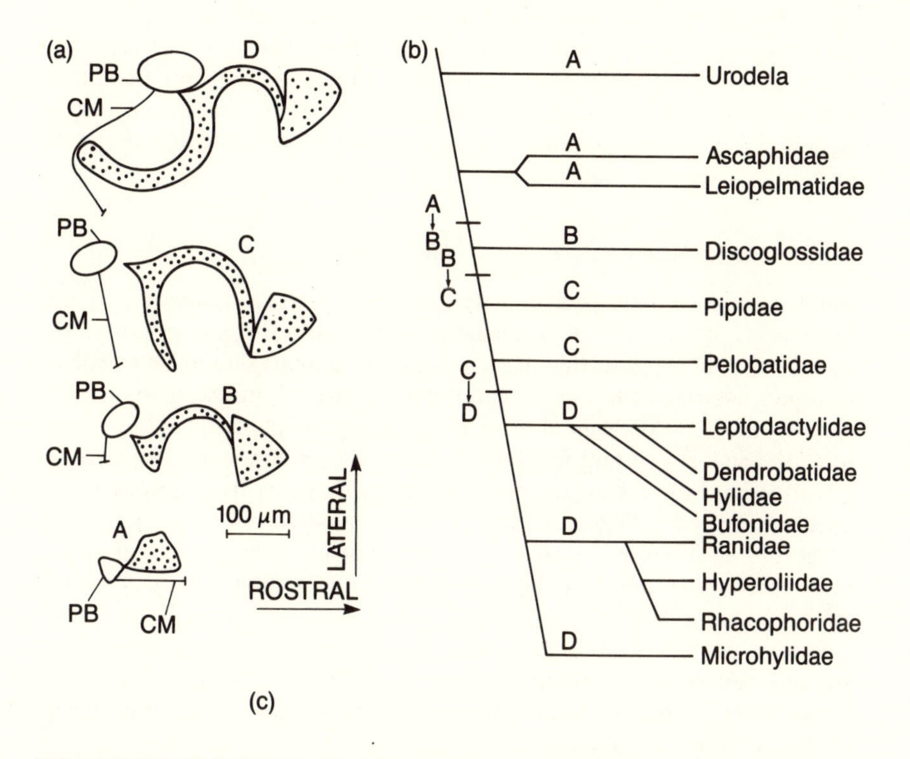

(c)

Character state	Family	Extinct genera	Extant		Species per character state
			Genera	Species	
A	Ascaphidae	2	1	1	
A	Leiopelmatidae	1	1	3	4
B	Discoglossidae	5	4	9	9
C	Pipidae	5	4	17	
C	Pelobatidae	3	8	49	66
D	Leptodactylidae	2	41	635	
D	Dendrobatidae	0	3	60	
D	Hylidae	3	33	560	
D	Bufonidae	1	19	277	2489
D	Ranidae	2	45	586	
D	Hyperoliidae	0	14	54	
D	Rhacophoridae	0	14	96	
D	Microhylidae	0	58	221	

Fig. 2. (a) The size and complexity of various character states of the amphibian papilla. (b) The distribution of character states of the amphibian papilla across anuran taxa with hypothesized evolutionary transitions. (c) The number of taxa as a function of the character state of the amphibian papilla.

of confounding cause and effect, and because the sample sizes are too small for rigorous statistical evaluation, e.g. there are only four character states of the amphibian papilla and thus only four data points. But if sensory and morphological constraints on the divergence of male courtship signals is an important phenomenon, it should be found in other taxa using other sensory modalities for mate recognition.

3.4 Stabilizing sexual selection on male traits

Many studies of conspecific mate recognition were motivated by the knowledge that the cost to a female of mating with the wrong species can be severe. This suggests that mate recognition systems should have evolved to reject heterospecifics as potential mates, even if, in doing so, females were to make an occasional error of rejecting a conspecific mate. Therefore, the sensory system involved in mate recognition should be strongly biased toward the mean, and thus should exert strong stabilizing selection on the male trait. This assertion has been invoked by several authors to suggest that (directional) sexual selection is unlikely because it would favor male traits that deviated from the mean (Templeton 1979; Paterson 1982; Gerhardt 1982).

Several studies have shown that sensory biases in mate recognition systems can result in stabilizing selection. One of the better examples comes from Gerhardt's studies of mechanisms of mate choice in tree frogs of the genus *Hyla* (summarized in Gerhardt 1982).

The call of *H. cinerea* has a frequency spectrum with two energy peaks, each of which matches the tuning properties of either the amphibian papilla or basilar papilla. The most attractive synthetic stimulus for eliciting female phonotaxis is that which combines the two frequency peaks. Gerhardt showed that for the low-frequency peak, females preferred that frequency closer to the population average rather than one that was higher or lower. These results suggest that there should be stabilizing selection on the male trait in nature, and measures of male mating success in the field are consistent with this interpretation (Gerhardt *et al.* 1987).

Why do the females prefer males with call characters close to the population average? Or, asking the question at another level of analysis, why do females have auditory tuning properties that result in preference for average calls? Gerhardt suggests that it is in response to selection for conspecific mate recognition. The sympatric tree frogs *H. gratiosa* and *H. squirella* have calls similar to that of *H. cinerea* but are slightly lower and higher, respectively, in frequency. Thus, if female *H. cinerea* chose calls that deviated substantially from the population mean, they would risk mating with a heterospecific. This study suggests that the mate recognition system evolved to effect conspecific matings. However, are selection for species recognition and sexual selection mutually exclusive phenomena?

The dichotomy between species recognition and sexual selection is a false one, as is evident from considering Gerhardt's study. For the green tree frogs it seems clear that the sensory biases of the female's auditory system and, by extension, the female choice behavior that it guides, enhance the female's chances of mating with conspecifics. This does not mean that the female preferences do not generate sexual selection on the male trait. Again, sexual selection by female choice refers to males' differential ability to attract mates. If females are more attracted to some males than to others, and if this differential attraction generates variance in male mating success, then sexual selection operates.

Sexual selection need not be directional, it can be stabilizing as in the green tree frog system in which sexual selection favors males with average call frequencies. In this example, and in others demonstrating stabilizing selection, there can be debate as to why females evolved preferences for average males. For the green tree frog the most logical argument is that females that avoid mating with conspecifics are favored by selection. The evolutionary forces responsible for the preference need not determine its current effects on male traits; even if the preference evolved for the benefit of mating with conspecifics, that does not mean that it cannot generate sexual selection.

There are many examples of female mate choice generating stabilizing selection on male traits. Many of these studies were conducted as investigations of species recognition and, because of the false dichotomy between species recognition and sexual selection, many of these studies have been cited as evidence for the lack of sexual selection. Understanding the processes of both species recognition and sexual selection will be enhanced if we abandon the notion that they are mutually exclusive.

3.5 Directional selection on male traits

Female preference for extreme, rather than mean, traits has received the most attention in the field of sexual selection. This is not surprising. It was the extreme elaboration of male traits and its obvious survival cost that led Darwin (1859, 1871) to consider sexual selection theory, and that motivated Fisher's (1958) hypothesis of runaway sexual selection. This also led to the controversy of the lek (Bradbury 1981; Beehler and Foster 1988; Höngstrum 1989): Why should females choose a male from whom they would accrue no immediate benefit, and why should they prefer a male whose own courtship promotes his early demise? It is perhaps in these instances that understanding the mechanism of female choice reveals the most about the operation of sexual selection.

The number of studies demonstrating that female preferences exert directional selection on male traits has increased dramatically in the last decade. There has not been a systematic review of the findings of such studies, but a casual survey seems to reveal some pattern.

Among frogs alone, studies have shown that females prefer lower frequency calls (Ryan 1980, 1983; Forester and Czarnowsky 1985; Robertson 1986; Ryan and Wilczynski 1988; Morris and Yoon 1989), calls with more notes (Wells and Schwartz 1984; Rand and Ryan 1981; Littlejohn and Harrison 1985), faster call rates or more calls (Sullivan 1983; Forester and Czarnowsky 1985; Schwartz 1986; Klump and Gerhardt 1987; Wells 1988; Forester *et al.* 1989), longer calls (Straughn 1975; Klump and Gerhardt 1987) and more intense calls (Fellers 1979; Zelick and Narins 1983; many of these studies are reviewed in Gerhardt 1988). Female crickets prefer longer calling bouts (Hedrick 1986) and more intense calls (Latimer and Sippel 1987), and female birds prefer larger song repertoires (Catchpole 1980; Searcy and Marler 1981; McGregor and Krebs 1982; Payne 1983; Searcy 1984; Catchpole *et al.* 1984). There are analogous studies of systems in which courtship is primarily visual: female swordtails prefer males with longer swords (Basola 1990); female guppies prefer males with a greater area of brighter colors (Kodric-Brown 1985; Houde 1987) and larger tails (Bischoff *et al.* 1985); female sticklebacks prefer larger males (Moodie 1982; Rowland 1989*b*), with more color (Semler 1971) and higher display rates (Ridley 1986); male sticklebacks prefer more rotund females (Rowland 1989*a*); male butterflies prefer females with larger wings (Tinbergen *et al.* 1942; Rutowski 1982) and faster wing beat (Magnus 1958); and female widowbirds, swallows and pied flycatchers prefer longer tails on their males (Andersson 1982; Møller 1988; Lijfeld and Slagsvold 1988). There are fewer studies dealing with chemical communication. However, it has been shown that male newts prefer stronger concentrations of female pheromone (Verill 1985). Also, female moths prefer larger males, and size is correlated with the amount of pheromone (Phelan and Baker 1986, 1987; Conner *et al.* 1990; Conner, pers. comm.). Rowland (1989*a*) discusses many of these examples.

In some studies, the female preference extends beyond the population's phenotypic range; that is, there is preference for a supernormal stimulus (Tinbergen 1948). For example, Sullivan (1983) showed that female toads prefer call rates that exceed those produced by any males. Rowland (1989*a*) showed a male preference for female sticklebacks who were distended far beyond the normal range, and a female preference for males that were 25 per cent larger than normal (Rowland 1989*b*). Also, Magnus (1958) reported that male mate choice in a moth favored a wing beat rate that greatly exceeded that exhibited by any females. Andersson's (1982) study is particularly interesting because not only did female widowbirds prefer artificial tails that exceeded the maximum length in the population, but there was no evidence of female preference when it was based on the population's normal variation. However, in most studies, the traits tested do not extend into the supernormal range, and thus the generality of preference for a supernormal stimulus cannot be evaluated.

Species recognition might sometimes generate directional selection. If two species are similar in a trait, females might be selected to prefer traits of conspecifics that differ most from those of heterospecifics (Trivers 1972). An example of just such an effect might be seen in the variation in dewlap color in the lizard *Anolis brevirostris* (Webster and Burns 1973). Dewlap color can be important in mate choice (Crews 1974), and populations of *A. brevirostris* vary in dewlap color. Those *A. brevirostris* that are nearest to a northern congeneric with light dewlaps have darker dewlaps, whereas those nearer to a southern congeneric characterized by darker dewlaps have lighter colored dewlaps. Unfortunately, the influence of dewlap-color variation on mate choice in this species has not been investigated.

As I emphasized in the previous section, there are many studies suggesting stabilizing selection. The disparity between the number of studies cited demonstrating stabilizing versus directional selection is not meant to reflect the relative frequency of these phenomena. However, the above review shows that in many cases females do not prefer the mean trait; females often exert directional rather than stabilizing selection.

When female choice generates directional selection, the direction is usually consistent, and it usually favors the more elaborate or greater quantity: longer tails, stronger odors, more intense calls, more complicated calls, larger song repertoires, brighter colors and more of them. It is unusual to find female preference for shorter tails, softer calls, simpler songs and duller colors. Thus there appears to be an inherent directionality in female preferences.

In Section 4.1, I will discuss in more detail competing hypotheses for the evolution of female preferences. One such hypothesis is Fisher's theory of runaway sexual selection. Unfortunately, there seems to be a common misconception among some behavioral ecologists that the continued exaggeration of a male trait is the necessary outcome of Fisher's hypothesis of runaway sexual selection (e.g. Alcock 1989). The essence of the runaway process is the genetic correlation of a male trait and a female preference through linkage disequilibrium. This causes an ever-accelerating increase in the frequency of the trait, due to selection by female choice, and the preference, as a correlated response to selection on the trait (see also Kirkpatrick 1987*b*). There is no inherent direction to the evolution of male traits under the runaway process alone, as many mathematical models have demonstrated (e.g. O'Donald 1983; Lande 1981; Kirkpatrick 1982). It is important to note that the runaway process applies to a specific mode of correlated evolution of trait and preference; this theory can accommodate, but does not predict, the exaggeration of male traits under sexual selection.

If the notion of an inherent directionality in female choice is true (i.e.

more, not less, elaborate), the runaway process by itself cannot account for patterns of male traits observed in nature. Population geneticists acknowledge this, and have suggested that biases in perceptual mechanisms can influence the direction in which the runaway sexual selection will proceed (e.g. O'Donald 1983; Kirkpatrick 1987*a*); thus the data are consistent with Fisher's hypothesis. Also, the directionality in female choice is consistent with hypotheses that suggest female choice is based on traits that indicate male quality (e.g. Zahavi 1975; Hamilton and Zuk 1982; Kodric-Brown and Brown 1984; Andersson 1986; Pomiankowski 1988), if we assume that the more elaborate the trait, the better the male quality, and the stronger the preference. In fact, this directionality is consistent with several hypotheses. For example, there is evidence that male courtship might have such an important effect on the reproductive state of the female, that in an experimental situation the female's fecundity decreases if she is deprived of the mate she first chose (Bluhm 1985; Yamamoto *et al.* 1989). In Section 4.3, I will discuss in more detail competing theories of the evolution of female preferences, and ask how our knowledge of mechanisms can add to or even reject some of the competing hypotheses.

3.6 Sensory biases and directional selection

An understanding of any inherent property of sensory systems that causes preferences resulting in directional selection, and especially preferences for exaggerated stimuli, would be an important addition to sexual selection theory. Such an understanding could explain mechanisms underlying the evolution of elaborate male traits in many animals, and thus it would contribute to our understanding of patterns of organic diversity, which is one of the major goals of evolutionary biology. The beginnings of such a perspective is suggested in some of the following studies.

Magnus (1958) showed that male butterflies preferred females with faster rates of wing beat, even when it exceeded the normal rate of 8–10 Hz. Using a model, he showed that males preferred supernormal rates, and that this preference continued up to a wing beat rate of 140 Hz, which is also the flicker-fusion rate of the butterfly's eye (the rate at which the individual flaps blur into a single movement). Therefore, the preference increased as the rate of retinal stimulation increased, and then ceased when the maximum rate of stimulation was reached. Not only does this suggest the sensory mechanism of (male) mate preference, it also shows that if the female's wing beat rate were not constrained by morphology and physiology, it might eventually reach a point at which further elaboration would be limited by the male's perceptual abilities. This would produce a result similar to that proposed by Fisher (1958), in which the counter-selection force of predation halts the further elaboration of a male's trait. Cohen (1984) has also discussed how sensory constraints on perceptual discrimination might impose an upper limit on the evolution of elaborate male traits under sexual selection.

Another indication of how retinal physiology can affect mate choice comes from crabs. In the fiddler crab, *Uca beebei*, some males construct pillars at the entrance to their burrows. Burrows with pillars are more likely to attract females, and Christy (1988) has suggested that the pillars are used by males to exploit the female's tendency to detect objects that project above the horizon (Zeil *et al.* 1986).

Rowland (1989*a*) also used animal models to investigate male choice based on a female trait. He had previously shown that male sticklebacks preferred larger, more gravid females, and suggested that the benefit to such a preference was due to the positive correlation between female size and female fecundity (Rowland 1982). In the more recent study, he showed that the male preference is elicited by the total projection area of the female, not just her volume due to distension with eggs. Males preferred a projectional area that greatly exceeded the normal range, suggesting that this preference is dictated by the amount of sensory stimulation of the retina. In addition, he pointed out that although in many cases this mechanism would result in males choosing more fecund females, it could also lead to males preferring to mate with females that were grossly distended with parasites.

The importance of red color in eliciting female sexual response in sticklebacks has been well known since Tinbergen (1953) reported interactions between these fish and red mail trucks. Semler (1971) showed female preference for males with more intense red. Interestingly, it is known that the female's sensitivity to red increases in the breeding season (Crounly-Dillon and Shama 1968). Also, Reimchen (1989) showed that a significant amount of the variation in the number of males with red nuptial coloration could be explained by the light transmission qualities of the environment; red nuptial colors were less common in water in which red wavelengths did not transmit well. In these fish, not only is the eye tuned toward red males, this tuning increases during the time of breeding but is susceptible to environmental noise that masks the signal.

The bright, polymorphic color patterns of male guppies are also well known, as are the studies of how predation and sexual selection act on these color patterns (Endler 1980, 1982, 1983; Kodric-Brown 1985; Breden and Stoner 1987; Houde 1987). Recent studies of the guppy's retina suggest how the variation in male color might be maintained. Archer *et al.* (1987) found that guppies exhibit a degree of polymorphism in visual pigments exceeding that reported for most vertebrates. They suggested that the pigment polymorphism might give rise to a perceptual polymorphism in females that could account for the maintenance of variation in the male trait.

Anurans have emerged as an especially good model for examining the neural bases of species-specific call preferences, and this approach is now being extended into the area of sexual selection. As suggested above, among species there is a good match between the frequencies in the advertisement call that contain the most energy and the tuning properties of the amphibian papilla and basilar papilla in the peripheral auditory system. I have discussed how male *P. pustulosus* increase the attractiveness of their call to females by adding chucks to the initial whine of the advertisement call. The chuck has a fundamental frequency of 200 Hz with 15 harmonics, and thus a frequency range of 200–3000 Hz. Besides preferring calls with chucks, females also show a more subtle call preference. They prefer chucks having a lower fundamental frequency. Larger males produce lower-frequency chucks, and in nature larger males have greater mating success because they are more likely to be chosen as mates by females. Thus the female preference for lower-frequency chucks exerts directional selection on the male trait (Ryan 1980, 1983, 1985).

Ryan *et al.* (1990) investigated the sensory basis of sexual selection in *P. pustulosus*. Because the chuck is harmonically structured, a preference for a lower fundamental frequency could be due to a preference based on any of the 15 harmonics. In this species, the frog's amphibian papilla is most sensitive to 700 Hz at threshold intensities and the basilar papilla is most sensitive to 2200 Hz. Although Ryan and Rand (1990) have shown that either the low or high half of the chuck is as attractive as the full chuck if the total amount of energy is the same, in nature more than 90 per cent of the energy is above 1500 Hz, falling in the frequency range to which the basilar papilla is most sensitive. Thus it would appear that the preference for the lower-frequency chucks might reside in the basilar papilla. The average dominant frequency from a random sample of 54 calls was 2500 Hz, higher than the average tuning of the female's basilar papilla (Fig. 3). This alone would suggest that calls with lower than average frequencies elicit greater neural stimulation in the females.

The amount of neural stimulation was quantified by determining the Fourier spectrum (frequency versus energy) of the chuck, and using a computer model to measure how much of this energy would pass through the tuning filter of the basilar papilla. Ryan *et al.* (1990) then asked if most chucks in the population would be more stimulatory if they were of lower frequency. To do this they multiplied all the frequencies in a chuck by a series of numbers (frequency multipliers) from 0.95 to 1.05. Frequency multipliers less than 1.0 would result in chucks with lower frequencies and frequency multipliers greater than 1.0 would produce higher-frequency chucks. They then determined the 'optimal' frequency multiplier; that is, the frequency multiplier that would result in the chuck that elicited the greatest amount of neural stimulation. The hypothesis that females prefer lower-frequency chucks because they elicit greater

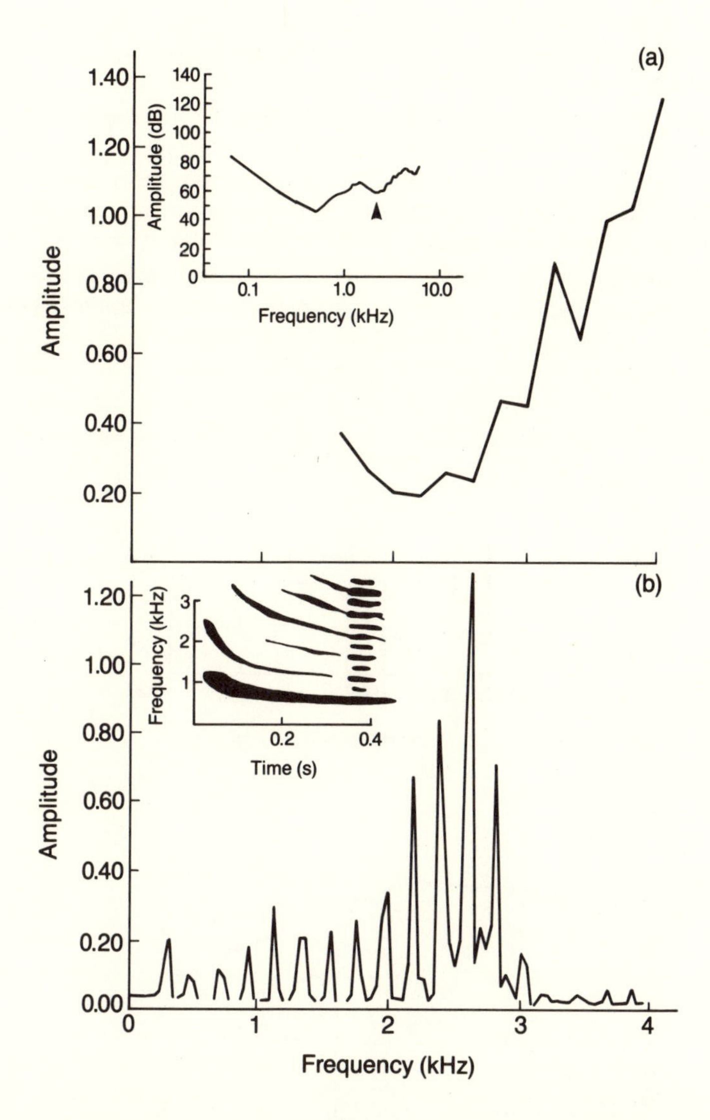

Fig. 3. (a) An average (N = 7) audiogram derived from the hindbrain of the frog *Physalaemus pustulosus* truncated to represent only those frequencies to which the basilar papilla is most sensitive. The inset shows the complete audiogram. (b) A Fourier spectrum, showing the distribution of call energy as a function of frequency for a single chuck. The inset shows a sonogram of whine plus one chuck.

neural stimulation predicts an optimal frequency multiplier significantly less than 1.0. The results support that hypothesis (Fig. 4).

A second property of the sensory system also results in directional selection on call frequency. Ryan *et al.* (1990) determined the average amount of neural stimulation for all of the 54 calls as a function of the frequency multiplier (Fig. 4). As expected from the previous results, the maximum was less than 1.0 (i.e. the unmanipulated call). However, the distribution of the amount of neural stimulation was asymmetric around 1.0. A chuck that was slightly lower in frequency than the optimum showed only a slight decrement in the amount of stimulation, whereas a

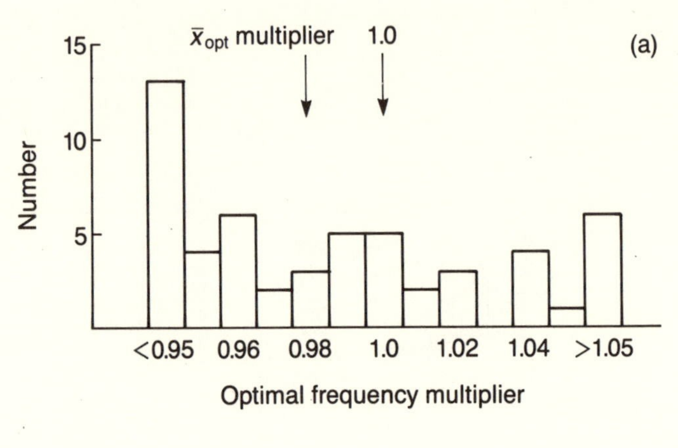

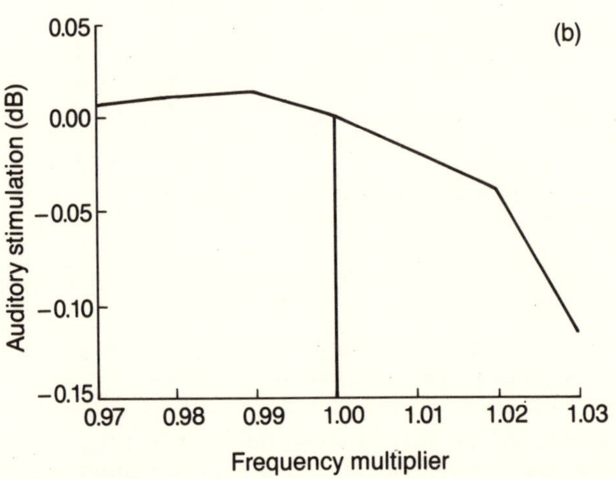

Fig. 4. (a) A frequency histogram of the optimal frequency multipliers of calls of 54 *Physalaemus pustulosus*. (b) The average amount of auditory stimulation elicited from the audiogram as a function of the frequency multiplier for all 54 calls.

chuck that was higher in frequency than the optimum showed a greater decrement in neural stimulation. Thus the disadvantage of producing a suboptimal chuck is more severe for higher-frequency chucks than lower-frequency chucks. Even if the peak tuning of the basilar papilla matched the mean dominant frequency of the chuck, the asymmetry in the tuning curve could still result in directional selection on chuck frequency by virtue of an asymmetric fitness function with respect to chuck frequency.

The cricket frog, *Acris crepitans*, offers another example of how female auditory tuning results in directional patterns of mate preferences. Ryan and Wilczynski (1988) showed that variation in the dominant frequency of calls among populations of cricket frogs (*c.* 3200–4000 Hz) resulted in different female preferences among populations. When presented with a choice between their own calls and calls from nearby Bastrop, female cricket frogs from Gill Ranch preferred the local call, because the tuning of the basilar papilla of the Gill Ranch females more closely matched the dominant frequency of the local call (3200 Hz) than the calls of the Bastrop males (3800 Hz).

The females did not always prefer the local call. For example, Bastrop females preferred calls from Gill Ranch to calls of their own males (Ryan and Wilczynski, unpublished data). In a series of studies of females from three populations in which the frequency differences were always *c.* 400 Hz, Ryan and Wilczynski (unpublished data) showed that over the range of frequencies tested, if there was a preference it was always for the call of the lower frequency, whether it was the local call or a foreign call. Large samples of tuning and calls from the same populations show that females usually are tuned slightly below the population's mean dominant call frequency. This can result in local mate preferences when the female is faced with the choice between the local call and a foreign call of higher frequency, or it can result in females preferring a foreign call of lower frequency to their own call. Furthermore, the slight mismatch between the mean female's tuning and the mean call's dominant frequency for the same population suggests the potential for sexual selection on call frequency. However, the 400 Hz differences that elicited preferences might be too large compared to the population variance in call frequency to have a significant effect within populations.

These studies of butterflies, crabs, fish and frogs all suggest that the total amount of stimulation received by the female is crucial in determining her mate preference. In the butterflies, it appears to be due to the rate at which the retina is stimulated, in the crabs by the orientation of the male's landmark, in the fish it is related to the area of the retina and photopigments that are stimulated, and in the frogs it is the result of the relationship between the call's dominant frequency and the tuning of the auditory system. Although these studies investigate very different signals in two sensory modalities, there seems to be a common theme: preference for greater stimulation (see also Morris 1956; Barlow 1977; Arak 1983; O'Donald 1983; West-Eberhard 1979; Burley 1985; Endler 1989).

Preference for a supernormal stimulus is a special case of females preferring more stimulating traits within the population's normal range, only more extreme. Staddon (1975) has offered a general explanation as to why some animals evolve preferential responses to supernormal stimuli. He gives an example of discrimination learning in which an animal is rewarded for a positive response to a light of 550 nm wavelength and receives no reward or a punishment for a response to a 500-nm light. When the animal is later tested with a series of stimuli, the most intense response has been shifted in the positive direction away from the stimulus for which positive reinforcement occurred, say 560 nm instead of 550 nm. This change in response is called a peak shift, and it occurs because of an asymmetry around the initially favored peak. If the animal responds to the positive side there may be no added advantage (reward), but if it responds to the negative side it has the risk of no reward or of punishment.

The asymmetry that causes the peak shift could have population effects similar to the asymmetric tuning of the basilar papilla in *P. pustulosus*; that is, it could result in an asymmetric fitness function of the male trait under selection. Staddon suggested that selection might be analogous to the positive and negative reinforcement in a discrimination paradigm, and O'Donald (1983) argued that Staddon's model could provide the direction for Fisher's runaway model. The mechanism proposed by Staddon, however, is crucially different than the mechanism implied by many of the studies reviewed above – selection for greater sensory stimulation. Depending on the conditioning paradigm, the peak shift could occur in either direction. Thus, like Fisher's hypothesis, Staddon's model could result in directional evolution but the direction is indeterminate.

Although many patterns of mate choice can be explained by differences in the amount of stimulation delivered by the male signal, this does not initially seem to explain another general pattern of mate choice, i.e. selection for novelty. However, I suggest that this is only a special case of preference for increased stimulation.

Several authors have discussed selection for novelty in different contexts. In a striking example, Burley (cited in Trivers 1985) fitted zebra finches with hats decorated with feathers and showed that females preferred these bizarre males to the normal, less ostentatious males (as long as the hats were white). Ehrman and Probber (1978) showed that in fruit flies the male phenotype in the minority had a mating advantage, although the generality of the rare male effect in fruit flies has been questioned (e.g. Peterson and Merrell 1983). Farr (1977) also demonstrated female preferences for novel male traits in guppies, and he suggested that preference for rare males is a special case of novelty.

The importance of novelty has been considered in some detail in the evolution of bird song, and discussions of the underlying mechanisms suggest that preference for novelty is a special case of preference for increased stimulation.

One of the most striking features of oscine birds is the diverse song repertoires of many species. The evolution of song repertoire has long been debated, and one of the seriously considered explanations is the monotony hypothesis (Hartshorne 1956). The crux of the argument is that animals habituate to repeated stimuli and novelty will be favored because it is most likely to 'draw attention', that is, to be perceived rather than merely detected.

Barlow (1977) pointed out that, in general, animal signals have to stand out against background noise, the source of which could originate in the environment or from conspecifics competing for the same communication channel. Novelty is one means by which an animal can become more apparent and thus increase its signal to noise ratio. If the receiver habituates, the novelty effect can increase the signal-to-noise ratio at the level of the receiver's sensory system. Hinde (1970) suggested that selection imposed by habituation might be an important force favoring both novelty and supernormal stimuli, and West-Eberhard (1979) argued that in sexual selection by female choice there might be a selective advantage to novelty *per se*, due to this release from habituation. If a selective advantage to novelty exists for the signaller, and if the mechanism responsible is an increase in the signal-to-noise ratio due to habituation, then selection for novelty is really only a special case of selection for greater sensory stimulation.

There are cases in which selection for novelty and familiarity might conflict. Bateson (1983) has shown that young Japanese quail imprint on birds with which they were raised, and then prefer mates that are similar to, but slightly deviant from, the model. He suggests that this results in an optimal balance between inbreeding and outbreeding. For example, first cousins are preferred over both siblings and more distantly related birds. ten Cate and Bateson (1988) suggested that within this preference there might be an asymmetry; among the mates that are slightly deviant from the model, individuals (in this case males) would prefer the mate that was more elaborately adorned. The reason for this asymmetry would be that more elaborate mates would also be more conspicuous. If this asymmetry existed, then mate selection would impose a directional bias on the evolution of male traits. ten Cate and Bateson (1989) provide evidence for such an effect. They showed that males imprinted to white birds with dots on the breast preferred white birds over birds with the wild, brown, dotless plumage. When given a choice among white birds, they preferred birds with the larger number of dots, regardless of the number of dots present on the model.

In summary, in this section I have reviewed studies showing that female choice generates directional selection on male traits. Usually, the direction is not random. Females tend to prefer traits that are more stimulating; the preferred traits are usually larger, brighter, more energetic and more complex. This bias in directionality is not predicted, although is easily accommodated, by the runaway sexual selection hypothesis. It is also consistent with predictions of the good genes hypothesis, and other hypotheses such as sexual selection for sensory exploitation, which will be discussed in the next section. In a few cases, knowledge of the mechanisms reveals the physiological locus at which the greater stimulation is generated, as well as constraints on signal elaboration. Finally, several researchers have discussed models or data that might result in asymmetric or directional mating preferences. These biases could impart a directionality on other processes, such as runaway sexual selection, and, as discussed below, might by themselves explain some patterns of the evolution of male traits under sexual selection.

4. EVOLUTION OF FEMALE PREFERENCES

In this section, I will discuss how mechanisms of mate choice can contribute to our understanding of the evolution of female preferences. Ever since Darwin, an important question in sexual selection has been if females choose males on the basis of male traits. That has now been well documented in a variety of studies. There is no doubt that female preferences can generate sexual selection. But why have these preferences evolved? This is now the most controversial question in sexual selection (e.g. Bradbury and Andersson 1987).

4.1 Hypotheses for the evolution of female preferences

Several hypotheses have been posited to explain the evolution of female preferences. Natural selection hyotheses, such as the good genes hyothesis (e.g. Zahavi 1975; Hamilton and Zuk 1982; Kodric-Brown and Brown 1984; Andersson 1986; Pomiankowski 1988), suggest that certain female preferences are favored by natural selection because they increase either the number or the quality (e.g. physical vigor) of offspring. There are many studies demonstrating adaptive mate choice. In most such cases, females receive an immediate natural selection advantage, such as more eggs fertilized and higher offspring survivorship due to superior territories or parental care (e.g. Thornhill 1976; Perrone 1978). However, there is currently little empirical evidence to suggest adaptive mate choice based on genetic benefits (Kirkpatrick 1987*a*), and genetic models have led to different interpretations of the internal validity of the good genes hypothesis (Kirkpatrick 1986; Andersson 1986; Pomiankowski 1988).

What is often viewed as the major alternative to natural selection hypotheses for the evolution of female preferences is runaway sexual selection (Fisher 1958). This hypothesis suggests that alleles for the female preference and the male trait genetically co-vary due to linkage disequilibrium. The preference evolves as a correlated response to selection on the trait. The interesting aspect of the runaway process is that it is the female preference that generates the selection on the male trait. Fisher refered to the runaway process as self-reinforcing choice, because when females prefer certain males, not only do their sons possess the trait but their daughters will mate with similarly endowed males. The stronger the preference, the stronger the selection on the trait, and thus the greater the correlated response in the preference.

In order for the runaway process to be initiated, the preference needs to be at a relatively high frequency. Fisher suggested that adaptive mate choice for good genes could confer an initial advantage on the preference by causing an increase in the frequency of the preference. Lande (1981) and Kirkpatrick (1982) have pointed out that stochastic processes such as drift could initiate the process. Another important factor could be the sensory biases discussed in the previous section. Sensory biases that might have evolved under selection pressures unrelated to mate choice could result in an increase in the frequency of a preference sufficient for the runaway process to be initiated.

The runaway hypothesis has been investigated extensively by population genetics models, its essential features validated, and some unexpected outcomes observed, but currently there is little empirical evidence to support the notion that runaway sexual selection has been an important process in the evolution of female preferences (see reviews by Kirkpatrick 1987*a*, 1987*b*).

Another hypothesis suggests that current preferences for traits have become established in the population for reasons not related to adaptive mate choice or sexual selection. In such instances, the current congruence between a preference and a trait is due to the male trait having evolved to exploit pre-existing biases in the female's sensory system. I refer to this hypothesis as sexual selection for sensory exploitation (Ryan and Rand 1990; Ryan *et al.* 1990). This hypothesis is evolutionary rather than mechanistic. As noted above, at some level the most preferred male traits exploit female sensory (including cognitive) biases. That fact says nothing about the evolutionary history of the preference and trait. Sensory exploitation states specifically that the male trait evolved to match a pre-existing sensory bias (see also Barlow 1977; West-Eberhard 1979; Arak 1983; Burley 1985; Kirkpatrick 1987*a*; Endler 1989).

I have discussed various patterns of mate choice and the underlying sensory mechanisms of mate choice that could favor the elaboration of male traits. Many evolutionary biologists might feel that these mechanistic studies are worthwhile only if they reveal the evolutionary history of sexual selection. (Of course, some comparative physiologists feel that studies of behavior are only important if they reveal something about sensor processing.) Studies of mechanisms of mate choice are important in their own right because they reveal how selection operates. Because sexual selection by female choice is a problem in communication, it cannot be understood completely without a knowledge of the interaction of the signal and the receiver. The knowledge of mechanisms alone might be very helpful in determining the currrent effects of a preference, but by itself if often tells us nothing about why the preference evolved. That is an historical question that requires techniques used to investigate history.

4.2 Passive and active mate choice

Not all researchers would agree with the assertion that knowledge of the mechanism alone cannot test evolutionary hypotheses. In fact, a great deal of attention has been given to just such an approach in sexual selection by female choice.

Parker (1983) made a distinction between two mechanistic categories of mate choice: active and passive. As an extreme example, consider two males vocalizing to attract mates. One male has a call of very low intensity, whereas the other has a much higher-intensity call. If the female is attracted to the call she first encounters, then she is likely to be attracted to the more intense call because of its larger active space. This is passive choice. In this case, there would be no need for conjecture as to the selective forces responsible for a female not choosing a male she never hears.

Passive choice also has been extended to cases in which the females perceive alternative signals but are attracted preferentially to the stimulus that is perceived as more intense; again, evidence of this type of passive choice is used to reject hypotheses of adaptive mate choice. The argument is that preference for the most intense signal reflects selection for females to mate with the closest male in order to decrease the costs (e.g. time and energy) of searching for a mate. If the male's signal indicated he was a better mate, then females would have been selected to endure these costs of searching. So, it is argued that because the female's behavior is not consistent with the adaptive hypothesis, that passive choice rejects such an explanation.

The more inclusive definition of passive choice seems parallel to the notion I attempted to document above; that is, females often are attracted to males with the most stimulating signals. However, it is important to understand that this is not the case. The passive/active dichotomy is not useful in cases in which the female actually perceives alternative stimuli, and in which the preference results from sensory biases enhancing the perception of one of the stimuli. Again invoking anuran vocal signals as an example, many studies have shown that the spectral tuning of the peripheral auditory system tends to match the spectral characteristics of the species-specific advertisement call. Thus the conspecific call will be perceived as more intense than a heterospecific call, and this difference in perceived intensity contributes to the preference. Is this conspecific preference passive attraction? And, if so, at what neurological level must the preference arise to be classified as active choice: if not the peripheral nervous system, then the central nervous sytem? Must it be in the forebrain rather than the hindbrain? Must it be cognitive rather than reflexive? Most neuroscientists assert that a preference among alternative stimuli results from enhanced neural stimulation at some level of the nervous system (e.g. Bullock 1986). Thus, except in some very obvious situations such as when the female is exposed only to the more intense stimuli, the classification of passive and active choice is difficult (see also Pomiankowski 1988; Sullivan 1989).

Besides operational problems, another difficulty arises when the mechanism of passive choice is used to reject hypotheses for the evolution of preferences. Returning to the example of peripheral tuning in a frog, many frequency preferences can be neutralized or reversed by changes in intensity (Gerhardt 1988). In fact, in some species preferences for conspecifics are intensity-dependent. This might be especially true for higher frequencies that overlap the most sensitive frequencies of the basilar papilla. This inner ear organ is thought not to be capable of pitch discrimination (Lewis and Lombard 1988), but the preference for different frequencies in this range is due to those that better match the peak tuning being perceived as more intense. In this case, should we then assume that a preference for conspecific calls is due to passive attraction, and conclude that selection for species recognition has had no role in the evolution of the preference? If calls are processed by only the basilar papilla, which is the case for many species, then the only option is intensity-dependent preferences, what Parker calls passive choice. Perhaps the intensity-dependence of the preference in this case tells us more about the sensory constraints on signal recognition in these animals than it does about the evolution of preferences.

The evolution of female preference is controversial because it has been difficult to gather the necessary empirical data to discriminate among hypotheses. The passive/active choice dichotomy seemed to offer a solution: if there is a preference, increase the intensity of the less preferred stimulus. If the preference is neutralized, then the curtain of time is torn and the evolutionary history of the preference is revealed. The controversy is solved. If only it were that simple! I suggest that as with the species recognition/sexual selection dichotomy, we also abandon the notion of passive versus active choice.

4.3 Sensory exploitation and history

Most studies of sexual selection by female choice address the current interactions and effects of the preference and the trait. It is often implied that if females prefer a certain trait, then the preference evolved under a (natural or sexual) selective advantage associated with that trait. Of course, most evolutionary biologists are aware of the critical difference between an evolved function and a current effect (Williams 1966), or an adaptation and an exaption (Gould and Vrba 1982). Both ideas remind us that current function, regardless of its fitness effects, does not always indicate the selective forces responsible for the evolution of the trait under consideration; this seems to have been forgotten in some research in behavioral ecology and sociobiology. Paraphrasing Kirkpatrick (1987*a*), the fact that females do not mate with dead males is not evidence for selection against necrophilia.

Depite the fact that animal behavior has its roots in a rich, comparative tradition (e.g. Lorenz 1950), the increased interest in behavioral ecology and sociobiology in the 1960s coincided with a decreased interest in phylogenetic (and mechanistic) aspects of animal behavior (Marler 1985). More recently, behaviorists have realized the necessity of using phylogenetic information to increase independence of data in testing hypotheses of adaptation (e.g. Harvey *et al.* 1978; Ridley 1983; Huey 1987; Pagel and Harvey 1988; Ryan 1988*a*). There are some cases in sexual selection when a knowledge of mechanisms, combined with meaningful phylogenetic information, can be used to examine hypotheses for the evolution of female preferences.

Although the natural selection and sexual selection hypotheses for the evolution of female choice make some critically different predictions, it has been difficult to collect empirical data to test these hypotheses (Kirkpatrick 1987*a*; Alcock 1989). However, there are some quite testable predictions when contrasting the natural and runaway sexual selection hypotheses with the sensory exploitation hypothesis. Both the natural selection and the runaway sexual selection hypotheses implicate the male trait as a causal agent in the evolution of the preference. The sensory exploitation hypothesis, on the other hand, predicts that the preference was established before the trait evolved. Therefore, appropriate phylogenetic data can produce a test of these hypotheses.

In Section 3.6, I reviewed data showing that female *Physalaemus pustulosus* preferred calls that contained lower-frequency chucks, that larger males produce lower-frequency chucks, and that this preference resulted in larger males having greater mating success in nature. The neurophysiological data suggest this preference might result from the tuning of the basilar papilla being lower than the mean dominant frequency of the population; that is, lower frequency calls are more stimulatory. The question of the evolution of the preference can be reduced to the evolution of this tuning bias.

The *P. pustulosus* species group is a well corroborated monophyletic unit consisting of two species pairs (Cannatella and Duellman 1984; Fig. 5). *P. pustulosus* and *P. petersi* are sister species, and both can produce calls with chucks. *P. coloradorum* and *P. pustulatus* comprise the other species pair. Neither of these species produces chucks, and chucks have not been reported in any of the other 40+ species in the genus (Barrio 1965). Thus parsimony dictates that the chuck is the derived condition that evolved in the ancestor of the *P. pustulosus–P. petersi* species pair (Ryan and Drewes, 1990).

Ryan *et al.* (1990) have determined the tuning of the basilar papilla of *P. coloradorum* and found that its best excitatory frequency is not significantly different from that of *P. pustulosus*. Thus the sensory bias responsible for the preference for lower-frequency chucks existed prior to the chuck (Fig. 5). This suggests that the distribution of frequencies in the chuck evolved to exploit the already existing biases of the female's basilar papilla. Both *P. coloradorum* and *P. pustulatus* produce whine-like advertisement calls that have very little energy in the range to which the basilar papilla is most sensitive. Therefore, *P. coloradorum* probably

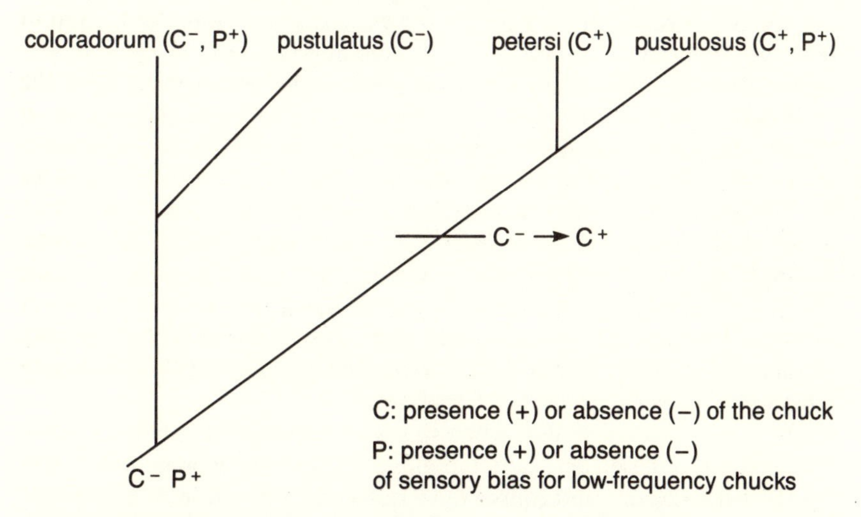

Fig. 5. The distribution of calls with chucks, the tuning of the basilar papilla that, in *P. pustulosus*, results in preference for low-frequency chucks, and the hypothesized evolutionary transitions among taxa of the *Physalaemus pustulosus* species group.

relies only, or at least primarily, on the amphibian papilla for call processing. However, even if the basilar papilla is not used in communication by *P. coloradorum*, all frogs have a basilar papilla and the basilar papilla is tuned to some frequency, often closely correlated to the frog's body size (Zakon and Wilczynski 1988). Thus it appears that the evolution of the chuck allowed stimulation of an inner ear receptor not used in communication in the earlier history of the species group.

An interesting aspect of the *P. pustulosus* study is that previous work has shown that females gain a reproductive advantage by choosing larger males, because they fertilize more of their eggs (Ryan 1983, 1985). These data could lead to the logical conclusion that this reproductive advantage generated selection for the preference. With the historical data now available, it seems that this is an incidental consequence, although certainly a selective advantage, and not the evolved function of the preference. Clearly a measure of current effects in this case does not accurately reflect historical processes.

One of the striking examples of an extremely sexually dimorphic male trait is the elaboration of the caudal fin in swordtails, as noted by Darwin (1871). Basolo (1990) demonstrated that the swordtail *Xiphophorus helleri* preferred males with longer tails. Swordtails constitute only one group of the genus *Xiphophorus*, the other group consists of the platyfish. Although the sword evolved in the swordtail clade, a recent study by Basolo (1989) suggests that the preference exists in the platy group as well. If artificial swords were attached to male platyfish, *X. maculatus*, females preferred the newly sworded males. As with the *Physalaemus* example, it appears that the preference was present in a common ancestor of species both with and without the male trait. This seems to be another case in which male swordtails evolved a trait that exploited already existing preferences.

Another example of pre-existing female preferences involves the species group of pygmy swordtails. *X. nigrensis* males span a size range of 18–40 mm in length, more than 90 per cent of the variation in body size is due to allelic variation at a single Y-linked locus (Kallman 1984, 1989), and only the larger males exhibit courtship behavior (Ryan and Causey 1989). Most populations of *X. pygmaeus*, however, consist almost entirely of males less than 25 mm in length, and never have males greater than 30 mm in length (Kallman 1984). Ryan and Wagner (1987) found that female *X. pygmaeus* preferred larger, courting *X. nigrensis* to their own smaller, non-courting males; when size was controlled, the heterospecific preference was based on courtship alone (Ryan and Wagner 1987). It is not clear if large size and courtship were secondarily lost in *X. pygmaeus*, or if *X. pygmaeus* retained the primitive condition and *X. nigrensis* evolved large size and courtship independent of the other swordtails; Kallman (1989) suggests the latter. Regardless, this study illustrates that the preference can exist independently of the trait. Furthermore, there is a preference residing in *X. pygmaeus* for both large size and courtship behavior, a preference that could be exploited if the appropriate male variation were to arise.

There are not many other examples that argue strongly for sexual selection for sensory exploitation, but some cases are suggestive. Arctiid moths have evolved the ability to perceive ultrasonics in order to detect bats and thus reduce predation risk. Some moths within this group have further evolved the ability to produce ultrasonics themselves to either jam the bat's echolocation signals or to advertise to the bats that they are distasteful. Conner (1987) has shown that at least one arctiid uses ultrasonics in courting the female. Although not yet conclusive, it appears that in males of this species the production of ultrasonic sound has become secondarily adapted to exploit a sensory modality of females that evolved under selection for avoidance of predators.

Previously, I discussed Christy's (1988) study that showed female fiddler crabs (*Uca beebei*) are attracted to pillars built by males near the openings of burrows, and that these structures exploit the female's visual system, which is adapted for detecting objects tht disrupt the horizon. Females of the closely related *U. stenodactylus* also prefer burrows with pillars even though their males do not build them (Christy, pers. comm.). This is yet another example of a preference existing independent of a trait. However, the necessary phylogenetic data are not available to make any conclusion about the historical relationships of preference and trait. Unfortunately, with the general lack of use of a phylogenetic framework by animal behaviorists, any understanding of history will be a slow process.

If a female preference for a male trait in any species is due to sensory exploitation, this need not imply that the sensory bias has never been under selection, even selection for mate choice. For instance, retinal sensitivities might evolve in response to available light, background colors in the environment or food colors. As an example, Ewert's (1980) classic studies of worm detectors in the toad's visual system shows that movements of a bar with the short edge perpendicular to the horizontal plane elicits approach, whereas movements of the same bar rotated 90 degrees elicits withdrawal. If these toads were to evolve visual courtship signals, the most effective form of the movement might be influenced by these motion detectors. Visual courtship signals are rare, but do exist, such as the foot-waving behavior of the Bornean frog (Harding 1982). An examination of the effectiveness of various types of movement in these species might be instructive. In another example, Fleischman (in press) pointed out that the eye of anoline lizards is exquisitely adapted for detecting fine movements of potential prey, and this sensitivity to movement is probably partly responsible for the evolution of the well-known push-up display of these lizards rather than a more static form of advertisement.

A sensory bias also could have resulted from selection for mate choice. Consider the preference for swords by the swordless platyfish (Basolo 1989). There might be a general preference in fish for larger males for adaptive reasons. The sword is one means by which a male could increase his apparent size, and the platyfish preference for the sworded males could result from a general preference for larger males. On the other hand, the pigmentation of the sword enhances the contrast of the fish against its background. There are several other possibilities and, encouragingly, most of them are testable.

Those of us who study female choice should be more explicit about the specificity of preferences. Is the preference for the sword directed only toward that specific morphological trait, or is it a more general, less-defined preference? And, clearly, we need to pay more attention to the phylogenetic distribution of traits and preferences, especially considering if the preference exists independent of the trait (e.g. Höngstrum 1989).

Sexual selection for sensory exploitation has been criticized because adaptationist theory predicts that female sensory biases should evolve to favor those male traits that are beneficial. Certainly, there are many examples of adaptive mate choice that bear out this prediciton. However, there are also many examples of sensory exploitation in other social contexts, besides mate choice, that clearly are disadvantageous for the signal recipient. Inter- and intraspecific deception in communication is a general example (Lloyd 1984), and brood parasites in birds a specific one. Parents of the host species preferentially feed the brood parasites while their own young starve to death or are thrown from their nest and die. The parents respond preferentially to the parasite because the large young with its increased gape response and enlarged associated markings acts as a supernormal stimulus that releases the parent's feeding response (Lack 1968). If ever selection should override a sensory constraint this would be the case.

5. CONCLUSIONS

Sexual selection by female choice involves communication, and the intent of this chapter has been to suggest the importance of understanding the role of the receiver in this process. Such an approach can only increase our understanding of the fascinating process of sexual selection by showing how properties of the receiver exert selection on male traits, and how they are responsible for some of the most bizarre morphologies and behaviors in the animal kingdom. It certainly will inform us about how evolution operates, and thus can contribute to hypotheses of the evolution of female preferences. When an understanding of mechanisms is combined with appropriate phylogenetic information, it might sometimes support a more parsimonious hypothesis of female preference – sexual selection for sensory exploitation. The main point of this chapter is that although behavioral ecology and population genetics have made important contributions to our understanding of sexual selection by female choice, we need to exploit the rich biology of this phenomenon and incorporate two additional levels of analysis, sensory mechanisms and phylogenetics. An integrative approach should enhance our understanding not only of sexual selection but of many problems in behavioral biology.

ACKNOWLEDGEMENTS

I thank D. Futuyma for suggesting that I write this chapter. Over the year I have especially benefitted from discussing these and related ideas with A. S. Rand, M. J. West-Eberhard and M. Kirkpatrick. I also appreciate discussions of these ideas, earlier drafts of the manuscript, or unpublished data with G. Barlow, A. Basolo, P. Bateson, D. Crews, R. Cocroft, W. Conner, D. Hews, J. Christy, J. Endler, M. Morris and W. Wilczynski.

REFERENCES

Alcock, J. (1989). *Animal behavior*, 4th edn. Sinauer Associates, Sunderland, Mass.

Alexander, R. A. (1975). Natural selection and specialized chorusing behavior in acoustical insects. In *Insects, science and society* (ed. D. Pimentel), pp. 35–77. Academic Press, New York.

Andersson, M. B. (1982). Female choice selects for extreme tail length in a widowbird. *Nature* **299,** 818–20.

—— (1986). Evolution of condition-dependent sex ornaments and mating preferences: Sexual selection based on viability differences. *Evolution* **40,** 804–816.

Arak, A. (1983). Mating behaviour of anuran amphibians: The roles of male–male competition and female choice. In *Mate choice* (ed. P. Bateson), pp. 181–210. Cambridge University Press, Cambridge.

Archer, S. N., Endler, J. A., Lythgoe, J. N. and Partidge, J. C. (1987). Visual pigment polymorphism in the guppy *Poecilia reticulata. Vision Res.* **8,** 1243–52.

Baker, M. C. and Cunningham, M. A. (1985). The biology of bird song dialects. *Behav. Brain Sci.* **8,** 85–100.

Baptista, L. F. (1985). Bird-song dialects: Social adaptation or assortative mating? *Behav. Brain Sci.* **8,** 100–101.

Barlow, G. W. (1977). Model action patterns. In *How animals communicate* (ed. T. A. Sebeok), pp. 98–134. Indiana University Press, Bloomington.

Barrio, A. (1965). El genero *Physalaemus* (Anura, Leptodactylidae) en la Argentina. *Physis* **25,** 421–8.

Basolo, A. (1989). Female preference for caudal fin elaboration in the platyfish, *Xiphophorus maculatus. Amer. Soc. Ich. Herp.*, San Francisco (Abstract).

Basolo, A. (1990). Female preference for male sword length in the green swordtail, *Xiphophorus helleri*. *Anim. Behav.* **40**, 332–8.

Bateson, P. B. (1983). Optimal outbreeding. In *Mate choice* (ed. P. Bateson), pp. 257–77. Cambridge University Press, Cambridge.

Beehler, B. M. and Foster, M. F. (1988). Hotshots, hotspots, and female preference in the organization of lek mating systems. *Amer. Nat.* **131,** 203–9.

Bischoff, J. A., Gould, J. L. and Rubenstein, D. I. (1985). Tail size and female choice in the guppy (*Poecilia reticulata*). *Behav. Ecol. Sociobiol,* **17,** 253–5.

Blair, W. F. (1964). Isolating mechanisms and interspecific interactions in anuran amphibians. *Quart. Rev. Biol.* **39,** 334–44.

Bluhm, C. (1985). Mate preferences and mating patterns of canvasbacks (*Aythya valisineria*). *Ornith. Monogr.* **37**, 45–56.

Boppré, M. and Schneider, D. (1985). Pyrrolizidine alkaloids quantitatively regulate both scent organ morphogenesis and pheromone synthesis in male *Creatanotus* moths (Lepidoptera: Arctiidae). *J. Comp. Physiol.* **157,** 569–77.

Borgia, G. (1986). Sexual selection in bower birds. *Sci. Amer.* **254,** 92–100.

Bradbury, J. W. (1981). The evolution of leks. In *Natural selection and social behavior: Recent research and new theory* (ed. R. D. Alexander and D. W. Tinkle), pp. 138–69. Chiron Press, New York.

—— and M. B. Andersson (ed.) (1987). *Sexual selection: Testing the alternatives.* John Wiley, Chichester.

Breden, F. and Stoner, G. (1987). Male predation risk determines female preference in the Trinidad guppy. *Nature* **329,** 831–3.

Brenowitz, E. A. (1985). Bird-song dialects: Filling in the gaps. *Behav. Brain. Sci.* **8,** 101–102.

Bullock, T. H. (1986). Some principles in the brain analysis of important signals: Mapping and stimulus recognition. *Brain Behav. Evol.* **28**, 145–56.

Burley, N. (1985). The organization of behavior and the evolution of sexually selected traits. *Ornith. Monogr.* **37**, 22–44.

—— (1986). Comparison of the band color preferences of two species of estrildid finch. *Anim. Behav.* **34,** 1732–41.

Campbell, B. (ed.) (1972). *Sexual selection and the descent of man.* Aldine, Chicago.

Cannatella, D. C. and Duellman, W. E. (1984). Leptodactylid frogs of the *Physalaemus pustulosus* species group. *Copeia* **1982,** 902–921.

Capranica, R. R. (1976). The auditory system. In *Physiology of the amphibia* (ed. B. Lofts), pp. 552–75. Academic Press, New York.

Catchpole, C. K. (1980). Sexual selection and the evolution of complex songs among European warblers of the genus *Acrocephalus*. *Behaviour* **74,** 149–66.

——, Dittami, J. and Leisler, B. (1984). Differential response to male song repertoires in female songbirds implanted with oestrodial. *Nature* **312,** 563–4.

Chambers, J. K. (1985). Social adaptiveness in human and songbird dialects. *Behav. Brain Sci.* **8,** 85–100.

Christy, J. G. (1988). Pillar function in the fiddler crab *Uca beebei* (II): Competitive courtship signalling. *Ethology* **78,** 113–28.

Cohen, J. (1984). Sexual selection and the psychophysics of mate choice. *Z. Tierpsychol.* **64,** 1–8.

Conner, W. E. (1987). Ultrasound: Its role in the courtship of the arctiid moth, *Cycnia tenera*. *Experentia* **43,** 1029–31.

——, Eisner, T., Vander Meer, R. K., Guerrero, A. and Meinwald, J. (1981). Precopulatory sexual interaction in an arctiid moth (*Utetheisa ornatrix*): Role of a pheromone derived from dietary alkaloids. *Behav. Ecol. Sociobiol.* **9,** 227–35.

——, Roach, B., Benedict, E., Meinwald, J. and Eisner, T. (1990). Courtship pheromone production and body size as correlates of larval diet in males of the arctiid moth. *Utetheisa ornatrix*. *J. Chem. Ecol.* **16**, 543–52.

Crapon de Caprona, M.-D. and Ryan, M. J. (1990). Conspecific mate recognition in swordtails, *Xiphophorus nigrensis* and *X. pygmaeus* (Poeciliidae): Olfactory and visual cues. *Anim. Behav.* **39,** 290–6.

Crews, D. (1974). Effects of group stability, male–male aggression, and male courtship behaviour on environmentally-induced ovarian recrudescens in the lizard *Anolis carolinensis*. *J. Zool. Lond.* **172,** 419–41.

Crounly-Dillon, J. and Sharma, S. C. (1968). Effect of season and sex on the photopic spectral sensitivity of the threespined stickleback. *J. Exp. Biol.* **49,** 679–87.

Darwin, C. (1859). *The origin of species.* (reprint of original). Random House, New York.

—— (1871). *The descent of man and selection in relation to sex* (reprint of original). Random House, New York.

—— (1882). Forward. On the modification of a race of Syrian street dogs. *Proc. Zool. Soc. Lond.* **25,** 367–70.

Dobzhansky, T. (1937). *Genetics and the origin of species.* Columbia University Press, New York.

Ehrman, L. and Probber, J. (1978). Rare *Drosophila* males: The mysterious matter of choice. *Amer. Sci.* **66,** 216–22.

Endler, J. (1980). Natural selection on color patterns in *Poecilia reticulata. Evolution* **31,** 76–91.

—— (1982). Convergent and divergent effects of natural selection on color patterns in two fish faunas. *Evolution* **36**, 178–88.

—— (1983). Natural and sexual selection on color patterns in peociliid fishes. *Environ. Biol. Fish.* **9,** 173–90.

—— (1986). *Natural selection in the wild.* Princeton University Press, Princeton.

—— (1989). Conceptual and other problems in speciation. In *Speciation and its consequences* (ed. D. Otte and J. A. Endler), pp. 625–48. Sinauer Asociates, Sunderland, Mass.

Ewert, J. P. (1980). *Neuroethology*. Springer-Verlag, Berlin.

Farr, J. A. (1977). Male rarity or novelty, female choice behavior, and sexual selection in the guppy, *Poecilia reticulata*. *Evolution* **31,** 162–8.

Fellers, G. M. (1979). Mate selection in gray treefrogs. *Copeia* **1979,** 286–90.

Fisher, R. A. (1958). *The genetical theory of natural selection*, 2nd edn. Dover Publications, New York.

Fitzpatrick, J. W. (1988). Why so many passerine birds? A response to Raikow. *Syst. Zool.* **37,** 71–76.

Fleischman, L. J. (in press). Design features of the displays of anoline lizards. *4th Anolis Newsletter*.

Forester, D. C. and Czarnowsky, R. (1985). Sexual selection in the spring peeper *Hyla crucifer* (Anura: Hylidae): The role of the advertisement call. *Behaviour* **92,** 112–28.

——, Lykens, D. V. and Harrison, W. K. (1989). The significance of persistent vocalisation by the spring peeper *Pseudacris crucifer* (Anura: Hylidae). *Behaviour* **108,** 197–208.

Fritzsch, B., Ryan, M., Wilczynski, W., Hetherington, T. and Walkowiak, W. (ed.) (1988). *The evolution of the amphibian auditory system.* John Wiley, New York.

Fuzessery, Z. M. (1988). Frequency tuning in the anuran central auditory system. In *The evolution of the amphibian auditory system* (ed. B. Fritzsch, M. Ryan, W. Wilczynski, T. Hetherington and W. Walkowiak), pp. 253–73. John Wiley, New York.

Gerhardt, H. C. (1982). Sound pattern recognition in some North American treefrogs (Anura: Hylidae): Implications for mate choice. *Amer. Zool.* **22,** 581–95.

—— (1988). Acoustic properties used in call recognition by frogs and toads. In *The evolution of the amphibian auditory system* (ed. B. Fritzsch, M. Ryan, W. Wilczynski, T. Hetherington and W. Walkowiak), pp. 455–83. John Wiley, New York.

——, Daniel, R. E., Perrill, S. A. and Schramm, S. (1987). Mating behaviour and male mating success in the green tree frog. *Anim. Behav.* **35,** 1490–1503.

Gibson, R. M. and Bradbury, J. W. (1985). Sexual selection in lekking sage grouse: Phenotypic correlates of male mating success. *Behav. Ecol. Sociobiol.* **18,** 117–23.

Gilliard, E. T. (1956). Bower ornamentation versus plumage characteristics in bower birds. *Auk* **56,** 450–51.

Gould, S. J. and Vrba, E. (1982). Exaptation – a missing term in the science of form. *Paleobiology* **8,** 4–15.

Graffen, A. (1987). Measuring sexual selection: Why bother? In *Sexual selection: Testing the alternatives* (ed. J. W. Bradbury and M. B. Andersson), pp. 221–33. John Wiley, Chichester.

Hamilton, W. D. and Zuk, M. (1982). Heritable true fitness and bright birds: A role for parasites? *Science* **218,** 384–7.

Harding, K. A. (1982). Courtship display in a Bornean frog. *Proc. Biol. Soc. Wash.* **95,** 621–4.

Hartshorne, C. (1956). The monotony threshold in singing birds. *Auk* **95,** 758–60.

Harvey, P. H., Kavanagh, M. and Clutton Brock, T. H. (1978). Sexual dimorphism in primate teeth. *J. Zool. Lond.* **186,** 475–86.

Hedrick, A. V. (1986). Female preferences for male calling bout duration in a field cricket. *Behav. Ecol. Sociobiol.* **19**, 73–7.

Hildebrand, M., Bramble, D. M., Liem, K. F. and Wake, D. B. (ed.) (1985). *Functional vertebrate morphology.* Belknap Press, Cambridge, Mass.

Hinde, R. A. (1970). *Animal behaviour, a synthesis of ethology and comparative psychology.* McGraw Hill, New York.

Höngstrum, J. (1989). Size and plumage dimorphism in lek-breeding birds: A comparative analysis. *Amer. Nat.* **134,** 72–87.

Houde, A. E. (1987). Mate choice based upon naturally occurring color-pattern variation in a guppy population. *Evolution* **41,** 1–10.

—— (1988). Genetic differences in female choice between two guppy populations. *Anim. Behav.* **36,** 510–16.

Hoy, R. R. (1978). Acoustic communication in crickets: A model system for the study of feature detection. *Fed. Proc.* **37,** 2316–23.

Huber, F. (1978). The insect nervous system and insect behaviour. *Anim. Behav.* **26,** 969–81.

Huey, R. B. (1987). Phylogeny, history, and the comparative method. In *New directions in ecological physiology* (ed. M. E. Feder, A. F. Bennett, W. W. Burggren and R. B. Huey), pp. 76–98. Cambridge University Press, Cambridge.

Huxley, J. S. (1938). Darwin's theory of sexual selection and the data subsumed by it, in light of recent research. *Amer. Nat.* **72,** 416–33.

Kallman, K. D. (1984). A new look at sex determination in poeciliid fishes. In *Evolutionary genetics of fishes* (ed. B. J. Turner), pp. 95–171. Plenum Press, New York.

—— (1989). Genetic control of size at maturity in *Xiphophorus.* In *Ecology and evolution of live-bearing fishes (Poeciliidae)* (ed. G. K. Meffe and F. F. Snelson), pp. 163–200. Prentice-Hall, Englewood Cliffs, N.J.

Keeton, W. T. (1979). Pigeon navigation. In *Neural mechanisms of behavior in the pigeon* (ed. A. M. Granda and J. H. Maxwell). Plenum Press, New York.

Kendrick, K. M. and Baldwin, B. A. (1987). Cells in temporal cortex of conscious sheep can respond preferentially to the sight of faces. *Science* **236,** 448–50.

Kirkpatrick, M. (1982). Sexual selection and the evolution of female choice. *Evolution* **36,** 1–12.

—— (1986). The handicap mechanism of sexual selection does not work. *Amer. Nat.* **127,** 223–40.

—— (1987*a*). The evolution of female preferences in polygynous animals. In *Sexual selection: Testing the alternatives* (ed. J. W. Bradbury and M. B. Andersson), pp. 67–82. John Wiley, Chichester.

—— (1987*b*). Sexual selection and female choice in polygynous animals. *Ann. Rev. Ecol. Syst.* **18,** 43–70.

Klump, G. M. and Gerhardt, H. C. (1987). Use of non-arbitrary acoustic criteria in mate choice by female gray tree frogs. *Nature* **326,** 286–8.

Kodric-Brown, A. (1985). Female preference and sexual selection for male coloration in the guppy *Poecilia reticulata. Behav. Ecol. Sociobiol.* **17,** 199–205.

—— and Brown, J. H. (1984). Truth in advertising: The kinds of traits favored by sexual selection. *Amer. Nat.* **124,** 309–323.

Lack, D. (1968). *Ecological adaptations for breeding in birds.* Methuen, London.

Lande, R. (1981). Models of speciation by sexual selection on polygenic traits. *Proc. Natl. Acad. Sci USA* **78,** 3721–5.

Latimer, W. and Sippel, M. (1987). Acoustic cues for female choice and male competition in *Tettigonia cantans. Anim. Behav.* **35,** 887–900.

Lewis, E. R. and Lombard, R. E. (1988). The amphibian inner ear. In *The evolution of the amphibian auditory system* (ed. B. Fritzsch, M. Ryan, W. Wilczynski, T. Hetherington and W. Walkowiak), pp. 92–123. John Wiley, New York.

Lijfeld, J. T. and Slagsvold, T. (1988). Female pied flycatchers *Ficedula hyopleuca* choose male characteristics in homogeneous habitats. *Behav. Ecol. Sociobiol.* **22,** 27–36.

Littlejohn, M. J. (1965). Premating isolating mechanisms in the *Hyla ewingi* complex (Anura: Hylidae). *Evolution* **19,** 234–43.

—— and Harrison, P. A. (1985). The functional significance of the diphasic advertisement call of *Geocrinia victoriana* (Anura: Leptodactylidae). *Behav. Ecol. Socio. Biol.* **16,** 363–73.

Lloyd, J. E. (1984). On deception, a way of all flesh, and firefly signalling and systematics. *Oxford Surv. Evol. Biol.* **1,** 48–84.

Lorenz, K. (1950). The comparative method in studying innate behavior patterns. *Symp. Soc. Exp. Biol. IV*, Cambridge.

Magnus, D. (1958). Experimentelle Untersuchungen zur Bionomie und Ethologie des aisermantels *Argynnis paphia* Girard (Lep. Nymph.). *Z. Tierpsychol.* **15,** 397–426.

Marler, P. (1985). Foreword. In *The túngara frog, a study in sexual selection and communication* (M. J. Ryan), pp. ix–xi. University of Chicago Press, Chicago.

Mayr, E. (1942). *Systematics and the origin of species.* Columbia University Press, New York.

—— (1963). *Animal species and evolution.* Belknap Press, Cambridge, Mass.

—— (1982). *The growth of biological thought.* Harvard University Press, Cambridge, Mass.

McGregor, P. K. and Krebs, J. R. (1982). Mating and song types in the great tit. *Nature* **297,** 60–61.

Møller, A. P. (1988). Female choice selects for male sexual tail ornaments in the monogamous swallow. *Nature* **332,** 640–42.

Moodie, G. E. E. (1982). Why asymmetric mating preferences may not show the direction of evolution. *Evolution* **36,** 1096–7.

Morris, D. (1956). The function and causation of courtship ceremonies. *Foundation Singer-Polignac: Colloque Internat. sur L'Instinct*, June 1954.

Morris, M. R. and Yoon, S. L. (1989). A mechanism of female choice for larger males in the treefrog *Hyla chrysoscelis. Behav. Ecol. Sociobiol.* **25,** 65–71.

Myrberg, A. A. Jr, Mohler, M. and Catala, J. D. (1986). Sound production by males of a coral reef fish (*Pomacentris partitus*): Its significance to females. *Anim. Behav.* **34,** 913–23.

O'Donald, P. (1983). Sexual selection by female choice. In *Mate choice* (ed. P. Bateson), pp. 53–66. Cambridge University Press, Cambridge.

Pagel, M. D. and Harvey, P. H. (1988). Recent developments in the analysis of comparative data. *Quart. Rev. Biol.* **63,** 413–40.

Parker, G. A. (1983). Mate quality and mating decisions. In *Mate choice* (ed. P. Bateson), pp. 141–66. Cambridge University Press, Cambridge.

Paterson, H. E. H. (1982). Perspectives on speciation by reinforcement. *S. Afr. J. Sci.* **78,** 53–7.

Payne, R. B. (1983). Bird songs, sexual selection, and female mating strategies. In *Female social strategies* (ed. S. E. Wasser), pp. 55–90. Academic Press, New York.

Perrone, M., Jr (1978). Mate size and breeding success in a monogamous cichlid fish. *Env. Biol. Fish.* **3,** 193–201.

Peterson, J. R. and Merrell, D. J. (1983). Rare male mating disadvantage in *Drosophila melanogaster. Evolution* **37,** 1306–1316.
Phelan, P. L. and Baker, T. C. (1986). Male size-related courtship success and intersexual selection in the tobacco moth *Ephestia elutella. Experentia* **42,** 1291–3.
—— and Baker, T. C. (1987). Evolution of male pheromones in moths: Reproductive isolation through sexual selection. *Science* **235,** 205–7.
Pomiankowski, A. N. (1988). The evolution of female mate preferences for male genetic quality. *Oxford Surv. Evol. Biol.* **5,** 136–84.
Radinsky, L. B. (1987). *The evolution of vertebrate design.* University of Chicago Press, Chicago.
Raikow, R. J. (1986). Why are there so many kinds of passerine birds? *Syst. Zool.* **35,** 255–9.
Rand, A. S. and Ryan, M. J. (1981). The adaptive significance of a complex vocal repertoire in a neotropical frog. *Z. Tierpsychol.* **57,** 209–214.
Reimchen, T. E. (1989). Loss of nuptial color in threespine sticklebacks (*Gasterosteus aculeatus*). *Evolution* **43,** 450–60.
Ridley, M. (1983). *The explanation of organic diversity, the comparative method and adaptations for mating.* Oxford University Press, Oxford.
—— (1986). *Animal behaviour: A concise introduction.* Blackwell Scientific, Oxford.
Robertson, J. G. M. (1986). Female choice, male strategies and the role of vocalizations in the frog *Uperolia rugosa. Anim. Behav.* **34,** 773–84.
Rose, G. J. and Capranica, R. R. (1983). Temporal selectivity in the central auditory system of the leopard frog. *Science* **219,** 1087–9.
Rowland, W.J. (1982). Mate choice by male sticklebacks *Gasterosteus aculeatus. Anim. Behav.* **30,** 1093–7.
—— (1989*a*). The ethological basis of mate choice in male threespine sticklebacks, *Gasterosteus aculeatus. Anim. Behav.* **38,** 112–20.
—— (1989*b*). Mate choice and the supernormality effect in female sticklebacks (*Gasterosteus aculeatus*). *Behav. Ecol. Sociobiol.* **24,** 433–8.
Rutowski, R. R. (1982). Epigamic selection by males as evidenced by courtship partner preference in the checkered white butterfly. *Pieris protodice. Anim. Behav.* **30,** 108–112.
Ryan, M. J. (1980). Female mate choice in a neotropical frog. *Science* **209,** 523–5.
—— (1983). Sexual selection and communication in a neotropical frog, *Physalaemus pustulosus. Evolution* **37,** 261–72.
—— (1985). *The túngara frog, a study in sexual selection and communication.* University of Chicago Press, Chicago.
—— (1986). Neuroanatomy influences speciation rates among anuran. *Proc. Natl. Acad. Sci, USA* **83,** 1379–82.
—— (1988*a*). Constraints and patterns in the evolution of anuran acoustic communication. In *The evolution of the amphibian auditory system* (ed. B. Fritzsch, M. Ryan, W. Wilczynski, T. Hetherington and W. Walkowiak), pp. 637–77. John Wiley, New York.
—— (1988*b*). Energy, calling, and selection. *Amer. Zool.* **58,** 885–98.
—— and Causey, B. A. (1989). 'Alternative' mating behavior in the swordtails

Xiphophorus nigrensis and *Xiphophorus pygmaeus* (Pisces: Poeciliidae). *Behav. Ecol. Sociobiol.* **24,** 341–8.

—— and Drewes, R. (1990). Vocal morphology of the *Physalaemus pustulosus* species group (Family Leptodactylidae): Morphological response to sexual selection for complex calls. *Biol. J. Linn. Soc.* **40**, 37–52.

—— and Rand, A.S. (1990). The sensory basis of sexual selection for complex calls in the túngara frog, *Physalaemus pustulosus* (sexual selection for sensory exploitation). *Evolution* **44**, 305–14.

—— and Tuttle, M. D. (1987). The role of prey-generated sound, vison, and echolocation in prey localization by the African bat *Cardioderma cor. J. Comp. Physiol.* **161**, 59–66.

—— and Wagner, W. E., Jr (1987). Asymmetries in mating preferences between species: female swordtails prefer heterospecific mates. *Science* **236,** 595–7.

—— and Wilczynski, W. (1988). Coevolution of sender and receiver: Effect on local mate preference in cricket frogs. *Science* **240,** 1786–8.

——, Fox, J. H., Wilczynski, W. and Rand, A. S. (1990). Sexual selection by sensory exploitation in the frog *Physalaemus pustulosus. Nature* **343**, 66–7.

Schwartz, J. J. (1986). Male calling behavior and female choice in the neotropical treefrog *Hyla microcephala. Ethology* **73,** 116–27.

Searcy, W. A. (1984). Song repertoire size and female preference in song sparrows. *Behav. Ecol. Sociobiol.* **14,** 281–6.

—— and Andersson, M. B. (1986). Sexual selection and the evolution of song. *Ann. Rev. Ecol. Syst.* **17,** 507–533.

—— and Marler, P. (1981). A test for responsiveness to song structure and programming in female sparrows. *Science* **213,** 926–8.

Semler, D. E. (1971). Some aspects of adaptation in a polymorphism for breeding colours in the threespine stickleback (*Gasterosteus aculeatus*). *J. Zool. Lond.* **165,** 291–301.

Staddon, J. E. R. (1975). A note on the evolutionary significance of 'supernormal' stimuli. *Amer. Nat.* **109,** 541–5.

Straughn, I. R. (1975). An analysis of the mechanisms of mating call discrimination in *Hyla regilla* and *Hyla cadaverina. Copeia* **1975,** 415–24.

Sullivan, B. K. (1983). Sexual selection in Woodhouse's toad (*Bufo woodhousei*). II. Female choice. *Anim. Behav.* **31,** 1011–17.

—— (1989). Passive and active choice: A comment. *Anim. Behav.* **37,** 692–4.

Templeton, A. (1979). Once again, why 300 species of Hawaiian *Drosophila*? *Evolution* **33,** 513–17.

ten Cate, C. and Bateson, P. (1988). Sexual selection: The evolution of conspicuous characteristics in birds by means of imprinting. *Evolution* **42,** 1355–8.

—— and Bateson, P. (1989). Sexual imprinting and a preference for 'supernormal' partners in Japanese quail. *Anim. Behav.* **38,** 356–8.

Thornhill, R. (1976). Sexual selection and parental investment in insects. *Amer. Nat.* **110,** 153–61.

Tinbergen, N. (1948). Social releasers and the experimental method required for their study. *Wilson Bull.* **60,** 6–52.

—— (1951). *The study of instinct.* Oxford University Press, Oxford.

—— (1953). *The social behavior of animals.* Butler and Tanner, London.

——, Meeuse, B. J. D., Boerma, L. K. and Varossieau, W. W. (1942). Die Balz des Samtfalters, *Eumenis* (=*Satyrus*) *semele* (L.). *Z. Tierpsychol.* **3,** 37–60.

Trivers, R. L. (1972). Parental investment and sexual selection. In *Sexual selection and the descent of man* (ed. B. Campbell), pp. 136–79. Aldine, Chicago.

—— (1985). *Social evolution.* Benjamen/Cummings, Menlo Park, Calif.

Vermeij, G. J. (1988). The evolutionary success of passerines: A question of semantics? *Syst. Zool.* **37,** 69–71.

Verrill, P. A. (1985). Male mate choice for large fecund females in the red spotted newt, *Notophthalmus viridescens:* How is size assessed? *Herpetologica* **41,** 382–6.

Wade, M. (1987). Measuring sexual selection. In *Sexual selection: Testing the alternatives* (ed. J. W. Bradbury and M. B. Andersson), pp. 197–207. John Wiley, Chichester.

Walker, T. J. (1974). Character displacement in acoustic insects. *Amer. Zool.* **14,** 1137–50.

Walkowiak, W. (1988). Central temporal coding. In *The evolution of the amphibian auditory system* (ed. B. Fritzsch, M. Ryan, W. Wilczynski, T. Hetherington and W. Walkowiak), pp. 275–294. John Wiley and Sons, New York.

Wallace, A. R. (1905). *Darwinism*, 3rd edn. Macmillan, London.

Webster, T. P. and Burns, J. M. (1973). Dewlap color variation and electrophoretically detected sibling species in a Haitian lizard, *Anolis brevirostris. Evolution* **27,** 368–77.

Wells, K. D. (1988). The effect of social interactions on anuran vocal behavior. In *The evolution of the amphibian auditory system* (ed. B. Fritzsch, M. Ryan, W. Wilczynski, T. Hetherington and W. Walkowiak), pp. 433–54. John Wiley, New York.

—— and Schwartz, J. J. (1984). Vocal communication in a neotropical treefrog, *Hyla ebraccata*: Advertisement calls. *Anim. Behav.* **32,** 405–420.

West-Eberhard, M. J. (1979). Sexual selection, social competition, and evolution. *Proc. Amer. Phil. Soc.* **123,** 222–34.

Williams, G. C. (1966). *Adaptation and natural selection, a critique of some current evolutionary thought.* Princeton University Press, Princeton.

Yamamoto, J. T., Shields, K. M., Millam, J. R., Roudybush, T. E. and Grau, C. R. (1989). Reproductive activity of force-paired cockatiels (*Nymphicus hollandicus*). *Auk* **106,** 86–93.

Zahavi, A. (1975). Mate selection: A selection for a handicap. *J. Theor. Biol.* **53,** 205–214.

Zakon, H. and Wilczynski, W. (1988). The physiology of the VIIIth nerve. In *The evolution of the amphibian auditory system* (ed. B. Fritzsch, M. Ryan, W. Wilczynski, T. Hetherington and W. Walkowiak), pp. 125–55. John Wiley, New York.

Zeil, J., Nalbach, G. and Nalbach, H. O. (1986). Eyes, eye stalks and the visual world of semi-terrestrial crabs. *J. Comp. Physiol.* **159,** 801–811.

Zelick, R. D. and Narins, P. M. (1983). Intensity discrimination and the precision of call timing in two species of neotropical treefrogs. *J. Comp. Physiol.* **153,** 403–12.

B chromosomes, selfish DNA and theoretical models: where next?

M. W. SHAW and G. M. HEWITT

1. INTRODUCTION

The phrase 'selfish DNA' imbues our hereditary material with the most hated of human vices. It gives authors the opportunity to grab a reader's attention instantly, offering either to expose the nature of original sin as inherent in the physical world, or to explain the transcendence of their base nature by chromosomes, by cells, and then by organisms, transmuting selfishness into cooperation. The scientific ideas embodied in the phrase, however, though not called by the same name, are as old as the knowledge that DNA encodes the genome of an organism. Östergren (1945) proposed that B chromosomes might be 'parasitic' nuclear elements, maintained because of their behaviour within a cell nucleus, rather than because they carried genetic information that aided an organism's survival or reproduction. B chromosomes, our primary focus in this chapter, are chromosomes whose presence or absence is polymorphic in a population. Their defining feature is their dispensibility: complete absence of other types of chromosomes would be lethal. In view of this dispensibility, it is almost inevitable that their effects on phenotype should not be severe. In practice, the effects of B chromosomes on external phenotype are almost exclusively restricted to alterations in viability, fecundity and similar quantitative characters: discrete phenotypic differences of a Mendelian kind determined by B chromosomes are extremely rare. Recent reviews of the literature pertaining to B chromosomes may be found in Jones and Rees (1982), Jones (1985) and Werren *et al.* (1988). Our present aim is to examine critically the theory relating to B chromosomes, and to speculate on likely fruitful areas of research in the light of this theory.

Because the number of B chromosomes within a nucleus is a polymorphic character, including odd numbers, their distribution at meiosis cannot be completely regular; in fact, pairing and disjunction in individuals carrying even numbers of B chromosomes is often irregular. The majority of B chromosome systems sufficiently analysed appear to involve 'selfishness' in the sense that B chromosomes tend to accumulate regardless of their effect on the fitness of the phenotype. As such, they are a paradigm of selfish DNA, and the earliest example to be widely studied; their behaviour within populations also has much in common with certain types of disease and parasite, as Östergren recognized. The study of B chromosomes, though interesting in its own right, may also shed light on how the integrity and organization of genomes evolved and is maintained (Eshel 1985). For example, one can ask how much of the excess DNA in genomes is actually the relic of the production of selfish elements and their subsequent neutralization by selection at some level, between cells, between individuals or between groups. Or one can ask to what extent evolutionary novelty can result from non-Mendelian elements holding populations far from optimal phenotypes, forcing shifts between adaptive peaks or constraining those peaks that are attainable.

Although an intuitively very useful metaphor, the notion of selfishness or parasitism is not well-defined for inherited elements like B chromosomes, transposable elements or certain viruses that do not have a mechanism to ensure exact equipartition to the products of meiosis. If such a mechanism exists, or if all products of meiosis are equally likely to participate in producing the next generation, the average frequency of the element in an infinite population or collection of finite populations cannot change unless the presence of the element alters fitness through survival or fecundity. But if such a mechanism does not exist, neither deleteriousness (i.e. zygotic fitness < 1) nor advantageousness (i.e. zygotic fitness > 1) of the element determine the long-term fate of the element: an advantageous element may disappear or a deleterious one persist. The persistence of a deleterious element is obviously selfishness, but an advantageous element can only continue in a population if its advantage outweighs any inherent tendency to disappear. So fitness changes produced by an element not subject to Mendelian inheritance do not solely determine its representation in a population; inheritance and fitness are inextricably bound together. A physical metaphor for such a system is illustrated in Fig. 1. This consideration must apply to cellular organelles like mitochondria and plastids; if distributed completely randomly to the products of mitosis, some lineages would accumulate enormous numbers, whereas others would die out for lack of any. Thus mechanisms must exist for regulating the numbers of organelles within a cell.

2. MODELS OF B CHROMOSOME POPULATION DYNAMICS

Modelling is the exploration of the consequences of hypotheses about the behaviour of a system. It may be a purely verbal activity, but mathematics frequently helps to extend chains of reasoning that would otherwise be impossible to grasp, to expose implicit hypotheses that might otherwise pass unnoticed, and to ensure that the reasoning in a model is internally

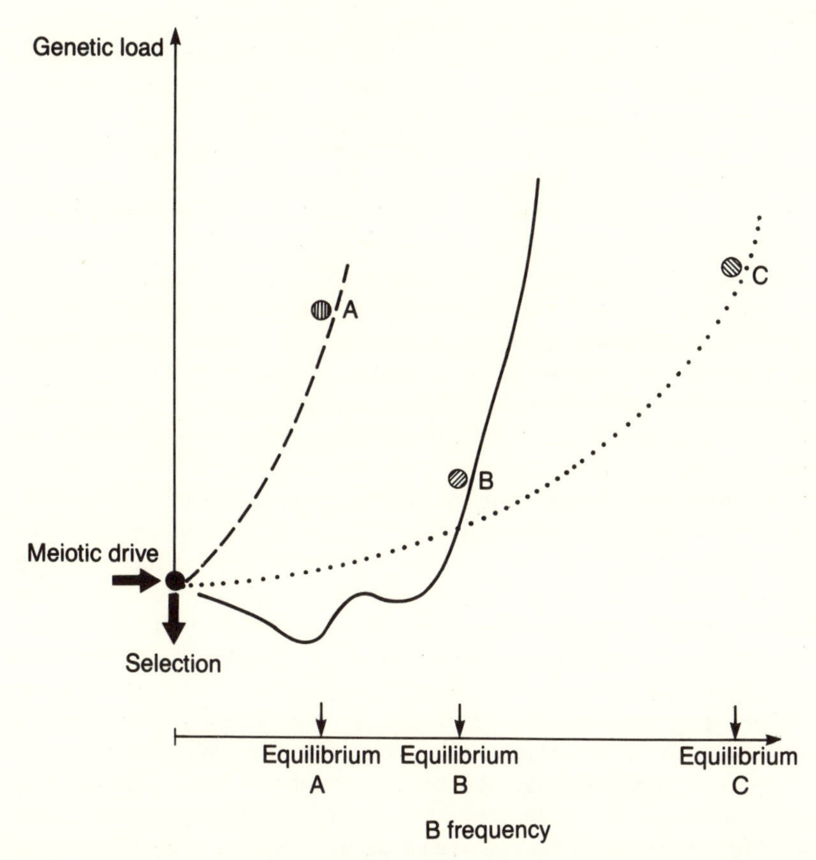

Fig. 1. A mechanical metaphor for the forces determining average B frequency. Gravity (selection) tends to pull the ball (the population) downwards. A force parallel to the ground (meiotic drive) also acts. The ball stops when gravity balances the force along the slope. Unless the slope goes uphill somewhere (increasing genetic load with more B's) the ball will never stop. Three systems – A, B and C – are shown. In A and C, small numbers of B's are disadvantageous; they differ in the balance between meiotic drive and selection. In B, small numbers of B's are advantageous; none the less the equilibrium frequency is below that in C.

consistent and logical. Modelling has a number of important uses. It can describe the *kinds* of behaviour a system supposed to be governed by certain rules may show, both revealing unexpected consequences and showing where pattern or regularity needs no explanation beyond the hypotheses of the model. It can suggest the strength and timing of interactions within the system, providing estimates of how likely it is that particular observations will be made, or positive or negative results obtained from experiments. Where mathematics is used, it may be possible to explore more complex systems and subtler interactions than would otherwise be possible. At the same time, the abstractness of mathematical language makes it easier to see when the essentials of two biological systems are the same, or, if they differ, where the essential differences lie. Like experiments, modelling may be done well or badly; bad models, like bad experiments, may tell us nothing new or even be positively misleading.

Sufficient data have been accumulated in a number of cases for workers to attempt mathematical or numerical models of the maintenance and evolution of B chromosomes within populations. It is useful to introduce first a model of a rather different genetic element, in order to indicate how general the processes involved are (we shall return to the analogies between B chromosomes and other non-Mendelian genetic elements later).

The model we want to consider is one designed to study transposable elements, residing on A chromosomes but capable of replicative transposition to new sites (Ajioka and Hartl 1988). Because of the very large number of copies of these sequences that can accumulate, no attempt was made to include parameters for transposition and fitness for every possible number of elements. Instead, the shape of the equilibrium distribution of numbers of elements per chromosome was deduced for a variety of different shapes of curve relating transposition probability to copy number and fitness to copy number. Assuming that the data they possessed represented equilibrium distributions – a point to which we shall return later – they were able to show that several different transposable elements had different relationships between transposition and copy number and fitness and copy number. In particular, they showed that it was not plausible in several of the cases to assume no effect on fitness.

The first model of a B chromosome system was elaborated for the wild Japanese lily, *Lilium callosum*, by Kimura and Kayano (1961). They set up a system of recursion equations, relating the frequency of each B karyotype in one generation to that in the next. Essentially, these equations involved a series of steps: (1) the generation of the array of seedling karyotypes produced by each adult karyotype; (2) the weighting of each array by the fecundity of the parental karyotype; (3) the adding up of these arrays across parental karyotypes; (4) the adding in of the appropriate proportion of clonal progeny; and (5) the weighting of the resulting array by the probabilities of survival of the different karyotypes. Kimura and Kayano (1961) had available to them numerical estimates for the fecundity and inheritance parameters in these equations; they assumed the natural population was at equilibrium and estimated survival probabilities for each karyotype. The plant was self-incompatible and therefore random mating was assumed. The rate of asexual reproduction via bulb formation was estimated crudely by assuming that spatial correlations in B frequency were caused solely through bulb formation. Their conclusion was that plants containing B chromosomes were always less fit than those without: the B chromosome was essentially 'selfish' or 'parasitic'.

Matthews and Jones (1982) analysed the rye (*Secale cereale*) system, modelling the passage of B chromosomes through meiosis and gametophyte development in some detail. This gave them many parameters, and they did not explore the parameter space fully. Their conclusions were limited to estimates of the sensitivity of the equilibrium number of B chromosomes to variations in individual parameters singly, all others being held at what were believed to be good, or at least typical, estimates. This is a dangerous procedure, and can obscure the underlying logic of what is going on. For example, Matthews and Jones comment that the removal of male gametes with large numbers of B chromosomes after gamete formation but before fertilization has 'remarkably little effect on the polymorphism' (Matthews and Jones 1982, p. 362). However, the figure supporting this conclusion was generated with all other parameters being held constant. In particular, at meiosis, all B chromosomes paired if possible and were otherwise distributed at random to meiotic products. At first gametophyte mitosis in both pollen grains and megaspores, 90 per cent of B chromosomes non-disjoin and 90 per cent of the non-disjoined products move into the functional nucleus. Thus the overall bias in the meiosis is 0.81, on both the male and female sides. At low B frequencies, therefore, the B frequency would increase by a factor of 1.6 in each generation. The loss in fitness of male gametes introduced by Matthews and Jones was linear with the number of B chromosomes; plants with more than four B chromosomes were assumed to die. So the main factor producing equilibrium here is the rapid generation of plants containing very large numbers of B chromosomes, producing mainly 4B gametes, mating with others to produce large numbers of dead offspring. If the effective drive rate is slightly reduced, by killing male gametes with 4B's, halving the effectiveness of 2B pollen etc., the main effect is to reduce the number of plants with B complements grossly in excess of four; equilibrium is still reached at a very high level, but the meiotic drive is slightly reduced at that equilibrium. At lower levels of meiotic drive, with more reasonable equilibria, the effect of reducing pollen fitness could be much larger; for some parameter values it could turn the fate of the B chromosome from long-term persistence into rapid loss. Matthews and Jones' main conclusions are, of course, correct: strong meiotic drive will cause a rapid increase of B frequency from initially low values to an equilibrium set by opposing natural selection. They are, however, wrong to say that 'selection against gametes and plants containing B's, even at levels much higher than observed in nature, cannot prevent the accumulation of B's within populations, provided that high rates of directed non-disjunction also occur' (Matthews and Jones 1982, p. 345). To take an extreme example with which to make the point, if karyotypes other than 0B and 2B were lethal, pairing and disjunction at meiosis normal, and the non-disjunction rate at pollen grain mitosis was u with all non-disjoined chromatids going to the functional pole, the fitness of the 2B karyotype must exceed $1/2u$ if the B is to be retained in the population. The parameters chosen are extreme, so that not all the details of the model need be explained, but the point is general.

Gregg *et al.* (1984) studied the equilibria possible with different fitnesses assigned to 1B and 2B individuals of the Australian plague locust *Chortoicetes terminifera* using point values for meiotic transmission rates derived from experiments. They showed that if the fitness of 1B's was less than a critical value, extinction of the B was certain unless 2B's were very much fitter than 0B's: on the face of it, an unlikely fitness pattern. (Note that fig. 4 in Gregg *et al.* (1984) is wrongly labelled.) Close to the critical fitness, the equilibrium B frequency was extremely sensitive to the fitness of 2B's. Shaw (1985) studied a very similar model of the British grasshopper *Myrmeleotettix maculatus*, and investigated the effect of varying meiotic transmission parameters and fitnesses on the final equilibrium reached. He pointed out that for an equilibrium B frequency to exist at all, drive must compensate for loss in fitness due to the B's: thus, Gregg *et al.*'s critical fitness is simply the inverse of the net excess transmission in their inheritance data. This feature, obvious as it seems once pointed out, seems to have been noticed independently in the context of B chromosomes, autosomal meiotic drive systems (Bruck 1957) and inherited disease (Fine 1975).

Shaw (1984) also looked at the magnitude of genetic drift around the equilibrium distribution of B karyotypes. It turns out that populations are not tightly constrained to the equilibria. For example, with an effective population size as large as 1000, a population of *M. maculatus* with plausible transmission rates and fitnesses would exist with a B frequency between 0.45 and 0.55 for 95 per cent of the time. This feature will be quite general in such models, because selection close to the equilibrium is relatively weak. Very large variances must therefore be attached to estimates of fitnesses, which assume natural populations are observed at equilibrium and use inheritance patterns themselves subject to errors of the order of at least 5–10 per cent (Fig. 2).

Although these models differ in detail, their essentials are necessarily the same, i.e. B's change in frequency for two reasons: individuals carrying them may have altered fitness, or the average number in the offspring may differ from that in the parents. In general, B's do not disjoin completely regularly. Even if they do, the products of non-disjunction will not die as is the case for A chromosomes, precisely because of their dispensibility: thus numerically variant individuals will inevitably enter the population.

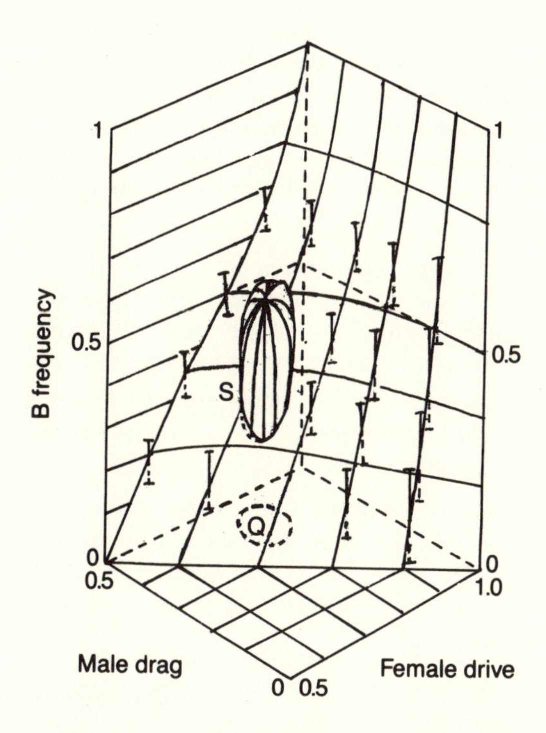

Fig. 2. Equilibrium values and the variation around equilibrium in the model of the B system in *Myrmeleotettix maculatus* proposed by Shaw (1985). The space is formed by axes representing the strength of female meiotic drive, the extent of meiotic 'drag' (loss of B's) in males, and the resulting equilibrium B frequency if 1B individuals are equal in fitness to 0B's, but 2B's are nearly sterile or inviable. Equilibrium B frequencies lie on the surface E. Variation around these equilibria because of genetic drift is indicated by the vertical bars: 95 per cent of the time a population with a genetically effective size of 1000 will have a B frequency between the upper and lower limits of a bar.

As an example of the implications of this for observations of real populations, suppose the average drive and drag rates in a population were measured to be 0.8 and 0.4 respectively, using 250 embryos from independent crosses, then Q is a 95 per cent confidence region for the true average transmission rates. Further suppose that equilibrium B frequency were measured using 200 individuals from a population with a genetically effective population size of 1000 (which is very large for this species). Then, S represents the joint 95 per cent confidence region for the three observations. B frequencies between the upper and lower extremes of frequency in this region are compatible with selective neutrality of the B, despite the large sample and population sizes assumed. Studies measuring only B frequency and meiotic transmission parameters will clearly yield estimates of fitness with very large variance. Although the model is a specific one, its conclusions will be qualitatively true for other B systems.

If an odd number of B's is present in a cell, there is no necessary mechanism to ensure that all products of meiosis have an equal probability of containing each possible number of B's. This means that the number of B's more or less must tend to increase or decrease. If the intrinsic tendency is to increase, it must at some stage become disadvantageous to have too many in an organism, because the number of B's present is in fact limited. So we can distinguish two types of situation:

1. B's tend to increase: equilibrium is reached when the unfitness of individuals with many B's – whether many be one or tens – balances the overall drive. Note that here the fitness of individuals with small numbers of B's can be irrelevant to the long-term fate of the B.
2. B's tend to decrease but are maintained by selection in favour of them. In this case, of course, a B is selfish only in the sense (if any) that any genetic element is.

In case (1), a genetic load is imposed on the population in which the B is polymorphic because at equilibrium individuals with B's are unfit on average. In case (2), a genetic load is imposed for the reverse reason. The analogy with inherited parasites is extremely close (L'Heritier 1970; Fleuriet 1988; Fine 1975).

3. SELFISH OR NOT?

The effects of B chromosomes on fitness have been estimated both directly and indirectly. The indirect method depends on knowing the inheritance patterns of the B and the equilibrium pattern of B karyotypes in some natural population. A matrix is set up to describe the progeny distribution of B's resulting from each possible mating in the population (or, if random mating is assumed, the distribution of B's in the gametes from each possible parent); a matrix of fitnesses for each genotype is also set up. The combination of the inheritance and fitness patterns gives an equation relating one generation's average B frequency to the previous generation's actual frequency. The study of these equations can reveal the long-term possible fates of the B: published examples only produce long-term extinction, indefinite increase, or stable equilibria. If it is assumed that natural populations are observed close to equilibrium and that the equilibrium is accurately estimated, then knowledge of inheritance patterns – relatively easier to obtain than fitness data – can be used to estimate the fitnesses of individuals possessing varying numbers of B's: the 'parent–progeny matrix method'. Kimura and Kayano (1961) were the first to apply this technique, studying a population of *Lilium callosum*, as mentioned before. They estimated all fitnesses as less than 1, with an accumulation mechanism in the pollen leading to a stable equilibrium.

Many papers have since used the method with data of varying precision to try to estimate fitness changes associated with the possession of B's.

Nur (1969*b*) demonstrated the rapidly declining fitness of male *Pseudococcus obscurus* with increasing numbers of B's coupled with strong meiotic drive of the elements; in females, there was no evidence for fitness differences associated with the B. Nur (1977) used a variant of the parent–progeny matrix method that avoids the artificiality of laboratory matings. Eggs were taken from gravid females caught in the wild and karyotyped, so that a sample of the egg population could be compared directly with the adult population. The changes between eggs and adults must be due to selection; however, the changes between adults and eggs are due to the combined effects of non-random meiotic transmission and sexual selection, assortative mating, inbreeding and fecundity differences. Assuming these sources of selection were negligible, Nur (1977) demonstrated strong meiotic drive in female *Melanoplus femur-rubrum* coupled with substantial selection against both 1B and 2B individuals of both sexes. Hewitt (1973*a*) found meiotic drive in female *M. maculatus*, coupled with meiotic loss ('drag') in males. In two populations, single B's appeared to be disadvantageous, but in another population the data suggested that individuals with a single B might be at an advantage to those without. Robinson and Hewitt (1976) studied the B frequency in this population at several times during the year starting with a newly laid population of eggs. They found some evidence for an increase in B frequency between adults and eggs, with a subsequent decline in frequency to the next generation of adults. However, the cycle was barely statistically significant (Shaw *et al.* 1985). (Both Robinson and Hewitt (1976) and Nur (1977) overestimated the effective egg sample size because correlations between the B content of eggs within a single pod or progeny were neglected.) Fuller data summarized in Shaw *et al.* (1985) suggest that meiotic drive in females in this population is stronger on average than estimated from the individuals studied by Hewitt (1973*a*), so no strong inferences can be made as to the fitness of 1B individuals with respect to 0B's. Gregg *et al.* (1984) concluded that the B chromosome in *Chortoicetes terminifera* was maintained by net meiotic drive with both 1B and 2B individuals being less fit than 0B individuals; they tempered this conclusion by noting the extreme sensitivity of their results to the exact equilibrium B frequency in the population.

Holmes and Bourgourd (1989) demonstrated directly that individuals of *Allium schoenoprasum* carrying small numbers of B's survived from seed to seedling better than individuals without B's; the numbers of B's in seeds seemed to be less than in their parents, suggesting meiotic loss rather than meiotic drive. By contrast, Thomson (1984) found direct evidence of a loss of B-carrying individuals of *Rattus fuscipes* between birth and maturity, but with transmission data consistent with an accumulation mechanism. The possible complexities of such studies are emphasized by the results obtained in experimental populations of *Secale cereale* and *Lolium perenne.* In *L. perenne*, selection within one generation favouring plants with B's was demonstrated at high sowing density in pot experiments; at low density, the average B frequency among plants within a cohort declined as it aged. Thus plants with B's unequivocally survived less well (Teoh *et al.* 1976). In *S. cereale*, there was evidence that B's became more disadvantageous at high density (Teoh and Jones 1978). Unfortunately, of course, such studies only demonstrate what could happen; they cannot demonstrate the actual fitnesses prevailing in the agricultural populations from which the experimental material is drawn. This is, however, a general problem in biology.

4. INBREEDING AND SELFING POPULATIONS

Moss (1969) claimed that reports of B's were restricted to outbreeding plant species, and considered this to be a real biological phenomenon. It would be most interesting to have the full data behind this comment published. Müntzing (1954) reduced the frequency of B's in populations of rye (*Secale cereale*) by forcible inbreeding. A similar experiment was carried out by Puertas *et al.* (1987), who showed that if B's were introduced into the partially inbreeding rye *Secale vavilovii* at moderate or high frequencies, they tended to be lost from the population quite quickly. Jones and Rees (1982) interpret Müntzing's result as an increased ability to tolerate B's in more heterozygous plants. This explanation suggests something intrinsic to inbred genomes which either removes excess transmission or makes the B's more deleterious. However, inbreeding should in general skew the distribution of B's towards larger numbers. Larger numbers are more likely to have deleterious effects on fitness, and it is this which will make it hard for a B to invade an inbreeding population.

As a general example, consider a B arising in a 99 per cent inbreeding population and in an otherwise identical outbreeding population, with slight disadvantage – say 10 per cent in a single copy – rising to effective lethality with three copies. Suppose also that meiotic drive is strong. In the outbreeder, almost all individuals have only one copy in the early stages of spread, for many generations. In the inbreeder, many of the offspring have multiple copies immediately; of their offspring, almost all have a lethal number. The average fitness of individuals with B's within this lineage is therefore very low and the chances that drift will keep the B in the population for a long time very small. The B is therefore likely to be eliminated very rapidly from the lineage. At the same time, the drive rate into the population at large is effectively very low because of the rarity with which the plant pollinates others. The situation is analogous to an infectious disease with rather high lethality and low infectiousness, which therefore does not cause epidemics. This argument applies even if the B chromosome, immediately upon its origin, was somewhat favourable to its carriers in low doses, because of the reduction in effective drive produced by selfing. Consider a population in which an individual with one B is normal, while the 2B karyotype is lethal. Denote the transmission rate through male gametes as t_m and through female gametes at t_f. Then one can write down the recurrence equations relating frequency in one generation to that in the next and show that the B will only increase in frequency when rare if the rate of selfing s in the population obeys the condition

$$s < (t_f + t_m - 1)/2t_f\, t_m$$

Even if t_f and t_m are both 1 (that is meiotic drive of 100 per cent through both male and female lines, leading to a doubling in frequency of the element in each generation in a random mating population), s must be less than 50 per cent. For a typical inbreeding plant with outbreeding rates of 1–2 per cent, the condition cannot be fulfilled by any combination of t_f and t_m. Advantageousness of the B in low doses will scarcely alter the argument because of the strength of the meiotic drive. (A less precise argument by Puertas *et al.* (1987) reaches a weaker conclusion; the argument is incomplete because they do not calculate changes in the *proportion* of individuals with B's.)

There is no experimental data relating to the invasion of similar inbreeding and outbreeding populations by B chromosomes, but there is some evidence that suggests that it is not necessary to invoke the A genome to explain the loss of B's from inbreeding populations. Cruz-Pardilla *et al.* (1989) showed that outbreeding rates were higher in B-containing individuals in a population of rye; reductions in fertility were consistent with the increase being caused by death of zygotes containing many B's as a result of selfing. However, a test of this sort in only one species can never be convincing as an explanation of the general phenomenon: more examples are needed, preferably over a longer time-scale than the three generations used by Puertas *et al.*, so that the dynamics of changes are clearer.

5. SLOWER SELECTIVE PROCESSES ACTING ON B's

If in a population not all individuals are equally fit at equilibrium, there is said to be a genetic load on the population (as in, for example, the sickle cell anemia polymorphism). As we have argued above, at equilibrium in a B chromosome system the selective removal of individuals with large numbers of B's just counteracts the effects of meiotic drive or selection for individuals with small numbers. Therefore, a B imposes a genetic load on a population. Two sorts of genetic variation are then selectively favoured, with opposite effects. First, mutations that ameliorate the disadvantages encountered by individuals with B's will be favoured provided the B frequency is sufficient to offset any deleterious effects they have in individuals without B's. Secondly, otherwise neutral mutations that reduce the frequency of transmission of the B will always be favoured when the B is at an equilibrium set by transmission and deleterious effects of large numbers of B's. Genetic constraints presumably prevent the first process from outrunning the second and producing an indefinite increase in the numbers of B's. These processes were discussed by Shaw (1984). If genetic variation is present in the B, variants with increased drive or decreased deleterious phenotypic effects will obviously increase.

Charlesworth and Langley (1986) point out that for transposable elements it is conceivable that 'self-regulation' could evolve, as individuals with larger numbers of transposable elements were selectively removed, generating an advantage to elements that saddled their hosts with fewer copies of themselves. They concluded that the idea would probably work only in rather rare cases because the load probably changes rather slowly with increasing number, and is in any case believed by most workers to be rather small. The load in some B systems can be much larger. Kimura and Kayano (1961) estimate *c.* 20 per cent for *Lilium callosum*; in *M. maculatus* it must be of the order of 5 per cent (Shaw 1984). Also, the fitness loss due to B's can increase very rapidly with increasing numbers of B's in an individual (Harvey and Hewitt 1979; Hutchinson 1975; Teoh *et al.* 1976; Teoh and Jones 1978; Kean *et al.* 1982; Romera *et al.* 1989; Shaw 1984). However, this mechanism for the evolution of self-regulation works only if a decrease in net drive decreases the number of individuals with many B's without decreasing the overall equilibrium frequency. Otherwise, a higher drive variant could invade whenever the equilibrium frequency dropped sufficiently, resulting in an evolutionary race with no set equilibrium point. However, B's that simultaneously increased the fitness of individuals with small numbers of B's and decreased drive would tend to replace others. The crucial difference is that a suppressor transposon can remain in the population – what is suppressed is transposition not inheritance – while a suppressor B suppresses itself.

Does selection for reduced drive mean B chromosomes will be short-lived? Not necessarily. Shaw (1984) compared the rates of suppression by genetic elements of large or small effect located on the A chromosomes with the rate of selection of variant B's with enhanced transmission and concluded that suppression was inherently 10 times slower than enhancement for the case of *M. maculatus*. The intrinsic reason for this is that suppression acts on the genetic load produced by the B (it is a second-order effect), while replacement of one B by another with enhanced drive simply depends on the difference in drive between the two (it is a first-order effect). Only a limited supply of genetic variation in drive rates of the B could alter this picture. Partial evidence for such co-evolutionary races has been found in *M. maculatus* (Shaw and Hewitt 1985; Shaw *et al.* 1985) and in *P. affinis* (Nur and Brett 1985, 1987). The data are particularly complete in *P. affinis*. At least two distinct loci that reduce transmission rates are present in natural populations, with more or less additive effects. The mechanism by which B transmission is reduced is different in the two loci (Nur and Brett 1988). Neither is present at high frequency in natural populations, probably because of damaging pleiotropic effects. Such effects were definitely demonstrated for one of the loci. Both these studies have concerned principally variation in the A genome; there is little evidence as to the available variation within the B chromosome. Nur and Brett (1987) found in *P. affinis* that their suppressor loci appeared to have similar effects on B's of distinct geographic origin. Shaw and Hewitt (1985) found weak evidence for slight differences in drive rate ascribable to differences between B chromosomes. B's are very morphologically variable (Jones and Rees 1982), and even within a single visual class there may be much small-scale morphological variation (Shaw 1983*b*). However, if drive were an intrinsic result of the geometry of the meiotic cell as may be the case in *M. maculatus* (Hewitt 1976), then there might be very little that a genetically inactive B chromosome could do to alter its behaviour within the cell.

The evidence that B chromosomes are genetically inactive is far from complete, however. It has been assumed that heterochromaticity implies inactivity, but active genes are known to exist in heterochromatic regions of *Drosophila* chromosomes (Hillaker *et al.* 1980). Ruiz-Rejón *et al.* (1980) and Oliver *et al.* (1982) have shown that a locus on the B chromosome of *Scilla autumnalis* controls the expression of an esterase locus on one of the A chromosomes, while having no effect on other electrophoretic loci examined. This suggests that B chromosomes are not completely passive objects, and a co-evolutionary race of the type referred to above could take place. A particularly dramatic phenotypic effect produced in a qualitative way by a B has been shown in the parasitic wasp *Nasonia vitripennis* (Werren *et al.* 1987; Nur *et al.* 1988). This B chromosome changes the sex of its carriers by destroying the other paternal chromosomes after fertilization; *N. vitripennis* has a haplodiploid sex determination system, so the resulting haploid zygotes are male. The drive experienced by this B chromosome is dependent on the proportion of eggs that the mother allowed to be fertilized; the greater this proportion, the more the over-representation of the B in the next generation. If the sex ratio is 50 per cent, with half the eggs being fertilized, the B experiences no drive (Skinner 1987). In such a case, the selective forces acting on suppressors are very strong if the element becomes at all frequent in the population, because of the complete loss of alleles carried in males with B's. This could be a very favourable system in which to examine co-evolution, because of the strength of the forces acting; however, the involvement of the sex ratio would complicate matters. For example, there is positive feedback if females adjust the proportion of sons they produce in relation to local male abundance; the more they try to skew the sex ratio to females by fertilizing a larger proportion of eggs, the more the B is over-represented in the next generation!

6. HOW LONG-LIVED ARE B CHROMOSOMES?

It is clear that most published karyotypes must be based on relatively few individuals drawn from relatively small proportions of a species total range. B's must therefore be relatively common. This may be due to frequent origin and a short lifetime (because of the accumulation of suppressor elements), except in rare cases where an equilibrium is possible at a B frequency that gives an average advantage to the possessors of B's. This could arise either *de novo*, or because a few B's were advantageous, and inheritance was modified to eliminate any natural tendency to increase – that is, secondary loss of drive. Alternatively, the birth of a B could be a rare event, but when successful the B would persist for a long time: this might be because selection for modifiers to suppress the B expression would be slow (depending on the equilibrium load × the drive reduction), while the response of the B, in terms of the time taken for a variant of higher drive to replace the original, would be fast. On this view, the hard part would be to be born: staying alive would be no problem.

Particularly interesting here is the suggestion by Gadi *et al.* (1982) that, based on G-band morphology, *Bandicota indica* and *Rattus rattus* share a homologous B. If the homology is real, it must predate the divergence of the two species. This could mean either that the B itself was very ancient, or that the species lineage had a predisposition to produce this form of B. For example, one particular chromosome might be particularly prone to polysomy for structural or genetic reasons (cf. human chromosome 2) and, consequently, the occasional modification into a B chromosome. Also in *Rattus*, Thomson *et al.* (1984) noted that the B in *R. fuscipes* in Australia was very similar morphologically to one described by Yosida (1977) in *R. rattus* in Japan. DNA sequence studies would be needed to try to test whether the B had arisen recently from a common source in each species, or was genuinely ancient. In either case, the outcome would be of interest.

Recent evidence shows that chromosomal variants may arise frequently (e.g. 1/50 in *Rumex acetosa*: Parker and Wilby 1989; 1/1000 in humans: Hook *et al.* 1984). Most are presumably very unfit. Very few of the human cases reported by Hook *et al.* (1984) survived birth, but Parker and Wilby were surveying plants capable of more or less normal development. It is also common experience that far more cytological abnormalities are to be found in embryos (e.g. grasshoppers) than among adults (e.g. Hewitt and East 1978). If most abnormalities are very unfit, their expected lifetime in populations must be very short, but the distribution would have quite long tails. Therefore, two categories of rare events could give rise to B's: the production of an element with more than the critical fitness in relation to drive, or the improbable persistence of an element with an equilibrium frequency of 0 for long enough for the equilibrium frequency to rise by modification of the B or the rest of the genome.

The concept of critical fitness may need explanation. As an example, in animals without accumulation in the germ-line between the zygote and formation of gametes, the maximum possible average transmission rate is 1 per sperm. Without directed non-disjunction at second meiotic division, the maximum possible average transmission rate is also 1 per ooctye. The critical fitness permitting a newly arising B chromosome with these characteristics to persist in a population is 0.5. Directed non-disjunction at second meiosis could produce a maximum conceivable drive of 3, and a critical fitness of 0.33. These are, of course, extreme lower limits; the critical fitness in real systems will usually be much greater. Somatic variation in numbers by itself is insufficient to reduce the critical fitness. However, if the average frequency in the gonads rises somatically, the critical fitness would be reduced. Such somatic accumulation is known, for example, in *Calliptamus barbarus* (Nur 1969*a*).

The likelihoods of the two scenarios suggested above for the origin of B's, and of the two hypotheses to explain the commonness of B's, are partly set by the frequency of variation among B's in transmission parameters. This is open to experimental study, which would however involve a great deal of time and effort: suitable model systems with many offspring and a short lifetime are needed.

It is notable that in plants the apparent origin of B-like elements has been reported several times (in haploid doubled *Nocotiana sylvestris*: Lespinasse *et al.* 1987; in inter-specific hybrids of *Coix:* Sabre and Deshpande 1987), whereas in animals there is weak evidence that some B's are very old, as noted above for the genera *Rattus* and *Bandicota*. This could reflect the greater complexity of early development in animals and the more modular construction of plants. This hypothesis would carry the corollary that B's might be more common among plants than animals, being more easily originated or less likely to produce very unfit phenotypes. Jones and Rees (1982) record over 1000 plant species with B's but less than 300 animals. However, whether this reflects nature or cytologists is of course unclear. For example, a more detailed examination of the literature on Orthoptera, widely studied because they are favourable cytogenetic material, revealed some 133 records of B's out of 1311 species for which karyotypes were known at the time of review (Hewitt 1979).

Only for *M. maculatus* has it been possible to put limits on the age of a B chromosome. Hewitt (1973*b*) suggested from its geographic distribution that the B in this species probably arose after the island of Britain became isolated from the continent of Europe, about 8000 years ago. On the other hand, the distribution of this B is not necessarily stable (Shaw 1983*a*). The observed rate of movement in the distribution puts a lower limit of about 1000 years on the age of the B (Shaw 1983*b*). A single locality is known on the island of Öland off Sweden where a morphologically similar B is found (Ramel 1980). Three hypotheses are possible concerning this situation:

1. The B arose separately from the British one.
2. An exceptional migration event (e.g. by aeroplane, or, even less probably for this grasshopper, by wind) has taken place.
3. The B was formerly widespread over the whole range of *M. maculatus* but has been eliminated by selection of modifiers everywhere except in these two island populations.

Using a combination of breeding and molecular genetics approaches all these hypotheses should be testable. Assuming neutral evolution of coding sequences on the B (which is possibly a derivative of the X: Hewitt 1973*b*) the age of the B in Britain should also be estimable.

Could B chromosomes evolve into standard members of the complement, via a two-stage process? First, while the B frequency remains high, selection to eliminate the meiotic load resulting from large numbers of B's could lead to regular pairing and segregation in cells with 2B's. This would lead to equilibrium distributions with no more than two B's present in any individual. Then mutations or changes that allowed the expression of haplosufficient loci (possibly diverged from their ancestral loci on the A's) might occasionally be favourable, leading to a higher fitness of B-containing individuals, and the presence in almost all individuals of two B's. Then over a long time period 'hidden' loci might gradually be re-expressed, with or without changes, as mutation exposed them and they turned out to be favourable. If the changes leading to inactivation of the loci on a B during its origination are essentially irreversible (e.g. complete deletion), then this process could not occur. Thus the process could be shown to be impossible; to demonstrate that it occurred if it were possible would be more difficult, probably depending on the chance discovery of a half-way stage in the process. We are grateful to Dr Uzi Nur for pointing out that in grasses there are many species with $2n = 14 + $ B's, but none with $2n = 16 + $ B's (Jones and Rees 1982). Similarly, there are many grasshoppers with $2n = 24 + $ B's (in females) but none with $2n = 26 +$ B's (Jones and Rees 1982). Thus, in these groups the process does not seem to have occurred. There are analogies here with several other processes, most obviously with the process of duplication and divergence, which is assumed to give rise, for example, to the mammalian myoglobin and haemoglobin families of genes (Bishop and Cook 1981). Similarly, in their original 'selfish DNA' paper, Orgel and Crick (1980) proposed that some 'junk' sequence DNA on conventional autosomes might eventually acquire new functions through the activation of new sequences or the alteration of control of already functioning sequences. The twist in the hypothesis for B chromosomes is that the change would eventually build a new chromosome pair. Nur (1962) suggested that the non-functional L chromosomes of *Sciara* and E chromosomes of cecidomyid midges, which are eliminated from somatic tissues, might have arisen as stabilized B chromosomes. The evidence he cited was that these chromosomes appeared to remain present in the genome in constant numbers via an opposition of accumulation and elimination mechanisms. This suggests that the elimination mechanism evolved to counter the accumulation, as discussed earlier. The E chromosomes are known to be essential for normal gametogenesis (Bantock 1970), so the parallel with our speculations above is extremely close. Going further afield, it is possible to draw parallels with the way in which in some fungi at first sight parasitic on plants may confer such an advantage on the plant by deterring herbivory as to turn the association into a symbiosis (e.g. Ahmad *et al.* 1986).

7. ARE B'S ANY DIFFERENT FROM OTHER SORTS OF SELFISH DNA?

We must briefly review some other types of 'selfish' DNA that may shed light on the processes involved during the presence of a B chromosome in a genome. Meiotic drive involves elements on the standard chromosome set that lead to over-representation of the chromosome carrying the element in the offspring. The best studied systems of this type are the segregation distorter (SD) of *Drosophila melanogaster* (this system has an enormous literature; two useful recent references are Lyttle 1989; Brittnacher and Ganetsky 1989), the sex-ratio distorter (SR) of several dipteran spp. (e.g. Wood and Ouda 1987; Maffi and Jayakar 1981) and the t-T system in *Mus musculus* (Silver 1985). In these systems, it is clear that sufficient genetic variability exists to mitigate the genetic load otherwise imposed on populations by the presence of these elements. That is, individuals that suffer less from the presence of the element are strongly selected, and the variation in ill-effects is heritable. For example, after the discovery of sex-ratio distorter systems, it was hoped that they could be used either to transport conditionally lethal genes (e.g. for insecticide sensitivity) into a population, or to reduce a population by dramatically biassing the sex ratio. Both prospects now seem unlikely, because of the great rapidity with which the gene pools of species challenged with these elements evolve so as to eliminate the distortion, or produce a balanced polymorphism with the elements producing meiotic drive in only a small proportion of their carriers (Curtis *et al.* 1976). In the SD system, two categories of co-adaptation or neutralization of the meiotic drive have been shown. The system is complicated in detail, but involves at least two very closely linked major elements, *Sd* and *Rsp*. Chromosomes carrying the *Rsp* allele are insensitive to the action of *Sd*, whereas sperm containing Rsp^+ degenerate before fertilization. So when a population uniformly $Sd^+ Rsp^+$ is challenged with an *Sd Rsp* chromosome, there is strong selection to increase the frequency of any Sd^+ allele chromosomes that happen to contain *Rsp*. There is evidence that this has occurred naturally, and it has been shown to occur rapidly in laboratory populations, at the rates predicted from the fitnesses and meiotic drive observed in separate experiments (Charlesworth and Hartl 1978). Lyttle (1979) linked the SD gene to the Y chromosome, increasing the genetic load in the population massively. In the population cages he studied, no major gene response was observed. None the less, within 12 generations the drive on the Y was substantially reduced by polygenes present on all chromosomes. Isogenic lines responded as fast and as variably as populations deliberately set up with much genetic variance present. Thus in this system, not only is response to the maverick element rapid, it can occur by many alternative routes, including both major genes and many different polygenic responses.

More recently, the study of mobile DNA sequences has produced an embarrassment of nuclear genetic elements that are not strictly inherited according to Mendel's rules. In these cases, work has mostly concentrated on understanding what is actually happening, where the elements are, and how they transpose (reviews in Berg and Howe 1988). It is clear that in at least some cases such elements do impose a genetic load on the population carrying them (Ajioka and Hartl 1988). The generality of this loss in fitness due to excessive numbers (whatever excessive may be in a particular instance) seems to be generally agreed by theoreticians and disputed by experimentalists. Basically, this is because the theoreticians are quite happy with loads of perhaps 1 per cent and correspondingly small fitness differences, whereas few molecular biologists consider anything less than lethality under laboratory conditions to constitute a selective difference (e.g. John and Miklos 1988). Fitness must play some part in regulating copy number in order to explain the sort of detailed correlations of DNA amount with ecological conditions found by, for example, Grime and Mowforth (1982). The distribution of, for example, the human *Alu* sequence suggests that many elements are the stabilized relics of what amounted to an epidemic, with rapid spread over a wide range during a short period. Subsequently, the elements have been stable and effectively neutral. Many millions of years later it is, of course, impossible to tell what brought the epidemic to an end. The relevance of the general picture to B chromosomes is that, once again, we have an impression of a genome of cooperating DNA sequences in which from time to time mavericks arise, to be eventually brought under control by the rest of the genome – at least in those species that survive to be studied.

8. PHENOTYPIC EFFECTS OF B'S

Most reported phenotypic effects of B's are quantitative: the few cases where gene expression has been claimed are either circumstantial or have subsequently been shown not to involve structural loci on the B. Ruiz-Rejón *et al.* (1980) thought they had found an isozyme polymorphism determined by the B in *Scilla autumnalis*, but analysis showed that the effect of the B was to express an otherwise silent structural locus on the A chromosomes (Oliver *et al.* 1982). This is a fascinating phenomenon, but still means that no isozymes have been found that are expressed from loci on a B. Staub (1987) found a leaf stripe phenotype controlled by B's in maize, but this was clearly a genetically quantitative effect, becoming more and more probable the more B's were present, and never appearing unless at least five B's were present in a cell. In *Haplopappus gracilis* (Jackson and Newmark 1960), achene colour is changed in the presence of one or more B's of a particular type: even here there is no evidence that these are not cases like that quoted for *S. autumnalis*. By contrast, B's very commonly have physiological effects on bulk properties of cells containing them, with quantitative effects on many properties. Any of these properties could be subject to selection for obvious or obscure reasons. It should, perhaps, be expected that B chromosomes would tend not to have dramatic phenotypic effects. Variants with large phenotypic effects will usually be deleterious (Fisher 1958), and we have already argued that right from their origin carriers of B's cannot be very much less fit than non-carriers. As most people in large, well-co-ordinated organizations know, the easiest way not to make any big mistakes is to do very little: thus it would be surprising if this were not the category into which the majority of B's fell. A glaring exception is the *psr* gene of *Nasonia vitripennis* (Nur *et al.* 1988), which, as mentioned above, changes the sex of all the offspring of males possessing it by the crude expedient of destroying all the other paternal chromosomes. Were it not for the haplodiploid sex determination system of this wasp, this trait would of course be lethal.

Most attention, at least in terms of numbers of species, has focused on effects on chiasma frequency and recombination, and many authors (Jones and Rees 1982; recent examples include Confalonieri 1988; Remis and Vilardi 1986) have suggested it as a 'function' of B chromosomes in populations, typically adducing as evidence the restriction of B's to marginal or central geographic areas of the species range, according to which occurred in their case. To play a role in the maintenance of B's in a population, an increase in chiasma frequency would at least have to cause the element to increase in frequency when rare. Individuals possessing a B would have to leave more surviving offspring than those without, by virtue of an increase in recombination. Recombination will only release variability if there is linkage disequilibrium in the population. Linkage disequilibrium decays rapidly in random mating populations unless there is selection to maintain it (e.g. Clegg *et al.* 1980; Hedrick *et al.* 1978). The effect of a B would be to slightly increase the rate of decay of disequilibrium; this could slightly accelerate the rate of adaptation of the population by reducing hitch-hiking, but the effect on an individual would be minuscule. The maximum effect would be achieved in the improbable case where linkage disequilibrium is shifted by selection to the maximum possible in opposite directions in alternate generations.

Most authors have probably had in mind group selection arguments based on better long-term adaptation to intermittently altering conditions. However, the waiting time for a variant to replace a less fit form is only logarithmically related to its initial frequency, and the waiting time until a favourable combination appears at least once is much less than linearly related to recombination rate, provided at least some recombination occurs. Therefore, a 10 per cent increase in the rate of release of variation when climate changes will certainly not lead to a 10 per cent increase in the survival probability of the population. In any case, the argument has never been restricted to neutral or nearly neutral B's. If a B was favoured only during climatic or other environmental changes, how would it survive in the population during the intervals between changes?

9. WORK POINTS

What areas of study might be fruitful? It is, of course, impossible to say. What follows is a list of ideas that we would follow up given infinite time and money. They may be useful as a spark to discussion.

Cataloguing the occurrence of B's has been helpful, but suffers from the difficulty of disentangling the effect of the energy, ability and interests of investigators from true natural differences between groups. Jones and Rees (1982) quote a figure of 1 per cent of all published karyotypes as having B's. But without some sampling strategy the meaning of this is unclear: ideally, one would like a frequency distribution showing the proportion of species having a given proportion of populations with a given frequency of B chromosomes – preferably subdivided by size of element! If one could age B's – which would be possible using DNA sequencing techniques for those occurring in several populations – one could look for associations of age with mitotic instability, phenotypic deleteriousness and regular inheritance. However, the effort involved would be stupendous, and the interesting results painfully slow in coming.

Developmental mitotic instability could give clues as to what is going on during development when, for example, a cell decides it is going to make roots from now on. In a similar vein, one would predict that fungi would not have B's, being apparently very intolerant of excess DNA, and frequently depending on rapid growth to survive. Exceptions might be the wood-rotting basidiomycetes, with their very specialized niche. Now that fungal karyotypes can be determined with pulse-gel electrophoresis, it should be possible to search for B's in fungi and test this hypothesis. Once again, however, the work involved is considerable, and such a project could only really be carried along by a larger project with broader concerns.

Emerging possibilities in biotechnology might include the possibility of purifying B's by pulse-gel electrophoresis and injecting the B's directly into oocytes or megaspores of populations or related species not naturally carrying B's in order to look at the extent of co-adaptation between the A and B genomes. This, of course, is a high-technology variation of the sort of experiment carried out by Puertas *et al.* (1987) and first suggested by Kimura (1962): but the technology might enable the experiment to be carried out in a much wider range of taxa, permitting generalizations to be made with more security.

Where do B's come from? In principle, they may be produced as duplications of normal autosomes or sex chromosomes and then diverge in function and sequence; or they may start small – essentially as an isolated kinetochore – and then grow, accumulating inactivated selfish elements at one or both ends. To study this, it might be possible to look at the total proportion of single copy, preferably coding, DNA in B's from many different organisms. If B's are generally 'born' from A chromosomes, evolving by changes in that DNA, the single copy DNA should retain homology with A sequences for long periods; B's built *de novo* would have no such homology except for transposable elements and other families of mobile DNA. Of course, mixtures of these processes may occur, complicating the interpretation. The work of Amos and Dover (1981) on satellites in B's in three species of tsetse flies provides a starting point for such work, but the use of satellites, which show evidence of rapid change and relatively frequent transposition between chromosomes, meant that they could draw no firm conclusions. None the less, it is notable that there was no evidence of particular homology with any chromosome of the standard complement. The elegant detective work used by Nur *et al.* (1988) to find the sex-ratio distorting B chromosome *psr* of *Nasonia vitripennis*, is a nice example of how modern techniques in molecular genetics may hugely improve our knowledge of the composition of B's. Here, sequences present only in *psr* genotypes turned out to be satellites. This suggested that *psr* might be heterochromatin or located in heterochromatin. Because *psr* was known to be inherited extrachromosomally, a search was made for a B chromosome, which was duly found cytologically. Because much of the DNA in the B was absent from the normal complement, Nur *et al.* (1988) suggest that the B might have originated in a cross-species hybridization. As mentioned above, interspecific hybridization has also been suggested as a source of B's in plants (Sabre and Deshpande 1987).

Recent advances in molecular genetic techniques mean that questions as to the organization and affinities of the DNA on B chromosomes may be addressed much more specifically and with a fair chance of success. Thus it should be possible to clone the DNA of an organism and challenge the B chromosome with probes from the normal complement by *in situ* slide hybridization. These methods are advancing rapidly; it is possible to locate relatively small homologus DNA sequences – even 500 base pairs using biotin labelling (Garson *et al.* 1987). Such techniques have recently shown the presence of rDNA in B chromosomes of *Crepis capillaris* (Maluszynska and Schweizer 1989). One can envisage other molecular approaches, for example using restriction enzymes to cut the A and B chromosome DNA, then separating the pieces on gels and using these to distinguish B chromosome material. One might use amplification techniques based on the polymerase chain reaction, coupled with labelling to make *in situ* probes from likely candidates. The possibilities are enormous, as will be the effort and time required!

Selfish DNA accumulation within a B could differ from the corresponding process in the A chromosomes in a number of ways. There should be fewer constraints from function than in the A chromosome set, so the rate at which selfish DNA can accumulate is potentially greater. However, pairing in many B chromosome systems is relatively rare, so processes expanding chromosomes or introducing new sequences by transfer from homologues would also be rarer than among the A chromosomes. In certain systems, this would make it possible to test particular ideas about the way categories of mobile DNA move or are stabilized. In a system in which pairing of B's was rare, for example, the rate of 'concerted evolution' should be slower on B's than on A's. However, B's should not be seen as an altogether unique phenomenon; there are other chromosomes that can be regarded as composed essentially of a kinetochore accompanied by 'junk', e.g. the giant chromosome of *Drosophila nasuto-*

ides (John and Miklos 1988). A study of the analogies between such chromosomes and B's might well be rewarding.

10. CONCLUSIONS

In a genome there are two ways to be successful: transmit yourself non-randomly, and make sure there are plenty of copies of cells containing you. Group selection will suppress the first but it is not obvious that it will do so efficiently enough to ensure the ubiquity of Mendelian segregation that we observe. Nunney (1989) makes a powerful point that group selection may efficiently ensure that observed taxa contain constraints to prevent selfishness appearing in the first place. Eshel (1985) suggests that a major advantage of the organization of genomes into chromosomes is that it means that for any given locus there are many others freely recombining with it which will be selectively favoured as suppressors of meiotic drive. This is a possible escape from the dilemma pointed out by Liberman (1975), who showed that in a two-locus system (one locus liable to meiotic drive, the other a suppressor), Mendelian segregation was always unstable if any degree of linkage whatsoever existed between the two loci. Another possible escape is to suppose that in a multilocus system things are different, and there may be powerful selection to suppress neighbours' selfishness, or respond to it by selfishness if it appears, rather as conditional cooperation can be a way out of repeated games of Prisoners Dilemma. B chromosomes represent a frequently extremely antisocial form of DNA: their study could shed much light on how a genome functions to enforce cooperation among its various parts.

REFERENCES

Ahmad, S., Johnson-Cicalese, J. M., Dickson, W. K., Funk, C. R. (1986). Endophyte enhanced resistance in perennial ryegrass to the bluegrass billbug *Sphenophorus parvulus*. *Entomol. Exp. Appl.* **41,** 3–10.

Ajioka, J. W. and Hartl, D. L. (1988). Population dynamics of transposable elements. In *Mobile DNA* (ed. D. E. Berg and M. M. Howe), pp. 939–58. American Society for Microbiology, Washington, D.C.

Amos, A. and Dover, G. A. (1981). The distribution of repetitive DNA's between regular and supernumerary chromosomes in species of *Glossina* (Tsetse): A two-stage process in the origin of supernumeraries. *Chromosoma* **81,** 673–90.

Bantock, C. R. (1970). Experiments on chromosome elimination in the gall midge, *Mavetiola destructor*. *J. Embryol. Exp. Morph.* **24,** 257–86.

Berg, D. E. and Howe, M. M. (ed.) (1988). *Mobile DNA*. American Society for Microbiology, Washington, D.C.

Bishop, J. A. and Cook, L. M. (1981). Genes, phenotype and environment. In

Genetic consequences of man-made change (ed. J. A. Bishop and L. M. Cook), pp. 1–36. Academic Press, London.

Brittnacher, J. G. and Ganetsky, B. (1989). On the components of segregation distortion in *Drosophila melanogaster.* IV. Construction and analysis of free duplications for the *Responder* locus. *Genetics* **121,** 739–50.

Bruck, D. (1957). Male segregation ratio as a factor in maintaining lethal alleles in wild populations of house mice. *Proc. Natl Acad. Sci. USA* **43,** 152–8.

Charlesworth, B. and Hartl, D. L. (1978). Population dynamics of the segregation distorter polymorphism of *Drosophila melanogaster. Genetics* **89,** 171–92.

—— and Langley, C. H. (1986). The evolution of self-regulated transposition of transposable elements. *Genetics* **112,** 359–83.

Clegg, M. T., Kidwell, J. F. and Horch, C. R. (1980). Dynamics of correlated genetic systems. V. Rates of decay of linkage disequilibria in experimental populations of *Drosophila melanogaster. Genetics* **94,** 217–34.

Confalonieri, V. A. (1988). Effects of centric-shift polymorphisms on chiasma conditions in *Trimeroptropis pallidipennis* (Oedipodinae, Acrididae). *Genetics* **76,** 171–80.

Cruz-Pardilla, M., Vences, F. J., Garcia, P. and Pérez de la Vega, M. (1989). The effect of B chromosomes on outcrossing rate in a population of rye. *Secale cereale* L. *Heredity* **62,** 319–26.

Curtis, C. F., Grover, K. K., Suguna, S. G., Uppal, D. K., Dietz, K., Agarval, H. V. and Kazmi, S. J. (1976). Comparative field cage tests of the population suppressing efficiency of three genetic control systems for *Aedes aegypti. Heredity* **36,** 11–29.

Eshel, I. (1985). Evolutionary genetic stability of mendelian segregation and the role of free recombination in the chromosomal system. *Amer. Nat.* **125,** 412–20.

Fine, P. E. M. (1975). Vectors and vertical transmission: An epidemiologic perspective. *Ann. N.Y. Acad. Sci.* **266,** 173–94.

Fisher, R. A. (1958). *The genetical theory of natural selection*, 2nd edn. Dover Publications, New York.

Fleuriet, A. (1988). Maintenance of a hereditary virus: The sigma virus in populations of its host, *Drosophila melanogaster. Evol. Biol.* **23,** 1–30.

Gadi, I. K., Sharma, T. and Raman, R. (1982). Supernumerary chromosomes in *Bandicota indica nemorivaga* and a female individual with XX/XO mosaicism. *Genetica* **58,** 103–108.

Garson, J. A., van den Berghe, J. A. and Kemshead, J. T. (1987). Novel non-isotopic *in situ* hybridisation technique detects small (lkb) unique sequences in routinely G-banded human chromosomes: Fine mapping of N-*myc* and β-*NGF* genes. *Nucl. Acids Res.* **15,** 4761–70.

Gregg, P. C., Webb, G. C. and Adena M. A. (1984). The dynamics of B chromosomes in populations of the Australian plague locust, *Chortoicetes terminifera* (Walker). *Can. J. Genet. Cytol.* **26,** 194–208.

Grime, J. P. and Mowforth, M. A. (1982). Variation in genome size – an ecological interpretation. *Nature* **299,** 151–3.

Harvey, A. W. and Hewitt, G. M. (1979). B-chromosomes slow development in a grasshopper. *Heredity* **42**, 397–401.

Hedrick, P., Jain, S. and Holden, L. (1978). Multilocus systems in evolution. *Evol. Biol.* **11,** 101–184.

Hewitt, G. M. (1973*a*). Variable transmission rates of a B-chromosome in *Myrmeleotettix maculatus* (Thunb.) (Acrididae: Orthoptera). *Chromosoma* **40,** 83–106.
—— (1973*b*). The integration of supernumerary chromosomes into the orthopteran genome. *Gold Spring Harbour Symp. Quantitative Biology* **38,** 183–94.
—— (1976). Meiotic drive for B-chromosomes in the primary ooctyes of *Myrmeleotettix maculatus* (Orthoptera: Acrididae). *Chromosoma* **56,** 381–91.
——(1979). Orthoptera: Grasshoppers and crickets. *Animal cytogenetics 3: Insecta 1.* Gebrüder Bornträger, Berlin.
—— and East, T. M. (1978). Effects of B chromosomes on development in grasshopper embryos. *Heredity* **41,** 348–56.
Hillaker, A. J., Appels, R. and Schalet, A. (1980). The genetic analysis of *D. melanogaster* heterochromatin. *Cell* **21,** 607–619.
Holmes, D. S. and Bougourd, S. M. (1989). B-chromosome selection in *Allium schoenoprasum.* I. Natural populations. *Heredity* **63,** 83–8.
Hook, E. B., Schreinemachers, D. M., Willey, A. M. and Cross, P. K. (1984). Inherited structural cytogenetic abnormalities detected incidentally in fetuses diagnosed prenatally: Frequency, parental-age associations, sex-ratio trends and comparisons with rates of mutants. *Amer. J. Human Genet.* **36,** 422–51.
Hutchinson, J. (1975). Selection of B chromosomes in *Secale cereale* and *Lolium perenne. Heredity* **34,** 39–52.
Jackson, W. D. and Newmark, P. (1960). Effects of supernumerary chromosomes on production of pigment in *Haplopappus gracilis. Science* **132,** 1316–17.
John, B. and Miklos, G. G. (1988). *The eukaryote genome in development and evolution.* Unwin Hyman, London.
Jones, R. N. (1985). Are B chromosomes selfish? In *The evolution of genome size* (ed. T. Cavalier-Smith), pp. 397–425. John Wiley, Chichester.
—— and Rees, H. (1982). *B Chromosomes,* Academic Press, London.
Kean, V. M., Fox, D. P. and Faulkner, R. (1982). The accumulation mechanism of the supernumerary (B-) chromosome in *Picea sitchensis* and the effect of this chromosome on male and female flowering. *Silvae Genetica* **31,** 126–31.
Kimura, M. (1962). A suggestion on the experimental approach to the origin of supernumerary chromosomes. *Amer. Nat.* **96,** 319–20.
—— and Kayano, H. (1961). The maintenance of supernumerary chromosomes in wild populations of *Lilium calosum* by preferential segregation. *Genetics* **46,** 1699–1712.
L'Heritier, P. (1970). *Drosophila* viruses and their role as evolutionary factors. *Evol. Biol.* **4,** 185–209.
Lespinasse, R., de Paepe, R. and Koulou, A. (1987). Induction of B chromosomes formation in androgenetic lines of *Nicotiana sylvestris. Caryologia* **40,** 327–38.
Liberman, U. (1975). Modifier theory of meiotic drive: Is mendelian segregation stable? *Theoret. Popul. Biol.* **10,** 127–33.
Lyttle, T. W. (1979). Experimental population genetics of meiotic drive systems. II. Accumulation of genetic modifiers of segregation distorter (SD) in laboratory populations. *Genmetics* **91,** 339–57.
—— (1989). The effect of novel chromosome position and variable dose on the genetic behaviour of the responder (*Rsp*) element of the segregation distorter (SD) system of *Drosophila melanogaster. Genetics* **121,** 751–63.
Maffi, G. and Jayakar, S. D. (1981). A two-locus model for polymorphism for

sex-linked meiotic drive modifiers with possible applications to *Aedes aegypti*. *Theoret. Popul. Biol.* **19,** 19–36.

Maluszynska, J. and Schweizer, D. (1989). Ribosomal RNA genes in B chromosomes of *Crepis capillaris* detected by non-radioactive *in situ* hybridisation. *Heredity* **62,** 59–65.

Matthews, R. B. and Jones, R. N. (1982). Dynamics of the B-chromosome polymorphism in rye. 1. Simulated populations. *Heredity* **48**, 345–69.

Moss, J. P. (1969). B-chromosomes and breeding systems. *Chromosomes Today* **2,** 268 (abstract).

Müntzing, A. (1954). Cytogenetics of accessory chromosomes (B chromosomes). *Caryologia* **6,** 282–301 (suppl.).

Nunney, L. (1989). The maintenance of sex by group selection. *Evolution* **43,** 245–57.

Nur, U. (1962). A supernumerary chromosome with an accumulation mechanism in the lecanoid genetic system. *Chromosoma* **13,** 241–71.

—— (1969*a*). Mitotic instability leading to an accumulation of B chromosomes in grasshoppers. *Chromosoma* **25,** 198–214.

—— (1969*b*). Harmful B-chromosomes in a mealy bug: Additional evidence. *Chromosoma* **28,** 279–97.

—— (1977). Maintenance of a 'parasitic' B chromosome in the grasshopper *Melanoplus femur-rubrum*. *Genetics* **87,** 499–512.

—— and Brett, B. L. H. (1985). Genotypes suppressing meiotic drive of a B chromosome in the mealybug, *Pseudococcus obscurus*. *Genetics* **110,** 73–92.

—— and Brett, B. L. H. (1987). Control of meiotic drive of B chromosomes in the mealybug, *Pseudococcus affinis (obscurus)*. *Genetics* **115**, 499–510.

—— and Brett, B. L. H. (1988). Genotypes affecting the condensation and transmission of heterochromatic B chromosomes in the mealybug *Pseudococcus affinis*. *Chromosoma* **96,** 205–212.

——, Werren, J. H., Eickbush, D. G., Burke, W. D. and Eickbush, T. H. (1988). A 'selfish' B chromosome that enhances its transmission by eliminating the paternal genome. *Science* **240,** 512–14.

Oliver, J. L., Posse, F., Martinez-Zapater, J. M., Enriquez, A. M. and Ruiz-Rejón, M. (1982). B chromosomes and E-1 isozyme activity in mosaic bulbs of *Scilla autumnalis* (Liliaceae). *Chromosoma* **85,** 399–404.

Orgel, L. E. and Crick, F. H. C. (1980). Selfish DNA: The ultimate parasite. *Nature* **284,** 604–607.

Östergren, G. (1945). Parasitic nature of extra chromosome fragments. *Bot. Notiser (Lund)* **2**, 157–63.

Parker, J. S. and Wilby, A. S. (1989). Extreme chromosomal heterogeneity in a small-island population of *Rumex acetosa*. *Heredity* **62,** 133–40.

Puertas, M. J., Ramirez, A. and Baeza, F. (1987). The transmission of B chromosomes in *Secale cereale* and *Secale vavilovii* populations. II. Dynamics of populations. *Heredity* **58,** 81–6.

Ramel, C. (1980). A B-chromosome system of *Myrmeleotettix maculatus* (Thunb.) (Orthoptera: Acrididae) in Sweden. *Hereditas* **92,** 309–312.

Remis, M. I. and Vilardi, J. C. (1986). Meiotic behaviour and dosage effect of B chromosomes on recombination in *Dichroplus elongatus* (Orthoptera: Acrididae). *Caryologia* **39,** 287–302.

Robinson, P. M. and Hewitt, G. M. (1976). Annual cycles in the incidence of B chromosomes in the grasshopper, *Myrmeleotetix maculatus* (Acrididae: Orthoptera). *Heredity* **36,** 399–412.

Romera, F., Vega, J. M., Diez, M. and Puertas, M. J. (1989). B chromosome polymorphism in Korean rye populations. *Heredity* **62,** 117–22.

Ruiz-Rejón, M., Posse, F. and Oliver, J. L. (1980). The B-chromosome system of *Scilla autumnalis* Liliacae): Effects at the isozyme level. *Chromosoma* **79,** 341–8.

Sabre, A. B. and Deshpande, D. S. (1987). Origin of B chromosomes in *Coix* L. through spontaneous interspecific hybridisation. *J. Heredity* **78,** 191–6.

Shaw, M. W. (1983*a*). Rapid movement of a B chromosome frequency cline in *Myrmeleotettix maculatus* (Orthoptera: Acrididae). *Heredity* **50,** 1–14.

—— (1983*b*). Movement of a cline for supernumerary chromosomes. In *Kew Chromosome Conference II* (ed. P. E. Brandham and M. D. Bennett), pp. 217–23. George, Allen and Unwin, London.

—— (1984). The population genetics of the B chromosome of *Myrmeleotettix maculatus* (Thunb.) (Orthoptera: Acrididae). *Biol. J. Linn. Soc.* **23,** 77–100.

—— and Hewitt, G. M. (1985). The genetic control of meiotic drive acting on the B-chromosome of *Myrmeleotettix maculatus* (Orthoptera: Acrididae). *Heredity* **54,** 259–68.

——, Hewitt, G. M. and Anderson, D. A. (1985). Polymorphism in the rates of meiotic drive acting on the B-chromosome of *Myrmeleotettix maculatus. Heredity* **55,** 61–8.

Silver, L. M. (1985). Mouse *t* haplotypes. *Ann. Rev. Genet.* **19,** 179–208.

Skinner, S. W. (1987). Paternal transmission of an extrachromosomal factor in a wasp: evolutionary implications. *Heredity* **59**, 47–53.

Staub, R. W. (1987). Leaf striping correlated with the presence of B chromosomes in maize. *J. Heredity* **78,** 71–4.

Teoh, S. B. and Jones, R. N. (1978). B chromosome selection and fitness in rye. *Heredity* **41,** 35–48.

——, Rees, H. and Hutchinson, J. (1976). B chromosome selection in *Lolium. Heredity* **37,** 207–213.

Thomson, R. L. (1984). B chromosome in *Rattus fuscipes*. 2. The transmission of B chromosomes to offspring and population studies: Support for the parasitic model. *Heredity* **52,** 363–72.

——, Westerman, M. and Murray, N. D. (1984). B-chromosomes in *Rattus fuscipes*. I. Mitotic and meiotic chromosomes and the effects of B-chromosomes on chiasma frequency. *Heredity* **52,** 355–62.

Werren, J. H., Nur, U. and Eickbush, D. (1987). An extrachromosomal factor causing loss of paternal chromosomes. *Nature* **327,** 75–6.

——, Nur, U. and Wu, Chung-I (1988). Selfish genetic elements. *Trends Ecol. Evol.* **3,** 297–302.

Wood, R. J. and Ouda, N. A. (1987). The genetic basis of resistance and sensitivity to the meiotic drive gene *D* in the mosquito *Aedes aegypti* L. *Genetica* **72,** 69–79.

Yosida, T. H. (1977). Supernumerary chromosomes in the black rat (*Rattus rattus*) and their distribution in three geographic variants. *Cytogenet-Cell Genet* **18,** 149–59.

Co-evolution between two symbionts: the case of cytoplasmic male-sterility in higher plants

D. COUVET, A. ATLAN,
E. BELHASSEN, C. GLIDDON,
P. H GOUYON and F. KJELLBERG

1. INTRODUCTION

Male-sterility, defined as the occurrence of individuals with no male function, i.e. females, among hermaphrodites is a frequent phenomenon in higher plants. Species where females (or male-steriles) and hermaphrodites (or male-fertiles) co-exist are called gynodioecious, and represent as many as 10 per cent of species in Angiosperms according to certain estimates (Delannay 1978).

Morphological syndromes of male-sterility are highly variable (Laser and Lersten 1972), as are the putative physiological mechanisms associated with male-sterility (e.g. Ahokas 1979; Musgrave *et al.* 1986 and references therein; Liu *et al.* 1988). However, in all cases where male-sterility is genetically determined, it seems at first sight barely credible that such a defect, the absence of male function, could be maintained by natural selection, and even be predominant in some populations (e.g. Baker 1963; Dommée *et al.* 1978).

Different explanations have been proposed for male-sterility. It has been suggested that 'gynodioecy, unlike unisexuality, is a highly adaptable outbreeding mechanism' (Mather 1940), because females are obligate outbreeders so that the frequency of inbreeding will vary with changing frequencies of females. Alternatively, the fact that male-steriles are better seed producers, in terms of number and/or quality, has been seen as an illustration of a compensation law between the production of pollen and seeds (Darwin 1877). Finally, male-sterility has been seen as an illustration of a conflict between cytoplasmic and nuclear genomes that possess different modes of inheritance (Cosmides and Tooby 1981). We will examine the relevance of these different views to explanations of the general occurrence of male-sterility in higher plants and, conversely, we will explore how the study of male-sterility can provide insights into breeding system evolution, resource allocation and nucleocytoplasmic interactions.

2. MECHANISMS OF MAINTENANCE OF MALE-STERILITY IN NATURAL POPULATIONS

2.1 Populations: Theory and observations

As well as rediscovering Mendel's laws, Correns (1908) also demonstrated that male-sterility did not follow these laws, but rather showed maternal inheritance. In subsequent studies, Mendelian genes were also shown to affect male-sterility and since then there has been a controversy about whether the inheritance of this trait is cytoplasmic or nuclear. The mode of inheritance is critically important because the conditions for maintenance of male-sterility in natural populations are very different depending on whether nuclear and/or cytoplasmic genes are involved.

In the case of nuclear inheritance, for females to exist, they must be more than twice as fecund as hermaphrodites (Lewis 1941). At equilibrium, the frequency of females,

$$p = (w-2)/(2^{*}w-2) \qquad (1)$$

where w is the fecundity of females (seed production) relative to hermaphrodites assuming equal survival of both sexes. Therefore, a positive correlation between the frequency of females and their relative fecundity is predicted (see Fig. 1, curve 2) with an expected maximum frequency of females of 50 per cent (at this point the fecundity of hermaphrodites is negligible, i.e. they behave like males and one obtains the classical results for expected frequencies of males and females). Survival differences between females and hermaphrodites will, however, influence their equilibrium proportions and the frequency of females can be greater than 50 per cent (van Damme and van Damme 1986). The relationship in eqn (1) is quite robust: it holds regardless of the dominance of the male-sterility alleles (Lewis 1941), or the number of nuclear loci responsible for the male-sterility (Ross and Shaw 1971). Obviously, the result also holds regardless of the mechanism responsible for this phenotypic difference, whether it be reallocation of resources not spent in pollen production or the fact that females are obligate out-crossers.

In the case of cytoplasmic inheritance, females will spread as long as they are more fecund than hermaphrodites (see Fig. 1, curve 1). Females will increase in frequency in the population and, if hermaphrodites are self-incompatible, the population will eventually become all female and go extinct. However, if hermaphrodites are self-compatible they can be maintained at a low frequency (Lewis 1941).

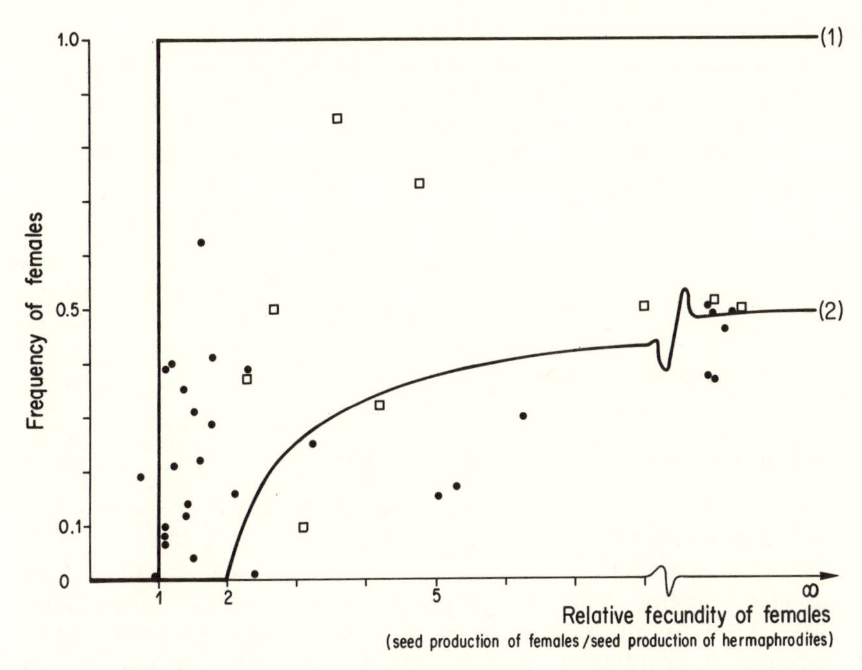

Fig. 1. Relationship between frequency of females and their relative fecundity: theoretical expectations and observations. Curve 1 is the expectation in the case of cytoplasmic inheritance of male-sterility; curve 2 is for nuclear or nucleocytoplasmic inheritance. The dots represent observations based on seed-set only; the squares represent observations that incorporate subsequent fitness components (see Appendix).

If both nuclear and cytoplasmic genes are responsible for male-sterility, i.e. the inheritance is nucleocytoplasmic (see Table 1 for an illustration of such a mechanism), the relationship between frequency of females and their relative fecundity is the same as in the case of nuclear inheritance (Fig. 1, curve 2, unpublished). This result can be intuitively understood by saying that at equilibrium, nuclear genes responsible for male-sterility will reach their equilibrium frequency in their own cytoplasm, so that the frequency of females among individuals carrying one particular cytoplasm is the same as in the case of nuclear inheritance.

We have compiled from the literature observations that have measured simultaneously the frequency of male-sterility and differences in fecundity between the two sexual forms (see Table 2 and Fig. 1). The observations do not nicely fit the theoretical expectations, represented by curves 1 and 2 in Fig. 1.

Table 1

An example of a nucleocytoplasmic genetic determinism of sexual phenotype depending on cytoplasm (S1 and S2) and genotype at restorer loci 1 and 2 (restorer alleles dominant)

locus 1 locus 2	R1R1 R2R2	R1R1 R2r2	R1R1 r2r2	R1r1 R2R2	R1r1 R2r2	R1r1 r2r2	r1r1 R2R2	r1r1 R2r2	r1r1 r2r2
S1	/	/	/	/	/	/	MS	MS	MS
S2	/	/	MS	/	/	MS	/	/	MS

Note:/, hermaphrodite; MS, female.

2.2 Attempts to reconcile observations and theory

The first explanation for such a discrepancy between theory and data was that measures of relative female fecundity were underestimates (Lloyd 1976). Such a measure should take into acount not only relative seed set but also other traits where female progeny might exhibit better performance such as germination rates, vigor at the different life-cycle stages, performance in different environments. For example, females in corn have a higher fecundity only in certain dry conditions (Vincent and Wooley 1972). However, the data for which differences subsequent to seed set (like germination rates or seedling vigor) were taken into account, are not closer to theoretical expectations (Fig. 1). Differences in survival between females and hermaphrodites could also be an explanation but few observations exist of such differences (but see van Damme and van Delden 1984). Moreover, if females lived longer than hermaphrodites, they should be more frequent when populations get older. Actually, the opposite is observed in several species: females are more frequent in more recently disturbed populations (Couvet *et al.* 1986; Belhassen *et al.* 1989). In any case, in view of the difficulty of measuring all possible differences between hermaphrodites and females (for example, longlasting effects of selfing in hermaphrodites), the possibility of a general underestimate of differences of fecundity and/or survival, by different investigators, on different continents and on different plant families cannot be ruled out. However, differences of performances between females and hermaphrodites would have to have been widely underestimated, at least by a factor of 2 in several cases, to account for the data in Fig. 1.

2.2.1 *The selfish gene approach*

The reconciliation of theory and observation is very dependent on the mode of inheritance of male-sterility: one needs to explain why females are so few in the case of cytoplasmic inheritance, while conversely why they are so numerous in the case of nuclear or nucleocytoplasmic inheritance. Most observations lie between curves 1 and 2 in Fig. 1, suggesting that selective pressures are antagonistic: cytoplasmic genes carried by females are transmitted more than those carried by hermaphrodites, whereas the reverse is true for nuclear genes. As a result, any cytoplasmic mutant determining male-sterility is favored, whereas the same is true for any nuclear mutant that restores male-fertility (Cosmides and Tooby 1981; Gouyon and Couvet 1985; Frank 1989). In fact, those cases of inheritance of male-sterility that were once thought to be nuclear have turned out to be nucleocytoplasmic after more extensive study (Couvet *et al.* 1986; and see Frank 1989 and below for the explanation), and this is confirmed by the fact that in the case of nuclear inheritance one does not predict the systematic excess of females given the phenotypic differences that have been measured between females and hermaphrodites (Fig. 1, but see the case of overdominance on a locus where heterozygozity determines male-sterility, e.g. Gregorius *et al.* 1982). In the case of nucleocytoplasmic inheritance of male-sterility, however, several explanations exist for such a systematic excess of females.

2.2.2 *Deleterious effects of restorer genes*

A common assumption of many of the models proposed to account for the maintenance of gynodioecy with a nucleocytoplasmic determination is that there are deleterious effects on fitness of the nuclear restorer alleles. A particular type of this fitness effect, the 'cost of restoration' (Delannay *et al.* 1981; Charlesworth 1981), involves the assumption that nuclear genes reduce fitness when they are present in individuals that do not carry the associated male-sterilizing cytoplasm. Frank (1989) pointed out that, in the absence of such negative pleiotropic effects on fitness of the restorer alleles, selection would drive the restorer alleles to fixation when the female advantage is less than 2, thus leading to a loss of the joint polymorphism.

There is a clear consensus arising both from models describing the maintenance of a stable joint polymorphism and also those giving dynamic nucleocytoplasmic polymorphisms (Gouyon and Couvet 1985; Frank 1989; Gouyon *et al.*, in press) that there should be some sort of 'cost of restoration' in plants with nucleocytoplasmic male-sterility. Unfortunately, the identification and measurement of such costs is difficult. Frank (1989) implemented the cost as a reduction in pollen fertility in hermaphrodites carrying restorer alleles but there is little, if any, empirical evidence to support such a mechanism. The situation is largely analogous to that in the area of the evolution of genetic resistance of hosts to pathogens or pests to pesticides with the cytoplasmic genes playing the rôle of pathogen/pesticide. It is usually *assumed* in these models that there is a cost of carrying resistance genes by the host in the absence of the pathogen and, conversely, that a cost exists to a pathogen carrying virulence genes in the absence of specific resistance genes in its host. However, in host–pathogen/pest–pesticide systems, the evidence for such effects being widespread is both limited and contradictory. The limited data on the fate of alleles conferring resistance following the removal of a particular pesticide sometimes indicates substantial selection against pest genotypes carrying those alleles (Greaves *et al.* 1977; Partridge 1979; McKenzie *et al.* 1982; Greaves 1986; Georgopoulos 1986), whereas in other cases there was no apparent selection against them (Whitehead *et al.* 1985; Georgopoulos 1986; Roush and Croft 1986).

Despite the absence of clear evidence favoring the assumption of either a 'cost of restoration' in nucleocytoplasmic male-sterility or a 'cost of resistance' in host–pathogen (pest–pesticide) studies, the idea of such costs existing remains intuitively attractive. Both restoration of male function and resistance to pathogens or pesticides are active processes (see later discussion), involving metabolic pathways whose cost of maintenance is unlikely to be zero. In addition, it is interesting to conceive of an 'arms race' between nuclear genes for restoration and their opposing cytoplasmic genes both because of the novel intracellular location of this conflict and because the different modes of inheritance of the two types of combatant present us with a possible asymmetry of costs.

2.2.3 *Disequilibrium within populations*

The previous section has dealt mainly with models that have focused on the conditions under which a joint nucleocytoplasmic polymorphism could be stably maintained. It is, therefore, appropriate to ask whether the existence of such stable joint polymorphisms is necessary to explain what is observed in nature. For example, could natural populations be in some type of dynamic equilibrium either measured at the level of the population (see Frank 1989; Gouyon *et al.*, in press) or measured at the level of a set of populations?

In natural populations, it is important to remember that many of the assumptions of theoretical models are not met. In particular, the assumptions of random mating are rarely appropriate for natural populations. In addition, when dealing with complex modes of inheritance such as are found determining gynodioecy, any given population will not necessarily contain all of the possible genetic variation relating to the trait being studied (e.g. a male-sterilizing cytoplasm may be found in the absence of its nuclear restorer alleles). Even if all genetic variation is present, it need not be distributed randomly over the whole area of the population. We shall use the term 'disequilibrium' to describe such situations.

Frank (1989) developed a model for a simple interbreeding population in which disequilibria were generated by mutation–migration rates of novel cytoplasmic male-sterility genes and nuclear restorer alleles. Under this mechanism, following the initial occurrence of a cytoplasmic male-sterilizing mutation, the proportion of females increases due to their higher relative fecundity. The female frequency continues to increase until either the females become pollen-limited or a mutation occurs for a nuclear allele restoring male fertility, after which the female frequency declines towards zero. Recurrent mutations of both cytoplasmic and nuclear genes thus produce episodic changes in female frequency and can be viewed as producing a dynamic polymorphism of nuclear and cytoplasmic genes if averaged over a suitable time period. The average periodicity of the episodes and the size of the fluctuations in female frequency are clearly strongly determined by the absolute magnitude of the combined mutation–migration parameters and their values relative to each other. Frank (1989) pointed out that population subdivision is an important part of the nucleocytoplasmic system, as the generation of new genetic variants by mutation and migration and their rate of loss by selection and/or drift determines the evolutionary dynamics (see also Gouyon and Couvet 1985; van Damme 1986).

The rôle played by population subdivision may be viewed in a slightly different manner by focusing on the fact that the spatial distribution of existing nucleocytoplasmic variation will not be expected to be random. Under conditions of restricted gene flow, local 'neighborhoods' may exist in which cytoplasmic male sterility genes can be found in the absence of nuclear restorer alleles, even though these restorer alleles may be present in other parts of the population. In such neighborhoods, the local frequency of females will continue to increase until the arrival, by migration, of the nuclear restorers. In essence, this argument is the same as that presented by Frank (1989), and it serves to emphasize the dramatic effect that restricted gene flow can have on slowing down the evolutionary dynamics of the nucleocytoplasmic sex-determining system. While the ultimate loss of females in subdivided populations is inevitable if nuclear restorer alleles are not lost from the total population, their rate of loss can be very slow. We have carried out many computer simulations of gynodioecy in subdivided populations, using values for dispersal and other life-history parameters derived from studies of *Thymus vulgaris*. If the populations are started in local disequilibrium, female frequency rises for approximately the first 10–20 generations (often reaching over 90 per cent), after which nuclear restorer alleles reach most local patches. However, the time for disappearance of the females is, on average, well over 500 generations after the foundation of the population. The modelling of such spatially structured populations serves to emphasize that their evolutionary dynamics are strongly dependent on the magnitude of local gene flow. The situation is complicated by an asymmetry of gene flow between cytoplasmic and nuclear genes. Nuclear genes in hermaphrodites may be dispersed both via pollen and seed while such genes in females are only dispersed by seed, as are cytoplasmic genes. On average, therefore, one should expect to find greater within-population structuring for cytoplasmic genes than for nuclear genes and greater structuring for those nuclear genes carried by females than those carried by hermaphrodites.

There is substantial evidence for restricted local gene flow in a number of seed plants including some gynodioecious species (e.g. van Damme 1986; Belhassen *et al.* 1987). The importance of such within-population structuring in determining nucleocytoplasmic polymorphisms in nature is still unclear and may well be species-dependent. However, the development of restricted migration hypotheses in the context of metapopulation models appears promising.

2.2.4 *Disequilibrium within a species*

As previously described, given nucleocytoplasmic inheritance of male-sterility, an excess of females can be due to the presence of a cytoplasmic male-sterility gene and absence of its corresponding nuclear restorer genes. This can be due to low migration rates of nuclear and cytoplasmic genes, as described above, but it can also depend on mutation rates of both type of genes (as formalized in Frank 1989): the more recent the occurrence of a new cytoplasmic male-sterility, the more females will be present. When cytoplasmic male-sterility mutants appear rarely (and/or restorer genes are present immediately after male-sterility has appeared) and if phenotypic differences between female and hermaphrodites are slight, male-sterile individuals will remain cryptic. In such a case, hermaphroditism would result from the accumulation of nuclear and cytoplasmic gene mutations when the female advantage is less than 2 and such an interpretation may apply to corn and tomatoes where more than 50 different nuclear restorer genes have been found (Kaul 1988, p. 15). In the following section, we will try to show that the dynamics of cytoplasmic genomes within an individual are such that deterministic processes can affect the rate of appearance of cytoplasmic male-sterility mutants, so that 'mutation' towards male-sterility is not solely a stochastic process. Where identified, the cytoplasmic genes responsible for male-sterility have turned out to be mitochondrial rather than chloroplast genes (Hanson and Conde 1985). This may be due to the fact that fully functional mitochondria are particularly necessary during male gametogenesis (Liu *et al.* 1988). In addition, the mitochondrial genome is genetically more variable than that of the chloroplast and therefore we will focus on the mutation potential of the mitochondrial genome.

Plant mitochondrial genomes are circular, consist largely of non-coding sequences (Sederoff 1987) and are numerous within a cell. The mitochondrial type that is inherited, which is the mitochondrial type of the germ cell lineage, is in fact a population of mitochondrial genomes and the heterogeneity of this population needs to be considered. On the one hand, due to uniparental inheritance of mitochondrial genes in higher plants and vegetative segregation during mitosis and meiosis, the fixation of one type of genome will depend in part on the number of mitochondrial genomes at each cell generation and may be rapid within a cell lineage (e.g. Birky *et al.* 1989). Therefore, low variability is expected within an individual that originates from one cell, the zygote. On the other hand, observations provide evidence for variability of mitochondrial genomes within an individual, especially in higher plants (Lonsdale *et al.* 1988; see Rand and Harrison 1989 for animals and Dujon and Belcour 1988 for yeast and fungi). A major type of variability in higher plants is in terms of genome structure, i.e. in sets and order of sequences. The plant mitochondrial genome can be present as a single molecule, the so-called master chromosome, or on different molecules that are smaller and represent subgenomic chromosomes that originate through recombination between repeated sequences that are present on the master chromosome (from Quétier and Vedel 1977; Lonsdale *et al.* 1988; Leaver *et al.* 1988). Recombination can lead to inversions, deletions (if, for example, the master circle splits into two circles, with one of them having no origin of replication) and also translocations and duplications of DNA fragments (see Small *et al.* 1989). With different sets of repeated sequences, the number of different mtDNA molecules that can appear can be quite large, and duplications yield new repeated sequences that can again give rise to another set of rearrangements through recombination. As a result of these recombinations, one finds variability of genome structure within a species (Palmer and Herbon 1988; Makaroff and Palmer 1988; Small *et al.* 1989). Moreover, occasional paternal leakage of mtDNA can lead to the appearance of new cytoplasmic types. Mixing together two different mitochondrial genomes, through mating in yeast (Oakley and Clark-Walker 1978) or protoplast fusion in plants (Belliard *et al.* 1979), has been shown to lead to new mtDNAs as a result of recombination.

Recombination between mtDNA molecules that change genome structure may not be just recombinational noise without phenotypic consequences but might generate phenotypic novelties like cytoplasmic male-sterility. Intramolecular recombination events have been shown to determine flagellar antigens in *Salmonella* (Zieg and Simon 1980), mating type in fungi (Klar *et al.* 1981) and, in higher plants, the expression of genes near recombination break-points (Makaroff and Palmer 1988). Besides changing gene order and hence their regulation, recombination can also create new genes. Cytoplasmic genes responsible for male-sterility have been shown to be the recombination product of several functional genes of the mitochondria (Dewey *et al.* 1986; Levings and Brown 1989).

As described above, what is transmitted over generations is a population of different mitochondrial molecules, and one needs to consider the nature of the differences between two cytoplasmic types: are they qualitative (some molecules being present in one cytoplasmic type and not in the other) or quantitative (differences of frequencies of these molecules). It has been shown that molecules prevalent in one cytoplasmic type, that were once thought to qualitatively characterize this type, were also present – albeit at very low frequencies – in an alternative cytoplasmic type (Small *et al.* 1989; Leaver *et al.* 1988), suggesting that differences between cytoplasms are more quantitative than qualitative. From this perspective, what would be called a cytoplasmic mutation at the level of the individual might only be a change of frequencies of the different mitochondrial molecules. Mutation in that case can be the result of 'population' processes that alter the frequencies of the different molecules.

Selective processes responsible for changes of cytoplasmic type can operate within and between cells. Not enough is known about the dynamics of mitochondrial genomes to describe what these selective processes are, if they act through replication rates of molecules (e.g. Blanc and Dujon 1980) or through other mechanisms such as segregation (e.g. Solignac *et al.* 1987), turnover and recombination. At the present time, the results of these within- and between-cell selective processes may well be 'a phenomenon in search of a molecular mechanism' (Backer and Birky 1985). In erythromycin resistance in yeast, two types of mitochondrial genome exist: that which confers resistance and that which does not. In the presence of erythromycin, the appearance of erythromycin-resistant yeasts is due to intracellular selection in favor of erythromycin-resistant mitochondria and not to selection between cells (see Backer and Birky 1985 and references therein for other cases, like in *Paramecium*).

The nuclear genome should influence which mitochondrial DNA molecules are present, especially as the enzymes that replicate the mitochondrial genomes are coded by nuclear genes (Attardi and Schatz 1988). Moreover, certain nuclear genes are known to induce specific deletions of the mitochondrial genome in certain tissues, such as in the case of muscular dystrophy (Zeviani *et al.* 1989). On the other hand, the artificial introduction of mitochondrial genomes of a species within cells of a foreign species can lead to preferential fixation during subsequent generations of the foreign mitochondrial genome (Niki *et al.* 1989). This suggests that the extant mitochondrial genome within a species may not be necessarily the one that possesses the best competitive abilities within its own cell lineage. It reminds one of the invasion of South America by North American mammals during the Pleistocene when the two continents were reunified (May 1978) and emphasizes that one should not think about nucleocytoplasmic interactions solely in terms of cooperation. With regard to male-sterility, one must distinguish reversion, where heritable changes of cytoplasmic types take place (so that all progeny of male-sterile individuals are hermaphrodites) implying an irreversible change in the cytoplasmic genome, from restoration of male-fertility, where nuclear genes modify the phenotype but not the cytoplasmic genotype. For example, reversion of *cms-T* (cytoplasmic male-sterility type T) in corn results from deletion of the gene *Urf-13*, whereas restoration is due to reduced expression of this same gene (Lonsdale 1987). The nuclear genotypes can also influence the rates of reversion (Small *et al.* 1988).

A comparable case is documented in the yeast, *Saccharomyces cerevisiae*, which is a facultative aerobe. The 'petite' mutation is the generic name of a wide variety of molecular rearrangements of mitochondrial genomes. MtDNA molecules in petite mutants are usually deletions of coding sequences but also an amplification of other sequences, notably replication origins (Dujon and Belcour 1988). Deletion–amplification events are supposed to happen through recombination across specific repeated sequences. The prevalence of wild-type mitochondrial DNA molecules of yeast in aerobic conditions is clearly the result of selection. If we consider selection among mtDNA molecules to represent 'individual' selection at the molecular level, then selections among mitochondria and among cells represent different hierarchical forms of 'group' selection. The possibility of selection among mitochondria within cells has been discussed by Lonsdale *et al.* (1988), who concludes that mitochondrial mutants can differ in their rates of replication or segregation. 'Suppressiveness' of these mutants measures the frequency of inheritance of the petite character versus the wild-type when a cross is made between these two phenotypes. Mutants in which replication origins of mtDNA are amplified are highly suppressive (Dujon and Belcour 1988).

As shown by these examples, several kinds of mitochondrial types are known that are not necessarily beneficial to nuclear genomes and may spread by selective processes. Several factors will determine the outcome of the selective processes through which a mitochondrial genome must go to produce a male-sterility 'mutation'. These factors can include recombination, replication and segregation rates whose effects will depend on the structuring of mitchondrial genomes between cells and possibly between mitochondria. The number of segregating DNA molecules at each cell generation will in turn determine the relative importance of selection acting between cells, between mitochondrial DNA molecules, and the rate of such selection will depend on the number of mitoses separating two reproductive events. The opportunity for frequent changes among mitochondrial genomes in a cell lineage may increase the probability of appearance of male-sterility and also of its reversion. As a result, one might propose that the frequent occurrence of male-sterility in particular taxa, for example the Labiatae (Delannay 1978), could be due to some special property of their mitochondrial genomes.

The fact that the presence of nuclear and mitochondrial genomes within the same cell is the result of a symbiosis of eukaryotic cells with bacteria is now well accepted and is evidenced by the phylogenetic proximity of mitochondria to other facultative symbionts, mutualists and parasites of the eukaryotic cell such as rhizobia, agrobacteria and rickettsia (see Weisburg *et al.* 1985; Yang *et al.* 1985). The partial breakdown of this endocellular symbiosis between the nuclear and the mitochondrial genome is exemplified by the petite mutation and male-sterility. Male-sterility results in a decrease of the reproductive ability of the other partner of the symbiosis, the nuclear genome, while there is no selective pressure whatsoever on cytoplasmic genomes to resume production of viable male gametes, through which they are not transmitted. The existence of endosymbionts that change the sex of their host, or of its progeny, according to their mode of inheritance, to maximize their own reproduction, has been shown in Crustacea where bacteria that exhibit maternal inheritance transform their male host into a female and where nuclear genes that confer resistance to such sex change are also known (Legrand *et al.* 1987); these nuclear genes can be seen as the equivalent of restorer genes in the case of male-sterility.

3. EVOLUTION OF PHENOTYPIC DIFFERENCES BETWEEN FEMALES AND HERMAPHRODITES

In populations exhibiting male-sterility, females usually have higher fecundity and/or survival than hermaphrodites (Lloyd 1976; van Damme 1984) and until now we have considered this phenotypic difference as a fixed parameter. We will now examine the two types of non-exclusive mechanisms that have been proposed to explain the phenotypic difference observed and to consider the evolution of these phenotypic differences between sexual forms. From now on, the term female advantage will be used to describe the superiority of performance of females in terms of fecundity and/or survival.

3.1 Resource reallocation in male-sterile individuals

In the 1980s, evolutionists progressively replaced their obsession with adaptation by an obsession with constraints, with the predominant questions being centered on the limits to adaptation rather than by adaptation itself. The genetic constraints involved in gynodioecy concern the mode of transmission of the genetic units determining the trait, and have been discussed above. With regard to physiological constraints, one of the first explanations given to the female advantage is the compensation law (Darwin 1877, p. 309). This can be interpreted now in terms of resource allocation. That is, some resource (e.g. carbon, nitrogen) is limiting organ production so that investment in male function decreases possible investment in female function or future survival. The way in which resource allocation influences evolutionary outcomes has been formalized using the fitness-set concept. This is the set of all possible phenotypes that can be realized, given the constraints acting on the general organization of the organism. Charnov *et al.* (1976) have shown that, with nuclear gender determination, a concave–convex fitness set favors gynodioecy (Fig. 2c.), whereas concave or convex fitness sets (see Figs 2a and 2b) lead to dioecy or hermaphroditism, respectively. The experimental study of such sets should be of great interest, as they define the limits of the evolution

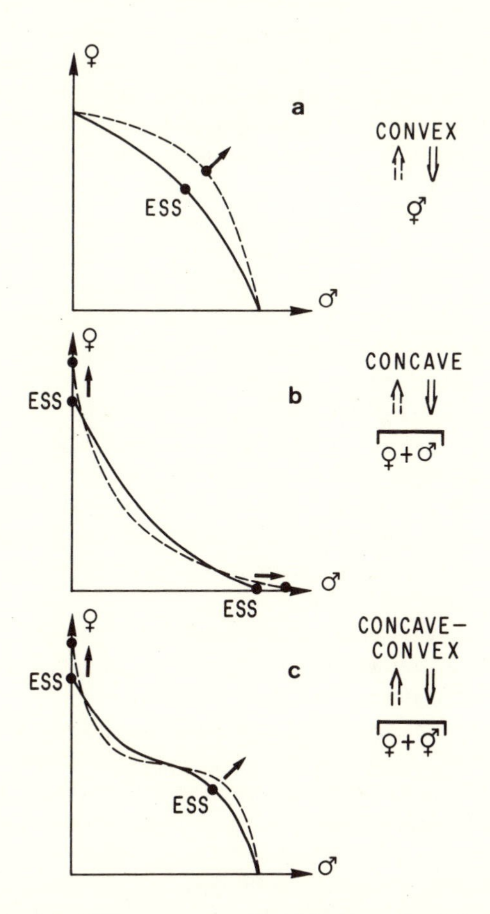

Fig. 2. The shape of the fitness-set as a determinant and as determined by the reproductive system, in the case of: (a) hermaphroditism; (b) dioecy; (c) gynodioecy (adapted from Charnov *et al.* 1976). The dotted lines show the expected direction of evolution, given a pre-existing fitness-set.

of reproductive systems. However, there are some severe practical and conceptual limitations to the value of fitness-sets.

There are three major practical problems with estimating fitness sets. First, the level of resource uptake by different individuals is often variable and this generates a positive relationship between male and female outputs, masking any compensation effect (van Noordwijk and de Jong 1986). This problem is resolvable by separating genetic from environmental components. Secondly, other functions may constitute sinks of variable strength for the same resources and also create a positive relationship between male and female functions. For example, if some individuals invest more in survival, their investment might be lower in both male and female sexual functions, compared to individuals that invest less in survival. Genetic variation in such 'alternative' non-sexual functions, if not recognized and factored out, may cause a spurious positive genetic correlation between the sexual functions. Thirdly, even if the preceding effects do not occur, it is unlikely that the extant genotypes will span the complete fitness set. For example, in a gynodioecious species where the fitness set has the shape given in Fig. 2c, the hermaphrodites should be found on the part of the compensation curve whose tangent intercepts the *y* axis at the all-female fitness point (Charnov *et al.* 1976). Such a result has actually been found in *Thymus vulgaris* (Atlan *et al.*, in prep). However, this can be interpreted either as indicating a linear compensation curve or as a concave–convex one, the concave part of which is 'invisible' because no individuals are represented there.

There are also conceptual limitations to the fitness-set concept because the fitness-set itself should evolve. The occurrence of hermaphrodites, by favoring any structure that improves both male and female functions should render the fitness-set more convex (Fig. 2a), whereas, in a dioecious species, gender differentiation can generate or increase the concavity of the fitness set (Fig. 2b) (Gouyon and Couvet 1985). From this perspective, the fitness-sets are self-reinforcing. The only way to escape from a concave or convex fitness-set is via a major modification of the basic rules defining fitness within a species. In some extreme cases, a change in ecological conditions can be sufficient (Kjellberg and Maurice 1988), but male-sterility mutants may also have such an effect.

Figure 2c shows a possible effect of cytoplasmic genes causing male sterility. By maintaining females that should not exist according to fitness-set rules, differentiation between females and hermaphrodites is favored and nucleocytoplasmic gynodioecy can therefore result in the concave–convex shape that was originally defined as the condition for the existence of nuclear gynodioecy. In this case, one could propose that the original male-sterility mutants did not possess a large advantage in fecundity relative to hermaphrodites, but that subsequent evolution may have led to a reallocation of resources such that now there is a large difference between females and hermaphrodites. The value of female advantage can then be viewed as the result of selection on sex allocation in hermaphrodites. Founder effects, deleteriousness of restorer genes and selection processes among mtDNA molecules may determine not only the occurrence and frequency of male-sterility but also the sex allocation of hermaphrodites and thereby the female advantage. For example, in *Thymus vulgaris*, where there are often populations with more than 50 per cent females (Dommée *et al.* 1978), the females produce usually 2–4 times more seeds than hermaphrodites (Assouad *et al.* 1978). Part of this large female advantage is due to specialization of hermaphroditic flowers toward male function (Couvet *et al.* 1985; see Arroyo and Raven 1975, for a more extreme case in *Fuchsia*). On the other hand, newly arisen male-sterile mutants only show a slight, if any, increase in seed production (e.g. Vincent and Wooley 1972 in corn; Vear 1984 in sunflower). Slight differences in fecundity between females and hermaphrodites may be present originally when male-sterility first appears in natural populations, but if the differences of fecundity then increase, they would further favor the maintenance of male-sterility.

Eventually, the shape of the fitness-set might be changed to become extremely concave, and the species may become dioecious. Gynodioecy is thought to have lead to dioecy in some families (e.g. Caryophyllaceae), whereas in other families gynodioecy seems to be a stable state, with an almost total absence of dioecy (e.g. Labiatae and Plantaginaceae) (Charlesworth 1981; and see Carlquist 1974, pp. 520–30). As discussed above, genetic constraints could explain this fact. However, morphodevelopmental constraints may equally be invoked. One can, for instance, think that the very rigid morphological determination of the number of ovules in the Labiatae might make it difficult to reduce female function and therefore produce an all-male flower. Other functional constraints like the necessity of high investment in pollen rewards as pollinator attractants might prevent unisexual forms from evolving (Charlesworth and Charlesworth 1987*b*).

3.2 Outbreeding in male-sterile individuals

The somewhat Panglossian view that prevailed in the 1970s led people to interpret the presence of every trait as a consequence of its beneficial effects. The most obvious effect of gynodioecy is the obligate outbreeding of females, and as a result many authors have tried to interpret the occurrence of females among hermaphrodites as a mechanism favoring allogamy. Following this idea, in 1963 H. G. Baker wrote that where male-sterility is present in self-incompatible species (e.g. *Plantago lanceolata* or *Hirschfeldia incana*), 'from the point of view of outcrossing, the gynodioecism appears to be superfluous'. If heterosis is included as a component of the female advantage, the conditions for the maintenance of male-sterility are the same as in the general case of nuclear inheritance (Valdeyron *et al.* 1973), as long as there is no overdominance associated with the locus that determines the sexual form itself (e.g. Gregorius *et al.* 1982). Several formalizations of the joint effect of resource allocation and heterosis have reached similar conclusions (e.g. Charlesworth 1981; Gouyon and Couvet 1987).

In a gynodioecious population with partly selfing hermaphrodites, females should on average be more heterozygous than hermaphrodites (Gouyon and Couvet 1987), because in gynodioecy (in contrast to dioecy) the sex of the mother is heritable, so that hermaphrodites are more likely to originate from selfing than females (Gouyon and Vernet 1982; Sun and Ganders 1986). The progeny of females will also be more heterozygous, so that due to heterosis both females and their progeny may perform better than hermaphrodites. Numerous studies have shown that indeed part of the higher fecundity of females is due to inbreeding depression of hermaphrodites (from Assouad *et al.* 1978 to Kohn 1988 and Jolls and Chenier 1989). The advantage of being outcrossed can help to maintain females in populations although phenotypic differences are usually not large enough to account for the frequency of females observed in natural populations, even when all the differences between females and hermaphrodites are taken into account (Sun and Ganders 1986; Jolls and Chenier 1989; and see data in Fig. 1 that concern self-compatible species; but see Kohn 1988).

In fact, avoidance of inbreeding will account for a substantial female advantage only when there is simultaneously a high selfing rate and high inbreeding depression (Charlesworth and Charlesworth 1987*a*). However, one must assume that hermaphrodites are unable to avoid inbreeding, although pollen from other individuals arrives regularly otherwise females would not be fertilized, as emphasized by Darwin (1877, p. 279).

Evidence that avoidance of inbreeding could be a proximate cause for the existence of females might be a positive correlation between the frequency of females and selfing rate, as observed in *Bidens* (Sun and Ganders 1986), but such a relationship may not be causal. The greater the frequency of females, the more isolated will be hermaphrodites and, as the selfing rate of hermaphrodites can depend on the relative amount of auto and allopollen, through late-acting self-incompatibility mechanisms (Seavey and Bawa 1986), one can predict that the more isolated are the hermaphrodites, the higher will be their selfing rate.Such a relationship has indeed been observed (Vaquero *et al.* 1989). More generally, correlations between proportions of females and population characteristics do not easily yield causal relationships. For example, in *Thymus vulgaris*, the proportion of females was first correlated with habitats where there was a high proportion of stones with diameters of 20–40 cm on the surface of the soil, which in Southern France are those areas that were once cultivated (Dommée *et al.* 1978), and in *Origanum vulgare* females were found to be more numerous in marginal populations than in the center of the distribution area of the species (Elena-Rossello *et al.* 1976). Krohne *et al.* (1980) in plantains and Gouyon *et al.* (1983) in thyme showed a relation between the proportion of females and the degree of disturbance of the environment and began to propose that the observed high proportions could correspond to non-equilibrium situations. Finally, Dommée *et al.* (1983) showed that, in successional systems, the proportion of females was higher in young populations than in old ones. This result has been confirmed in post-fire successions for the same species (Belhassen *et al.* 1989). It is particularly tempting to propose that the need for genetic variation generated by outcrossing is stronger in disturbed or recently colonized habitats than in old stable ones, given that in self-compatible species gynodioecy promotes outbreeding. However, in disturbed areas, populations are, on average, also younger than in stable ones, and in that case the association of founder effects and restricted gene flow (see above) may also result in high proportions of females. As such, gynodioecy almost certainly interacts with the whole biology of the species concerned, influencing in which habitats the species can live and the overall success of extant species (and perhaps the extinction of others).

Comparing families of Angiosperms yields some evidence of a negative association between dioecy and self-incompatibility (Charlesworth 1985), and a positive association is also suspected between gynodioecy and dioecy (Charlesworth and Charlesworth 1978). One is tempted to relate these results and to propose that avoidance of inbreeding is an 'emergent property' of gynodioecious species, in that gynodioecy decreases the probability of extinction of a species in phyla where self-incompatibility mechanisms do not exist. Thus, although outbreeding can no longer be accepted as a proximate cause of male-sterility in most well-studied cases, population and species correlations might agree with a vision of gynodioecy as an outbreeding mechanism (Gouyon and Couvet 1985). The maintenance of sex has also been proposed to be the result of species selection (Gouyon *et al.* 1988; Nunney, 1989), emphasizing that diversity of reproductive systems can also be due to selection at the population and/or species level.

4. CONCLUSIONS

Male-sterility is a useful tool for studying the interaction of selective processes at different hierarchical levels. We have seen that the presence of male-sterility might involve selection within cells between different mitochondrial DNA molecules in interaction with nuclear genomes, selection between different individuals due to genetic (outbreeding) and physiological effects (resource allocation), selection among populations (differences of deme reproductive output, due to variability of frequency of females), and perhaps even species selection processes. We propose that changes at the level of cytoplasmic genomes, coupled with the dynamics of nuclear and cytoplasmic gene pools, are the proximate cause of male-sterility, and that associated phenomena like sex reallocation of female and hermaphrodites and inbreeding depression may be evolutionary consequences that further explain population and species patterns.

ACKNOWLEDGEMENTS

This chapter is dedicated to Professor Georges Valdeyron who provided the inspiration for much of this work.

REFERENCES

Ahokas, H. (1979). Cytoplasmic male-sterility in barley. III. Maintenance of sterility and restoration of fertility in the *msml* cytoplasm. *Euphytica* **28,** 409–420.

Arroyo, M. T. K. and Raven, P. H. (1975). The evolution of subdioecy in morphologically gynodioecious species of *Fuchsia*. *Evolution* **29,** 500–511.

Assouad, M. W., Dommée, B., Lumaret, R. and Valdeyron, G. (1978). Reproductive capacities in the sexual forms of the gynodioecious species *Thymus vulgaris* L. *Bot. J. Linn. Soc.* **77,** 29–39.

Atlan, A., Gouyon, P. H. and Couvet, D. (in prep.). Sex allocation in a hermaphroditic plant: *Thymus vulgaris* and the case of gynodioecy.

Attardi, G. and Schatz, G. (1988). Biogenesis of mitochondria. *Ann. Rev. Cell. Biol.* **4,** 289–333.

Backer, J. S. and Birky, C.W. (1985). The origin of mutant cells: Mechanisms by which *Saccharomyces cerevisiae* produces cells homoplasmic for new mitochondrial mutations. *Current Genet.* **9,** 627–40.

Baker, H. G. (1963). Evolutionary mechanisms in pollination biology. *Science* **139,** 877–83.

Belhassen, E., Dockes, A.C., Gliddon, C. and Gouyon, P. H. (1987). Dissemination et voisinage chez une espece gynodioique: Le cas de *Thymus vulgaris* L. *Genetique, Selection et Evolution* **19,**(3), 307–320.

——, Trabaud, L., Couvet, D. and Gouyon, P. H. (1989). An example of nonequilibrium processes: gynodioecy of *Thymus vulgaris* in burned habitats. *Evolution* **43,** 662–7.

Belliard, G., Vedel, F. and Pelletier, G. (1979). Mitochondrial recombination in cytoplasmic hybrids of *Nicotiana tabacum* by protoplast fusion. *Nature* **281,** 401–403.

Birky, C. W., Fuerst, P. and Maruyama, T. (1989). Organelle gene diversity under migration, mutation, and drift: Equilibrium expectations, approach to equilibrium, effects of heteroplasmic cells, and comparison to nuclear genes. *Genetics* **121,** 613–27.

Blanc, H. and Dujon, B. (1980). Replicator regions of the yeast mitochondrial DNA responsible for suppressiveness. *PNAS USA* **77,** 3942–6.

Boutin, V. (1987). Selection sexuelle et dynamique de la sterilite male dans les populations de betteraves sauvages, *Beta maritima* L. Doctoral thesis, USTL, Lille.

Carlquist, S. (1974). *Island biology*. Columbia University Press, New York.

Charlesworth, D. (1981). A further study of the problem of the maintenance of females in gynodioecious species. *Heredity* **46,** 27–39.

—— (1985). Distribution of dioecy and self-incompatibility in angiosperms. In *Essays in honour of John Maynard Smith* (ed. J. J. Greenwood and M. Slatkin), pp. 237–68. Cambridge University Press, Cambridge.

—— and Charlesworth, B. (1978). A model for the evolution of dioecy and gynodioecy. *Amer. Nat.* **112,** 975–97.

—— and Charlesworth, B. (1987*a*). Inbreeding depression and its evolutionary consequences. *Ann. Rev. Ecol. Syst.* **18,** 237–68.

—— and Charlesworth, B. (1987*b*). The effect of investment in attractive structures on allocation to male and female functions in plants. *Evolution* **41,** 948–68.

Charnov, E. L., Maynard-Smith, J. and Bull, J. J. (1976) Why bean hermaphrodite? *Nature* **263,** 125–6.

Correns, C. (1908). Die rolle der mannlichen keimzellen bei der geschlechtsbeststimmung der gynodioecischen pflanzen. *Ber. Deutsch. Bot. Ges.* **26a,** 686–701.

Cosmides, L. M. and Tooby, J. (1981). Cytoplasmic inheritance and intragenomic conflict. *J. Theoret. Biol.* **89,** 83–129.

Couvet, D., Henry, J. P. and Gouyon, P. H. (1985). Sexual selection in hermaphroditic plants: The case of gynodioecy. *Amer. Nat.* **126,** 294–9.

——, Bonnemaison, F. and Gouyon, P. H. (1986). The maintenance of females among hermaphrodites: The importance of nuclear–cytoplasmic interactions. *Heredity* **57,** 325–30.

Darwin, C. R. (1877). *The different forms of flower on plants of the same species.* D. Appleton, New York.

Delannay, X. (1978). La gynodiocie chez les angiospermes. *Naturalistes Belges* **59,** 223–35.

——, Gouyon, P. H. and Valdeyron, G. (1981). Mathematical study of the evolution of gynodioecy with cytoplasmic inheritance under the effect of a nuclear restorer gene. *Genetics* **99,** 169–81.

Dewey, R. E., Levings, C. S. and Timothy, D. H. (1986). Novel recombinations in the maize mitochondrial genome produce a unique transcriptional unit in the Texas male-sterile cytoplasm. *Cell* **44,** 439–49.

Dommée, B., Assouad, M. W. and Valdeyron, G. (1978). Natural selection and gynodioecy in *Thymus vulgaris* L. *Bot. J. Linn. Soc.* **77,** 17–28.

——, Guillerm, J. L. and Valdeyron, G. (1983). Regime de reproduction et heterozygotie des populations de thym, *Thymus vulgaris* L. dans une succession post-culturale. *CR Acad. Sci. Paris, Serie III* **296,** 111–14.

Dujon, B. and Belcour, L. (1988). Mitochondrial DNA instabilities and rearrangements in yeast and fungi. In *Mobile DNA* (ed. D. E. Berg, and M. M. Howe), pp. 861–78. American Society for Microbiology, Washington, D.C.

Elena-Rosello, J. A., Kheyr-Pour, A. and Valdeyron, G. (1976). La structure genetique et le regime de la fecondation chez *Origanum vulgare* L. Repartition d'un marqueur enzymatique dans deux populations naturelles. *C.R. hebd. Seanc. Acad. Sci. Paris Ser. D* **283,** 1587–9.

Frank, S. A. (1989). The evolutionary dynamics of cytoplasmic male-sterility. *Amer. Nat.* **133,** 345–76.

Georgopoulos, S. G. (1986). Plant pathogens. In *Pesticide resistance – strategies and tactics for management*, pp. 100–110. National Research Council, National Academy Press, Washington, D.C.

Gouyon, P. H. and Couvet, D. (1985). Selfish cytoplasm and adaptation: Variations in the reproductive system of thyme. In *Structure and functioning of plant populations* (ed. J. Haeck and J. W. Woldendorp), Vol. 2. North-Holland, New York.

—— and Couvet, D. (1987). A conflict between two sexes, females and hermaphrodites. In *The evolution of sex and its consequences* (ed. S. C. Stearns), pp. 245—61. Birkhauser-Verlag, Berlin.

—— and Vernet, P. (1982). The consequences of gynodioecy in natural populations of *Thymus vulgaris* L. *Theoret. Appl. Genet.* **61,** 315–20.

——, Lumaret, R., Valdeyron, G. and Vernet, P. (1983). Reproductive strategy and disturbance by man. In *Ecosystems and disturbance* (ed. H. Mooney), pp. 214–25. Springer-Verlag, Berlin.

——, Gliddon, C. J. and Couvet, D. (1988). The evolution of reproductive systems: a hierarchy of causes. In *Plant population ecology* (ed. A. J. Davy, M. J. Hutchings and A. R. Watkinson), pp. 23–33. Blackwell Scientific, Oxford.

——, Vichot, F. and Van Damme, J. M. M. (in press). Nuclear-cytoplasmic male-sterility: Single-point equilibria versus limit cycles. *Amer. Nat.*

Greaves, J. H. (1986). Managing resistance to rodenticides. In *Pesticide resistance – strategies and tactics for management*, pp. 236–44. National Research Council, National Academy Press, Washington, D.C.

——, Redfern, R., Ayers, P. B. and Gill, J. E. (1977). Warfarin resistance: A balanced polymorphism in the Norway rat. *Genet. Res. Camb.* **89,** 295–301.

Gregorius, H. R., Ross, M. D. and Gillet, E. (1982). Selection in plant populations of effectively infinite size: III. The maintenance of females among hermaphrodites for a biallelic model. *Heredity* **48,** 329–43.

Hanson, M. R. and Conde, M. F. (1985). Functioning and variation of cytoplasmic genomes: Lessons from cytoplasmic–nuclear interactions affecting male fertility in plants. *Int. Rev. Cyt.* **94,** 213–67.

Horovitz, A. and Beiles. A. (1980). Gynodioecy as a possible population strategy for increasing reproductive output. *Theoret. Appl. Genet.* **57,** 11–15.

Jolls, C. L. and Chenier, T. C. (1989). Gynodioecy in *Silene vulgaris*: Progeny success, experimental design, and maternal effects. *Amer. J. Bot.* **76,** 1360–67.

Kaul, M. L. H. (1988). *Male-sterility in higher plants.* Monographs in theoretical and applied genetics Vol. 10. Springer-Verlag, Berlin.

Kjellberg, F. and Maurice, S. (1988). Seasonality in the reproductive phenology of *Ficus*: Its evolution and consequences. *Experientia* **45,** 653–60.

Klar, A. J. S., Strathern, J. N., Broach, J. R. and Hicks, J. B. (1981). Regulation of transcription in expressed and unexpressed mating type cassettes of yeast. *Nature* **289,** 239–44.

Kohn, J. R. (1988). Why be female? *Nature* **335,** 431–3.

Krohne, D. T., Baker, I. and Baker, H. G. (1980). The maintenance of the gynodioecious breeding system in *Plantago lanceolata* L. *Amer. Midl. Natur.* **103,** 269–79.

Laser, K. D. and Lersten, N. R. (1972). Anatomy and cytology of microsporogenesis in cytoplasmic male-sterile angiosperms. *Bot. Rev.* **38,** 425–54.

Leaver, C. J., Isaac, P. G., Small, I. D., Bailey-Serres, J., Liddell, A. D. and Hawkesford, M. J. (1988). Mitochondrial genome diversity and cytoplasmic male-sterility in higher plants. *Phil. Trans. Roy. Soc. Lond.* **B319,** 165–76.

Legrand, J. J., Legrand-Hamelin, E. and Juchault, P. (1987). Sex determination in Crustacea. *Biol. Rev.* **62,** 439–70.

Levings, C. S. and Brown, G. G. (1989). Molecular biology of plant mitochondria. *Cell* **56,** 171–9.

Lewis, D. (1941). Male-sterility in natural populations of hermaphrodite plants. *New Phytol.* **40,** 56–63.

Liu, X. C., Jones, K. and Dickinson, H. G. (1988). Cytoplasmic male sterility in *Petunia hybrida*: Factors affecting mitochondrial ATP export in normal and cytoplasmically male sterile plants. *Theoret. Appl. Genet.* **76,** 305–310.

Lloyd, D. G. (1976). The transmission of genes via pollen and ovules in gynodioecious angiosperms. *Theoret. Popul. Biol.* **9,** 299–316.

Lonsdale, D. M. (1987). Cytoplasmic male-sterility: A molecular perspective. *Plant Physiol. Biochem.* **25,** 265–71.

——, Brears, T., Hodge, T. P., Melville, S. A. and Rottmann, W. H. (1988). The plant mitochondrial genome: Homologous recombination as a mechanism for generating heterogeneity. *Phil. Trans. Roy. Soc. Lond.* **B319,** 149–63.

Makaroff, C. A. and Palmer, J. D. (1988). Mitochondrial DNA rearrangements and transcriptional alterations in the male-sterile cytoplasm of *Ogura* radish. *Mol. Cell. Biol.* **8,** 1474–80.

Mather, K. (1940). Outbreeding and separation of the sexes. *Nature* **145,** 484–6.

May, R. M. (1978). The evolution of ecological ecosystems. *Sci. Amer.* **239,** 118–35.

Mckenzie, J. A., Whitten, A. J. and Adena, M. A. (1982). The effect of genetic background on the fitness of diazinon resistance genotypes of the Australian blowfly, *Lucilia cuprina. Heredity* **19,** 1–19.

Musgrave, M. E., Antonovics, J. and Siedow, J. N. (1986). Is male-sterility in plants related to lack of cyanide-resistance respiration in tissues? *Plant Sci.* **44,** 7–11.

Niki, Y., Chigusa, S. I. and Matsuura, E. T. (1989). Complete replacement of mitochondrial DNA in *Drosophila. Nature* **341,** 551–2.

Nunney, L. (1989). The maintenance of sex by group selection. *Evolution* **43,** 245–57.

Oakley, K. M. and Clark-Walker, G. D. (1978). Abnormal mitochondrial genomes in yeast restored to respiratory competence. *Genetics* **90,** 517–30.

Palmer, J. D. and Herbon, L. A. (1988). Plant mitochondrial DNA evolves rapidly in structure, but slowly in sequence. *J. Mol. Evol.* **28,** 87–97.

Partridge, G. G. (1979). Relative fitness of genotypes in a population of *Rattus norvegicus* polymorphic for warfarin resistance. *Heredity* **43,** 239–46.

Philipp, M. (1980). Reproductive biology of *Stellaria longipes* Goldie as revealed by a cultivation experiment. *New Phytol.* **85,** 557–69.

Quétier, F. and Vedel, F. (1977). Heterologous populations of mitochondrial DNA molecules in higher plants. *Nature* **268,** 365–8.

Rand, D. M. and Harrison, R. G. (1989). Molecular population genetics of mtDNA size variation in crickets. *Genetics* **121,** 551–69.

Ross, M. D. and Shaw, R. F. (1971). Maintenance of male-sterility in plant populations. *Heredity* **26,** 1–8.

Roush, R. T. and Croft, B. A. (1986). Experimental population genetics and ecological studies of pesticide resistance in insects and mites. In *Pesticide resistance – strategies and tactics for management*, pp. 257–70. National Research Council, National Academy Press, Washington, D.C.

Seavey, S. R. and Bawa, K. S. (1986). Late-acting self-incompatibility in Angiosperms. *Bot. Rev.* **52,** 195–219.

Sederoff, R. R. (1987). Molecular mechanisms of mitochondrial-genome evolution in higher plants. *Amer. Nat.* **130,** S30–S45.

Small, I., Earle, E. D., Escote-Carlson, L. J., Gabay-Laughnan, S., Laughnan, J. R. and Leaver, C. J. (1988). A comparison of cytoplasmic revertants to fertility from different *cms-S* maize sources. *Theoret. Appl. Genet.* **76,** 609–618.

——, Suffolk, R. and Leaver, C. J. (1989). Evolution of plant mitochondrial genomes via substoichiometric intermediates. *Cell* **58,** 69–76.

Solignac, M., Genermont, J., Monnerot, M. and Mounolou, J. C. (1987). *Drosophila* mitochondrial genetics: Evolution of heteroplasmy through germ line cell divisions. *Genetics* **117,** 687–96.

Sun, M. and Ganders, F. R. (1986). Female frequency in gynodioecious species correlated with selfing rates in hermaphrodites. *Amer. J. Bot.* **73,** 1645–8.

Valdeyron, G., Dommée, B. and Valdeyron, A. (1973). Gynodioecy: Another computer simulation model. *Amer. Nat.* **107,** 454–9.

Van Damme, J. M. M. (1984). Gynodioecy in *Plantago lanceolata* L. III. Sexual reproduction and the maintenance of male steriles. *Heredity* **52,** 77–93.

—— (1986). Gynodioecy in *Plantago lanceolata* L. V. Frequencies and spatial distribution of nuclear and cytoplasmic genes. *Heredity* **56,** 355–64.

—— and Van Damme, R. (1986). On the maintenance of gynodioecy: Lewis' results extended. *J. Theoret. Biol.* **121,** 339–50.

—— and Van Delden, W. V. (1984). Gynodioecy in *Plantago lanceolata* L. IV. Fitness components of sex types in different life cycle states. *Evolution* **38,** 1326–36.

Van Noordwijk, A. J. and De Jong, G. (1986). Acquisition and allocation of resources: Their influence in variation in life history tactics. *Amer. Nat.* **128,** 137–42.

Vaquero, F., Vences, F.J., Garcia, P., Ramirez, L. and Perez de la Vega, M. (1989). Mating system in rye: Variability in relation to the population and plant density. *Heredity* **62,** 17–26.

Vear, S. A. (1984). The effect of male-sterility on seed yield and oil content in sunflower. *Agronomie* **4,** 901–904.

Vincent G. B. and Wooley, D. G. (1972). Effect of moisture stress at different stages of growth: Cytoplasmic male-sterile corn. *Agron. J.* **64,** 599–602.

Webb, C. J. (1981). Test of a model predicting equilibrium frequencies of females in populations of gynodioecious angiosperms. *Heredity* **46,** 397–405.

Weisburg, W. G., Woese, C. R., Dobson, M. E. and Weiss, E. (1985). A common origin of Rickettsiae and certain plant pathogens. *Science* **230,** 556–8.

Whitehead, J. R., Roush, R. T. and Norment, B. R. (1985). Resistance stability and co-adaptation in diazinon-resistant house flies (Diptera:Muscidae). *J. Econ. Entomol.* **78,** 25–9.

Yang, D., Oyaizu, Y., Olsen, G. J. and Woese, C. R. (1985). Mitochondrial origins. *PNAS* **82,** 4443–7.

Zeviani, M., Servidei, S., Gellera, C., Bertini, E., Dimauro, S. and Didonato, S. (1989). An autosomal dominant disorder with multiple deletions of mitochondrial DNA starting at the D-loop region. *Nature* **339,** 309–311.

Zieg, J. and Simon, M. (1980). Analysis of the nucleotide sequence of an invertible sequence controlling element. *PNAS* **77,** 4196–200.

APPENDIX

Data from the literature of the relationship between frequency of females and their relative fecundity[a]

Species	Frequency of females (%)	Relative fecundity of females	Reference
Rhus ovata	15	5.0	Lloyd (1976)
Rhus integrifolia	17	5.3	
Pimelea prostata	37	10.0	
Cortaderia richardii	38	2.3[b]	
Pimelea traversii	39	2.3	
Leucopogon melaleucoides	39	1.1	
Origanum vulgare	40	1.2	
Hebe subalpina	49	38.5	
Pimelea sericeo-villosa	49	21.3	
Pimelea oreophila	50	18.5	
Cirsium arvense	50	1916[b]	
Cortaderia selloana	51	21.3[b]	
Hirschfeldia incana	7	1.1	Horovitz and Beiles (1980)
	10	1.1	
Stellaria longipes	50	2.7[b]	Philipp (1980)
Gingidia montana	30	6.2	Webb (1981)
	35	1.4	
Gingidia decipiens	12	1.4	
	25	3.2	
Gingidia trifoliata	14	1.5	
Gingidia baxteri	16	2.1	
	31	1.6	
Gingidia enysii	4	2.4	
Gingidia flabellata	0.3	0.9	
	21	1.3	
Scandia rosaefolia	28	1.8	
	46	33.3	
Lignocarpa carnosula	41	1.8	
Lignocarpa diversifolia	36	25.0	

(*continued*)

APPENDIX *continued*

Species	Frequency of females (%)	Relative fecundity of females	Reference
Plantago lanceolata	4	1.5	van Damme
	8	1.1	(1984)
	22	1.6	
Thymus vulgaris	10	3.1[b]	Couvet et al.
	50	8.0[b]	(1986)
	73	4.8[b]	
	85	3.7[b]	
Beta vulgaris	19	0.7	Boutin
	62	1.6	(1987)
Cucurbiata foetidissima	32	4.2[b]	Kohn (1988)

[a]Adapted from *Gouyon and Couvet, (1987)*.
[b]Data that included measures of fitness components subsequent to seed-set (like germination rate or seedling vigor). These data are represented with a square in Fig. 1.

Studying the evolution of physiological performance

ALBERT F. BENNETT and
RAYMOND B. HUEY

1. INTRODUCTION

The study of physiology has largely developed in almost complete independence from the study of evolution. The practitioners, goals and philosophical bases of the fields have been different (Mayr 1982) such that little communication exists between them (see Futuyma 1986; Feder 1987). Nevertheless, the two fields have obvious importance and relevance to each other. The diversity and design of particular functional systems can be properly understood only from the selective, genetic and historical perspectives that evolution provides; and the evolutionary processes of selection and adaptation can be truly understood only when the mechanistic bases underlying functional systems are elucidated (Arnold 1983).

The field of physiological ecology (ecological physiology), a hybrid of comparative physiology and natural history, is a natural place for the interaction of physiology and evolutionary biology to occur. One of its major goals has been to document the role of the environment in shaping the diversity of physiological, morphological and behavioral features of organisms (Feder 1987). Its perspective is therefore fundamentally evolutionary to the extent that it considers how organisms came to be the way they are and how they might change in the future. In practice, however, physiological ecology has been most successful in discovering how various physiological systems work in different animals (e.g. nasal glands in birds and reptiles or the mechanisms of cellulose digestion in ruminants versus non-ruminants: Schmidt-Nielsen 1972) and in analyzing the complex biophysical exchanges of heat and mass between organisms and the environment (Porter 1989). It has been less successful in elucidating why and how particular systems or capacities have evolved. The literature of physiological ecology is, of course, replete with correlations between tolerance capacities of organisms and environmental gradients (e.g. of dehydration resistances of frogs with hydric gradients: Prosser 1986). Nevertheless, adaptation is often simply assumed in such correlative studies and not tested directly (Gould and Lewontin 1979). Because such evolutionary discussions are typically vague, physiology has – not surprisingly – made little impact on contemporary studies in evolution (Futuyma 1986).

Despite the historical indifference between the fields of evolution and physiology, we believe the potential for productive interchange is now very high. Physiological ecology is branching out into new areas (cf. Feder *et al.* 1987), many of which focus on historical, genetic and evolutionary issues. Several recent papers (e.g. Lauder 1981; Lande and Arnold 1983; Arnold 1983, 1987; Watt 1985; Bennett 1987; Huey 1987; Feder 1987; Koehn 1987; Powers 1987; Huey and Kingsolver 1989; Pough 1989) suggest methods for more rigorous evaluations of evolutionary hypotheses in the hope that evolution and adaptation cease to be bedeviling topics for physiological ecology. We believe that evolutionary biologists and functional biologists will increasingly be able to interact to their mutual profit and understanding, particularly on topics of genetics and natural selection. To this discussion, physiological ecologists can bring a thorough understanding of organismal function, an appreciation for the organism as an integrated unit, and the ability to analyze complex interactive effects of environmental factors on organismal capacities and performance. Most importantly, they can structure and execute experiments to test specific hypotheses about organismal function.

Here we discuss some approaches that we and our colleagues are currently undertaking to investigate aspects of functional capacities of systems and their evolution. To provide a coherent focus, we restrict our examples to studies of maximal locomotor performance (e.g. speed, endurance). Because of its great energetic cost and its intimate dependence on muscle and nervous system physiology, the physiological bases of locomotor capacity have received considerable study from mechanistically oriented physiologists during the past decade. Because of its importance to animal behavior in nature, the ecological and evolutionary consequences of locomotor performance have also been examined by ecologically and evolutionarily oriented physiologists. Indeed, locomotor performance may be the area of physiology in which evolutionarily relevant studies are best developed.

Good reasons exist for expecting that maximal locomotor performance may be of selective and hence evolutionary importance. Maximal athletic performance may well influence an animal's predation success (Elliott *et al.* 1977; Webb 1986), its success in social dominance and reproduction (Garland *et al.* 1990*b*) and its ability to escape from predators or other noxious factors (Christian and Tracy 1981; Huey and Hertz 1984*a*). Yet how generally valid are these associations? Is maximal locomotor performance routinely or periodically important to the biology of animals in nature such that the Olympic athlete is a suitable paradigm for animal locomotion

(Hertz *et al.* 1988)? Or, to paraphrase a statement from Ecclesiastes, is the race necessarily to the swift or the battle to the strong? Certainly, the image of a cheetah sprinting after a gazelle is firmly etched in one's mind. Yet, not all animals are cheetahs, and perhaps locomotor capacity is not a significant factor in the lives of most animals. The key issues concern the ecological and evolutionary importance to animals of maximal locomotor capacity. Is, for example, a faster individual more successful in capturing prey? Or is an individual with greater endurance able to spend more time foraging or is it more dominant socially? Is selection on performance constrained by mechanistic and genetic considerations?

To help answer these and related evolutionary questions, we and our colleagues have conducted a series of investigations of locomotion, principally using reptiles as study systems. These studies are designed to test explicit hypotheses involving the evolutionary significance of maximal performance. The general approach that we have used involves the integration of a two-part process. First, using test procedures that are ecologically realistic, we measure maximal performance of individuals in the laboratory. Secondly, relying on detailed information on the background ecology of our study organisms, we then analyze variation in performance (intraspecific, interspecific) from quantitative genetic, ontogenetic, demographic and phylogenetic perspectives. An understanding of the process and pattern of physiological evolution requires such a multi-level approach (Arnold 1983, 1987; Huey and Bennett 1986; Bennett 1987). Studies that focus narrowly on correlative patterns or that test relationships between performance and correlates of fitness only during a short part of an animal's life-cycle are incomplete at best and potentially misleading.

In this chapter, we begin with a short discussion of how we measure and interpret maximal performance. We then address a series of evolutionary issues. First, how variable, repeatable and heritable is maximal performance? In other words, can performance potentially respond to selection? Secondly, what information on the process of natural selection in 'real-time' can we glean from studies of maximal performance in the field? Is selection directional or stabilizing, and what might be the key selective agent(s)? Thirdly, what information on the patttern of selection over evolutionary time emerges from comparative studies? Are historical patterns and real-time processes congruent?

2. MEASURING LOCOMOTOR CAPACITY

In deriving methods to study the evolutionary significance of maximal locomotor performance, we have established and followed two important guidelines:

1. Although locomotion and its components could potentially be studied at a variety of physiological levels (enzyme kinetics to organismal performance *per se*), we concentrate on measures of organismal performance (e.g. speed, endurance) because such measures provide the most direct link between integrated physiological capacities and Darwinian fitness (Bartholomew 1958; Huey and Stevenson 1979; Arnold 1983; Bennett 1989). Lower-level systems, though they sometimes correlate with organismal-level performance (Bennett *et al.* 1984; Garland 1984; Taigen *et al.* 1985; Watt 1985; Koehn 1987; Powers 1987), are generally better suited for analyses of the mechanistic bases of variation in organismal performance, not its ecological consequences.
2. The type of performance being measured as well as the actual test conditions in the laboratory should be ecologically relevant to the natural history of a particular organism (Huey and Stevenson 1979; Bennett 1980; Arnold 1983; Huey *et al.* 1984; Pough 1989). Thus, knowledge of the biology of a given species might dictate the measurement of maximal speed for one species, but of relative agility on a narrow perch for another (Losos and Sinervo 1989). Locomotor tests should be done on an appropriate substrate and slope (Huey and Hertz 1984*a*; Losos and Sinervo 1989), at ecologically relevant body temperatures (Bennett 1980; Huey and Dunham 1987), during times of normal diel activity (Huey *et al.* 1989), and (when appropriate) in response to multiple environmental or physiological factors (Bennett 1980; Shine 1980; Garland and Arnold 1983; Huey and Hertz 1984*a,b*; Crowley 1985*b*; Wilson and Havel 1989; Pough 1989; Huey *et al.* 1990). Moreover, test conditions should allow animals to express normal behavior (e.g. defensive or aggressive responses) during the experiment (Hertz *et al.* 1982; Crowley and Pietruszka 1983; Arnold and Bennett 1984; Garland 1988; Pough 1989). Although it would be naive to think that any laboratory measures of performance will ever perfectly correspond to natural performances (Pough 1989), careful attention to the natural history of an organism should increase the probability that laboratory measurements are meaningful.

For our own studies we measure as many as three aspects of locomotor performance: maximal burst speed in a sprint, maximal capacity for exertion (time or distance to exhaustion while running at high speed) and endurance capacity (time to exhaustion while running at low speed). These variables should depend on relatively independent physiological traits (Table 1), and have generally been found to be independent of each other or have only a very weak positive association (Bennett 1980; Garland 1988; van Berkum *et al.* 1989; Jayne and Bennett 1990*a*; Huey *et al.* 1990; Shaffer *et al.*, in press). Taken together these variables define a

Table 1
Factors anticipated to limit locomotor performance at different levels of exertion (from Bennett 1989)

Speed	*Limiting factor*
Slow sustainable speed	Motivation, ultimately fuel
Fast sustainable speed (endurance)	Maximal oxygen consumption
Fast non sustainable speed (maximal exertion)	Anaerobic metabolism
Maximal burst speed	Structure/function of the musculoskeletal complex

'locomotor performance space' that bounds most of the potential capacities for behavior of which an animal is capable (Bennett 1989). [Note, however, that frogs achieve considerbly higher rates of oxygen consumption while calling than while engaged in vigorous locomotion (Taigen and Wells 1985; Taigen *et al.* 1985).] As mentioned above, however, the natural history of a particular organism may sometimes dictate that other measures of locomotion (e.g. agility, acceleration) might be relevant as well (Losos, 1990*a,b*).

In interpreting measures of locomotor performance, one must recognize that all behavior (not just locomotor behavior) depends on physical activity and that maximal levels of all behaviors are limited by the same physiology and morphology that limits locomotor activity (Table 1) (Bennett 1989). Therefore, our measurements of locomotor capacity define not only the limits of locomotor capacity *per se*, but also of behavioral 'work' of any type within a given metabolic mode. For example, maximal exertion may provide information on the ability to repel an intruder or perhaps to dig a burrow, not just on the ability to outrun a predator.

To measure maximal speed, we use electronic racetracks that have a series of photocell stations at set intervals along the track (Huey *et al.* 1981; Miles and Smith 1987). When an animal is chased down the track, it breaks the light beams in succession; a computer records the elapsed time until each beam is broken and then calculates the maximal speed over all intervals during a run. Typically, we race each individual several times and analyze the single fastest speed for each individual. Speed is positively correlated with initial acceleration (Huey and Hertz 1984*a*), so the data on individual variation in maximal speed also provide information on individual variation in acceleration.

To measure capacity for maximal exertion, we pursue an animal at high speed around a circular racetrack until it becomes exhausted (judged by loss of righting response) or assumes a static aggressive or defensive posture (Arnold and Bennett 1984; Garland 1984, 1988). The total distance travelled is a good index of effort, but speed or total time (or distance covered in the first minute or two: Bennett 1980) are also sometimes measured.

To measure endurance capacity, we place an animal on the slowly moving belt of a motor-driven treadmill. The elapsed time until the animal becomes exhausted, as indicated by its falling off the tread and losing its righting response, serves as an index of endurance. Detailed field data on movement rates in nature can suggest an ecologically relevant treadmill speed (Huey and Pianka 1981; Huey *et al.* 1984).

Each of these factors is relatively easy to determine experimentally on large numbers of individuals. Moreover, we can measure performance in each mode on two different days to test for short-term repeatability of the variable. Because performance measurements do not harm the animals, we can conduct release–recapture studies to measure long-term repeatability of individual performance.

A legitimate concern about such measures of locomotor capacity is the extent to which they are influenced by motivational factors. Perhaps speed or endurance measured under laboratory conditions depends more on the willingness of an animal to perform in an artificial situation than on physiological or morphological limits to locomotion, and thus does not represent a true maximal capacity. This possibility is always present when organismal traits are considered, and especially behavioral ones. Accordingly, we and our colleagues have made numerous attempts to determine whether our measurements of locomotor capacity in fact correlate with underlying physiological and morphological systems: if such correlations cannot be found, then performance measurements might reflect motivational differences rather than physiological limitations.

Individual endurance and maximal exertion are in fact generally linked with aerobic metabolic capacity and with anaerobic capacity, respectively (e.g. Garland 1984; Garland and Else 1987). In an important study, Garland (1984) showed that 89 per cent of the size-corrected inter-individual variability in endurance of the lizard *Ctenosaura similis* is explained by only four variables (maximal oxygen consumption, heart- and thigh-muscle mass, and hepatic aerobic enzyme activity) and that 58 per cent of the variability in maximal exertion is correlated with only two variables (maximal carbon dioxide production and anaerobic enzyme activity of skeletal muscle). [Garland's analysis has been criticized (Pough 1989) for analyzing overlapping physiological 'levels', but that criticism does not negate Garland's demonstration of links between physiology/morphology and performance.] Similarly, endurance and maximal exertion of different species of African lacertid lizards correlate with aerobic and anaerobic capacities, respectively (but not with muscle enzyme activities) (Bennett *et al.* 1984).

For burst speed, studies to date are less comprehensive, but physiological and morphological correlates of variation in speed are now documented for both interspecific (Losos 1990*a,b*; see Section 6) and some intraspecific comparisons (Miles 1987; Snell *et al.* 1988; Tsuji *et al.* 1989; Huey *et al.* 1990; but see Garland 1984, 1985). For example, individual burst speed is strongly correlated with muscle-fiber diameter in desert iguanas (*Dipsosaurus*: Gleeson and Harrison 1988) and with relative limb length in lizards of the genus *Urosaurus* (Miles 1987). However, hindlimb length or thigh-muscle mass has little or no effect on performance of individuals of several taxa (Garland 1984, 1985; Garland and Else 1987; Tsuji *et al.* 1989; Huey *et al.* 1990). Speed in garter snakes (*Thamnophis* spp.) is correlated with the ratio of body to tail vertebrae: animals departing from the mean ratio have lower speeds (Arnold and Bennett 1988; Jayne and Bennett 1989). However, experimental manipulation of tail length over the observed range of variation does not significantly alter speed, and therefore this association is not a simple biomechanical linkage (Jayne and Bennett 1989).

The role of body size on locomotor performance of neonate *Sceloporus occidentalis* has recently been studied using a novel manipulative experiment (Sinervo and Huey 1990). Neonates from a southern population are large at birth and have a high speed and stamina relative to the small neonates from northern populations. To determine whether the differences in performance are merely an allometric consequence of interpopulational differences in body size, Sinervo and Huey (1990) removed some yolk from eggs from the southern population, thereby producing miniaturized neonates equivalent in size to northern hatchlings. Miniaturized southern neonates no longer were relatively fast, but they still maintained high stamina relative to the northern neonates. Thus interpopulational differences in speed are an allometric consequence of differences in egg size, but differences in stamina must also reflect differences in physiology. Size manipulation (Sinervo 1988, 1990) adds an important experimental dimension to studies of the allometry of physiological performance.

Overall, our estimates of locomotor performance appear to have at least a partial mechanisitic basis and therefore represent legitimate meaures of maximal locomotor capacity. Presumably, measurements of locomotor capacity are also influenced by individual differences in behavior as well (e.g. motivation to sprint). Interestingly, such tendencies may run in families (van Berkum and Tsuiji 1987).

3. STUDY ORGANISMS

Reptiles have numerous attributes that make them suitable and popular systems for physiological and ecological studies (e.g. Milstead 1967; Huey *et al.* 1983; Seigel *et al.* 1987). They may occur in populations of very high density and have relatively low vagility, permitting repeated longitudinal sampling of individuals under field conditions. They are completely independent from parental care at birth, and so ecology and survivorship are uncomplicated by this factor. They can be easily captured and manipulated, and individuals can be marked and returned to field conditions or often kept successfully in captivity for long periods (e.g. Arnold, 1981). Important from the viewpoint of locomotion, they also seem relatively inflexible in adjustment of their locomotor capacities: they neither train physically nor lose capacity under laboratory conditions (Gleeson 1979; Garland *et al.* 1987; but see John-Alder and Joos, in press).

In collaboration with many colleagues, we have concentrated our studies on three taxa: garter snakes (*Thamnophis sirtalis*), fence lizards (*Sceloporus occidentalis*) and canyon lizards (*Sceloporus merriami*). We intentionally selected these species because the availability of background behavioral, ecological and demographic information makes these among the best known of reptiles. This wealth of background information facilitates our ability to pose and to interpret relevant functional hypotheses.

Garter snakes were studied by Jayne and Bennett between 1985 and 1988 in the vicinity of Eagle Lake in northern California. The ecology of this isolated montane population was previously studied extensively by S. J. Arnold and coworkers (e.g. Kephart 1982; Kephart and Arnold 1982). Moreover, the activity physiology of nearby populations and related species is also well known (Arnold and Bennett 1984; 1988; Garland 1988; Garland *et al.* 1990*a*; Garland and Bennett, in press).

Fence lizards were studied by R. B. Huey, T. Garland Jr, B. Sinervo, J. S. Tsuji and F. H. van Berkum in south-central Washington, near the northern limit of the species' range. Pilot studies were conducted by Tsuji and van Berkum in 1983. A full-scale study was started in 1985 but was terminated after one year when the study area was logged. Fence lizards are well known ecologically, behaviorally and physiologically.

Two populations of canyon lizards in Big Bend National Park, Texas have been studied since 1984 by R. B. Huey, A. E. Dunham, K. L. Overall and R. A. Newman. This study is continuing. These populations have been the focus of long-term ecological and demographic studies (Dunham 1981; Dunham *et al.* 1989). In terms of behavior, ecology and demography, these are the best known populations of lizards.

4. ATTRIBUTES OF LOCOMOTOR CAPACITY

For selection to act on a trait and for evolution to occur, the trait must possess two attributes, namely variability and heritability. Moreover, to facilitate the detection of selection with longitudinal investigations, traits should be repeatable over time, for a single measure of performance will then adequately characterize an individual throughout the interval of the study (Huey and Dunham 1987). Locomotor capacity possesses these attributes.

4.1 Variability

Traditionally, physiological ecologists have shown little interest in documenting inter-individual variability within populations, regarding variability only as an unfortunate feature obscuring central tendency (Bennett 1987). However, high levels of variability are seen in a variety of physiological variables: coefficients of variation are often 10–40 per cent for physiological traits (e.g. Garland 1984; Bennett *et al.* 1989). Locomotor performance capacity is also highly variable among different individuals (e.g. Bennett 1980; Huey and Hertz 1984*b*; Bennett *et al.* 1989; Tsuji *et al.* 1989; Huey *et al.* 1990). For example, size-corrected coefficients of variation for speed, exertion and endurance are, respectively, 27, 28 and 63 per cent for adult lizards (*Ctenosaura similis*: Garland 1984) and 19, 51 and 56 per cent for neonatal snakes (*Thamnophis sirtalis*: Jayne and Bennett 1990*b*). Among neonatal garter snakes, speed, maximal exertion and endurance vary by 10-fold, 20-fold and 100-fold, respectively (Arnold and Bennett 1988; Jayne and Bennett 1990*b*). Similarly, speed and endurance vary 5-fold and 14-fold in neonate fence lizards (*Sceloporus occidentalis*: Tsuji *et al.* 1989) and 3-fold and 14-fold, respectively, in adult canyon lizards (1- to 3-year-old *S. merriami*: Huey *et al.* 1990). Clearly, locomotor capacity shows considerable variability. Interestingly, a few individuals are truly exceptional. For instance, two female canyon lizards had endurance times that were greater than six standard deviations above the population mean (Fig. 1; Huey *et al.* 1990).

4.2 Repeatability

Is the locomotor capacity of an individual animal consistent over time? In other words, is an individual that is relatively fast at one time likely to be relatively fast in the future? Establishing the temporal repeatability of performance through time is a key step in any evolutionary analysis of individual variation, especially in studies that attempt to analyze phenotypic variation among individuals (Arnold 1986; Bennett 1987; Huey and Dunham 1987; Falconer 1989). Significant short-term (day-to-day or week-

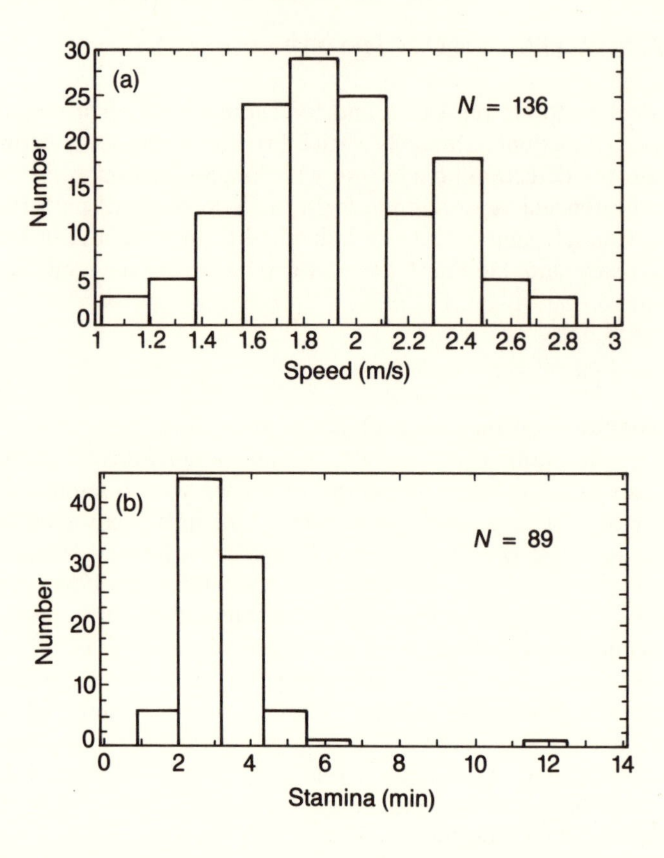

Fig. 1. Histograms of locomotor performance of adult female (aged 1–3 years) canyon lizards (*Sceloporus merriami*) from Boquillas Canyon, Big Bend National Park. (a) Maximal burst speed on a 2-m track, and (b) endurance on a treadmill at 0.5 km/h. Reprinted from Huey *et al.* (1990) with permission.

to-week) repeatability of locomotor performance capacity has been found in nearly every species in which it has been examined (amphibians: Putnam and Bennett 1981; Bennett *et al.* 1989; Shaffer *et al.*, in press; lizards: Bennett 1980; Crowley and Pietruszka 1983; Garland 1984, 1985; Huey and Hertz 1984*b*; John-Alder 1984; Kaufmann and Bennett 1989; van Berkum *et al.* 1989; Huey *et al.* 1990; mammals: Djawdan and Garland 1988; snakes: Garland and Arnold 1983; Arnold and Bennett 1988; Jayne and Bennett, 1990*a*). Correlation coefficients of size-corrected locomotor performance between different measurement days are generally 0.5–0.8 in these studies. Repeatability even remains significant among individuals across different physiological states, such as body temperature, again in every species of lizard examined (Bennett 1980; Huey and Hertz 1984*b*;

Huey and Dunham 1987; Kaufmann and Bennett 1989; van Berkum *et al.* 1989).

The ultimate test of repeatability involves individuals living under field (not laboratory) conditions. This involves measuring the performance of a large sample of individually marked animals, releasing them into the field, and then subsequently recapturing and remeasuring performance of the survivors. Huey and Dunham (1987) found that the speed of an individual *Sceloporus merriami* in two populations was positively correlated with its speed a full year later. Using data from additional years and also controlling for sex, population and age, Huey *et al.* (1990) recently confirmed that speeds of adults (aged 1–3 years) were significantly repeatable for at least one year (Fig. 2), and they demonstrated that endurance was similarly repeatable. One year is a substantial length of time for these small lizards (the cohort-generation time is about 1–1.5 years: A. E. Dunham, personal communication). Moreover, the between-year repeatabilities of speed of lizards are in fact slightly higher than those measured for thoroughbred racehorses and greyhounds, equivalent to those for some morphological traits of birds, and even higher than those for reproductive traits in some birds (see references in Huey and Dunham 1987).

Is locomotor performance stable even during periods of rapid growth? In other words, is a fast neonate likely to be a fast yearling or a fast

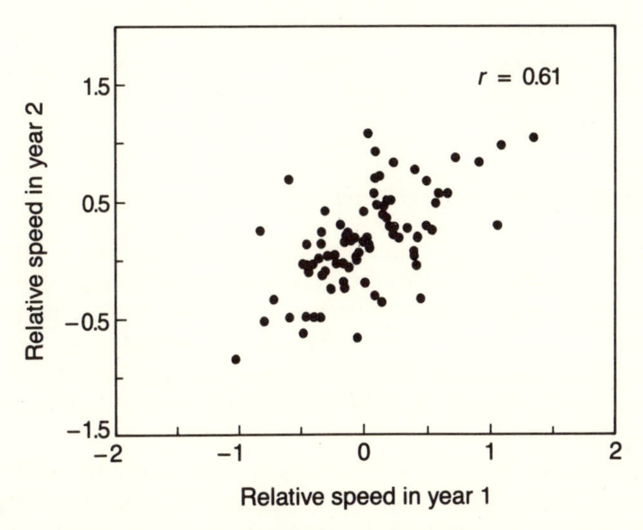

Fig. 2. Relative maximum speed of adult canyon lizards (*Sceloporus merriami*) is repeatable between years ($r = 0.61$, $N = 86$, $P < 0.001$). Relative speeds are actual speeds that have been standardized for population and sex. From data in Huey *et al.* (1990).

adult? van Berkum *et al.* (1989) measured the speed and endurance of almost 300 hatchling fence lizards, *Sceloporus occidentalis* (aged 14 days), released the animals into the field, and then remeasured the speed and endurance of surviving lizards over a year later. Although only a few hatchlings survived, size-corrected endurance (Pearson's $r = 0.77$, $n = 11$, $P < 0.005$) and possibly speed ($r = 0.38$, $n = 12$, $P = 0.06$) were repeatable, even though the lizards had grown by more than 10-fold in mass. Size-corrected speed and endurance were also repeatable over periods of one year (but not 2 years) in a field population of garter snakes, *Thamnophis sirtalis* ($r = 0.25$, $n = 185$, $P = 0.001$ for speed and $r = 0.22$, $n = 166$, $P = 0.005$ for endurance: Jayne and Bennett 1990*a*).

Recent studies with a salamander (*Ambystoma*: Shaffer *et al.*, in press) provide an interesting counterpoint with studies of reptiles. Although speed and endurance of these salamanders are repeatable within a metamorphic stage, locomotor performances are not repeatable across metamorphic stages. The lack of repeatability presumably reflects the radical shifts in morphology and ecology associated wtih metamorphosis (Shaffer *et al.*, in press).

In summary, our studies with reptiles demonstrate that individual differences in locomotor performance capacity are repeatable, even over long time intervals in nature, even at different body temperatures, and even in rapidly growing individuals. These demonstrations are important, for high repeatability greatly facilitates attempts to study natural selection on individual performance in nature. These results imply that relative performance capacities of animals do not change radically during the time intervals over which survival is monitored.

4.3 Ontogeny of locomotor performance

Although these former studies examine whether *relative* locomotor performance is stable over 1 or 2 years, they do not describe the stability of *absolute* locomotor performance. Is the average absolute performance of animals of a given age sensitive to between-year changes in the environment? Similarly, does absolute performance change drastically as an animal matures and then ages (Pough 1983, 1989; Taigen and Pough 1985; Huey *et al.* 1990)?

Determining the environmental sensitivity of absolute performance requires comparisons of performance of individuals of equivalent age and sex in different years. In an initial comparison, speeds of canyon lizards in two populations were virtually identical in two years (1984, 1985), suggesting that absolute performance is insensitive to environmental conditions (Huey and Dunham 1987). However, with data from additional years, significant between-year differences were detected (Huey *et al.* 1990). Interestingly, speed (but not endurance) in a given summer was positively correlated with the cumulative rainfall during the previous winter and spring.

Determining the ontogeny of absolute performance requires comparisons of individuals of different ages. Unfortunately, ontogenetic profiles of physiological performance have received little attention, except during early growth phases (Pough 1989; van Berkum *et al.* 1989; Jayne and Bennett 1990*a*). The reason is obvious: physiologists rarely know the exact ages of their study animals. This difficulty can be circumvented, however, by conducting physiological studies on populations with individuals marked from near birth because ages are automatically known from mark-recapture data (Huey *et al.* 1990).

The long-term demographic studies of A. E. Dunham and K. Overall made it possible to study the ontogeny of locomotor performance in the lizard *Sceloporus merriami* (Huey *et al.* 1990). In humans and greyhounds (Ryan 1975), maximal speed increases initially with age, reaches a maximum in early adulthood, and then eventually declines during senescence. Do locomotor capacities of *S. merriami* show a similar ontogenetic pattern? Performance might not decline in old adults if mortality rates are so high that few animals actually survive to old age in nature or if only the fastest lizards survive. Speed and endurance are initally low in hatchling *S. merriami* but high in 1-year-olds (young adults). Our sample size of older adults is still limited; however, in regression analyses for lizards 1–3 years old, age actually has a significant positive effect on endurance ($P = 0.003$) but may have a negative effect on speed ($P = 0.053$). In both cases, however, the effects are very weak. Moreover, because the adult age structure of these populations is heavily biased towards 1-year-olds (Dunham 1981), the performance of adult *S. merriami* must be essentially unaffected by age.

We wish to emphasize the importance of considering demography in studies that examine the relationship between performance and fitness. Imagine an investigator who attempts to test the hypothesis that the relative calling (or hopping) ability of male frogs influences their mating ability, but who does not know either the age of individual males or the age structure of the population. If age (independent of size; e.g. see Smith *et al.* 1986) influences performance, then any attempt to analyze a correlation (or the lack thereof) between performance and reproductive success will necessarily be confounded both by age and by the age structure. Knowledge of individual age and of population age structure should be a prerequisite to analyses of performance and fitness (Charlesworth 1980; Huey *et al.* 1990).

4.4 Heritability

Finally, we ask whether locomotor performance capacity has a genetic basis. Or, more properly, is there heritable variation in maximal speed, exertion and endurance, such that these traits have the genetic potential to respond to selection? Knowing the heritability of a trait is required for predictions of the dynamics of evolutionary change (Arnold 1987, 1988; Falconer 1989). However, heritability has been measured for few traits (ecological, physiological or morphological), at least in natural populations. Quantitative geneticists measure heritability in several ways, but the approach used so far for studies of locomotion in natural populations is the most basic. It involves raising full-sib families and then analyzing the patterns of variation within versus among families. Significant heritabilities have been found for speed in horses (Langlois 1980; Gaffney and Cunningham 1988), dogs (Ryan 1975) and humans (Bouchard and Malina 1983*a,b*). Speed, exertion and endurance, as well as defensive behaviors, all have significant heritabilities in garter snakes (genus *Thamnophis*) (Arnold and Bennett 1984; Garland 1988; Jayne and Bennett, 1990*a*, *b*; Arnold and Bennett, unpublished data). Speed and endurance are heritable in fence lizards (*Sceloporus occidentalis*) (van Berkum and Tsuji 1987; Tsuji *et al.* 1989). Standard and resting, but not maximum, oxygen consumption appear heritable in skinks (*Chalcides ocellatus*) (Pough and Andrews 1984). In all of the listed studies (Table 2), locomotor capacity shows moderate and significant heritability. The estimates based on full-sib breeding designs are, however, only first approximations of heritability; they do not exclude maternal and dominance effects. Experiments using more sophisticated breeding designs are encouraged.

These same full-sib data can address a related genetic issue, specifically the genetic correlation between traits, such as between speed and endurance. Evolutionary ecologists have recently become very interested in genetic correlations, because such correlations can profoundly affect evolutionary responses to selection (Lande and Arnold 1983; Arnold 1987, 1988). If two traits are genetically correlated, then direct selection on only one of the traits will lead to simultaneous evolutionary changes in the correlated trait. Two studies of locomotor performance have thus far addressed this issue. *A priori* might expect a negative genetic correlation between speed and endurance, as they emphasize different types of skeletal muscle fibers and limb proportions. However, Garland (1988) found a positive, rather than a negative, genetic correlation between speed and endurance in neonatal garter snakes. Tsuji *et al.* (1989) found no significant genetic correlation among these variables in hatchling fence lizards. On this very limited basis, the evolution of speed and endurance are predicted to be positively coupled in garter snakes but independent in fence lizards.

Table 2
Broad-sense heritabilities of locomotor performance capacity

Species	Performance	Heritability	Reference
Greyhounds	speed	0.23	Ryan (1975)
Racehorses	speed	0.02–0.65	Langlois (1980)
Thamnophis radix	time-form	0.39–0.76	Gaffney and Cunningham (1988)
	speed	0.59–0.76	Arnold and Bennett (unpublished data)
Thamnophis sirtalis	speed	0.58	Garland (1988)
	speed	0.28	Jayne and Bennett (1990*a*)
	exertion	0.26	Jayne and Bennett (1990*b*)
	endurance	0.70	Garland (1988)
	endurance	0.68	Jayne and Bennett (1990*a*)
Sceloporus occidentalis	speed	0.59	van Berkum and Tsuji (1987)
	speed	0.36	Tsuji *et al.* (1989)
	endurance	0.36	Tsuji *et al.* (1989)
Drosophila melanogaster	wing-beat frequency	0.39	Curtsinger and Laurie-Ahlberg (1981)
	power output	0.33	Curtsinger and Laurie-Ahlberg (1981)
Lygaeus kalmii	flight duration	0.20	Caldwell and Hegmann (1969)

5. ANALYZING EVOLUTIONARY PROCESS: NATURAL SELECTION

Several different strategies are used to study the evolutionary adaptation of physiology or other traits (Endler 1986*a*; Huey and Kingsolver 1989). The one traditionally used by physiological ecologists involves examining comparative data on contemporaneous organisms (e.g. species in different environments) and then deriving *post-hoc* reconstructions of *patterns* that have evolved over historical time (Huey and Bennett 1986, 1987; Bartholomew 1987; Feder 1987); we review examples of this comparative approach later in this chapter. An alternative approach involves studying selection on phenotypic traits in nature (Lande and Arnold 1983; Endler 1986*a*; Mitchell-Olds and Shaw 1987; Schluter 1989): here one analyzes evolutionary *processes* in real time. This approach has been used very elegantly in several recent studies, especially those by Peter Grant and his colleagues (e.g. Boag and Grant 1981; Grant and Grant 1989) on morphological variation in Darwin's finches. However, the specific protocol we use to study selection on performance was conceived by Arnold and Bennett (see Arnold 1983). Specifically, we first measure locomotor performance in a cohort of known-age individuals, release them into the field, and then recapture the survivors sometime later. Using a variety of statistical procedures (e.g. multivariate analyses: Lande and Arnold 1983; fitness functions: Schluter 1989; randomization tests: Jayne and Bennett 1990*b*), we test hypotheses concerning whether maximal locomotor capacity of an individual animal is, for example, correlated with growth or survivorship. By monitoring the behavior (e.g. movement rates, social dominance) of these same individuals, one can also test supplementary hypotheses relating maximal locomotor capacities to foraging or social behaviors. In combination, these approaches enable us to determine whether and how maximal performance influences correlates of fitness in real time.

The basic statistical approach can be depicted for the case of direct selection on a single phenotypic trait such as speed. The frequency distribution of speed in an initial cohort of individuals is compared to the distribution of those individuals known to survive the interval of selection (e.g. Fig. 3). The speeds compared are those of the initial cohort before release, on the assumption that speed is repeatable (see above) during the interval of selection. [If the factor, such as speed, is correlated with body size, multivariate techniques are required (Lande and Arnold 1983; Endler 1986*a*; Crespi and Bookstein 1989; Jayne and Bennett 1990*b*).] These initial distributions may be altered by selection in several ways. If directional selection favored faster individuals between the time of release and recapture, then the survivors should be a relatively fast subset of the original cohort and thus have a significantly greater mean performance. This might be the case if faster individuals were more adept at capturing prey or at avoiding predators. At least theoretically, directional selection could alternatively favor slow individuals, because fast individuals might have a higher risk of injury, as in thoroughbred racehorses, or incur higher energetic costs associated with maintaining 'high-performance' muscles (Goldspink 1981; Taigen 1983). If stabilizing or normalizing selection favored individuals of average speed, then the distribution for the survivors should have the same mean as the initial distribution but a reduced variance. Here we describe results from three such field studies, one with snakes and two with lizards.

5.1 Locomotor capacity of garter snakes

Between 1985 and 1988, Jayne and Bennett (1990*a,b*) followed the effect of locomotor capacity and body size on the survivorship of individual animals in a local population of garter snakes (*Thamnophis sirtalis*) in northern California. In 1985, 40 gravid females were collected, and these later gave birth to 275 offspring in the laboratory. Speed, exertion and endurance were measured within one week of birth, and all of the animals

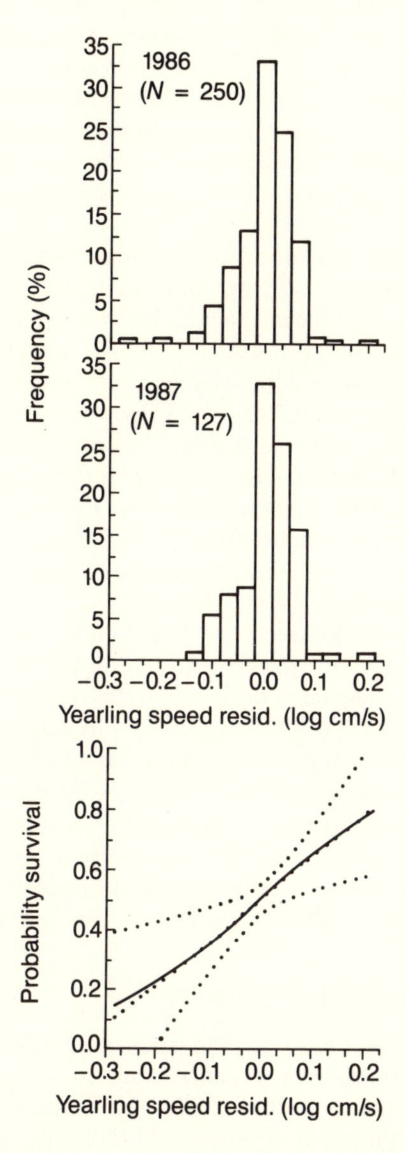

Fig. 3. The effect of speed on survivorship in yearling garter snakes (*Thamnophis sirtalis*). The top panel indicates the distribution of size-corrected residuals of speed of animals released in 1986, and the center panel of the subset of these animals recaptured in 1987. The probability of survival (lower panel) is estimated with both a regression (dotted line with 95 per cent confidence limits) (Lande and Arnold 1983) and a cubic spline function (solid line) (Schluter 1989). Data from Jayne and Bennett (1990*b*).

were released into their original population within 2 weeks of birth. During the following year, some of these animals were recaptured, remeasured and re-released, along with nearly 400 additional animals resident in the population. Survivorship and locomotor performance were followed for all individuals over a total of 3 years. Because all locomotor parameters were found to be size-dependent and because size positively influenced survivorship of neonates, all of the analyses were carried out on size-corrected residuals.

Locomotor capacity significantly predicted survivorship, but only after the first year of life; all probabilities reported are for one-tailed randomization tests comparing means between years. During the first year, neonate size (length, but not mass) was directly correlated ($P = 0.022$) with survivorship, as has been found in the lizard *Uta stansburiana* (Ferguson and Fox 1984). However, no measure of locomotor performance was directionally associated with survival ($P = 0.19$–0.73). During the second year of life, however, both speed ($P = 0.007$; Fig. 3) and exertion ($P = 0.008$) were positively related to survival, and endurance was nearly so ($P = 0.06$). For snakes older than 2 years, speed continued to be an important correlate of survivorship ($P = 0.001$), and exertion ($P = 0.08$) and endurance ($P = 0.10$) were marginally significant. As snakes in this population do not reach adult size for 3–4 years after birth, this differential mortality associated with size and locomotor capacity occur prior to reproduction. Locomotor capacity is thus under natural selection in this population. In addition, the mass residual on length was found to be under stabilizing selection in yearling snakes ($P = 0.006$): both relatively thin and heavy animals were at a selective disadvantage. Selection intensities (Schluter 1989) on these functions are similar in magnitude to those measured for morphological characters in natural populations of small birds (Bumpus 1899; Boag and Grant 1981; Schluter and Smith 1986).

5.2 Locomotor capacity of fence lizards

To examine selection on speed and endurance, Huey, Garland, Tsuji and van Berkum (unpublished data) captured 49 gravid fence lizards (*Sceloporus occidentalis*) in June 1985. The females were taken to the laboratory, where they laid eggs within 1–2 weeks. When the resulting hatchlings ($N = 296$) were 2 weeks old, we measured their speed, endurance and body size and then released them (age = 3 weeks) on our study area in early August. Six weeks later, shortly before the hatchlings began their first winter dormancy, we recaptured as many of the survivors as possible, remeasured their size and performance, and re-released them. We staged a second recapture the following May, and a third and final recapture in August, when the subadult lizards were about 13 months of age (these lizards require 2 years to mature). During each recapture we measured the size and performance not only of the lab-raised hatchlings, but also of field-raised hatchlings of the same age (*c.* 400). Because our recaptures were more frequent than those of the garter snake (above) or canyon lizard (below) studies, we analyze selection over short time intervals (first summer, first winter and first spring through summer), for which performance is highly repeatable (van Berkum *et al.* 1989). For each interval, we can determine whether speed, endurance or body size at the beginning of a time period influenced survivorship as well as growth rate (survivors only).

Based on preliminary analyses, the patterns appear consistent across all three time intervals, for both lab- and field-raised hatchlings. We present here only data on the effect of speed on survivorship for the first interval (from age 3 weeks to 9 weeks, lab-born hatchlings). Comparisons of the distributions of speeds at release for the entire original cohort and for the survivors demonstrate that the survivors were a random subset of the original cohort, at least with respect to speed (randomization test on average performance, $P = 0.35$). Endurance similarly had no effect on survival ($P = 0.17$). Locomotor performance was also uncorrelated with short-term feeding success (indexed by size of fecal pellets from recaptured lizards) or with the growth rates of the survivors. Nevertheless, we did detect two cases of selection. First, survival of the hatchlings during the first winter dormancy was sharply reduced if their tail was broken near the base, suggesting that complete tails may confer survival advantages in dormancy in ways removed from those associated to defense against predators (Bauwens and Thoen 1981). Secondly, the effects of body size on survival were contrary to those observed for garter snakes (above) or for *Uta* (Ferguson and Fox 1984): hatchlings from large eggs had slightly lower survivorship than did hatchlings from small eggs (Sinervo, Huey, Tsuji and van Berkum, unpublished data). In contrast to Lack's (1954) hypothesis of stabilizing selection on egg size, selection in this population is strongly directional and favors small eggs: females that make many small eggs not only produce more hatchlings, but also produce hatchlings with an increased probability of survival.

Because logging of the habitat forced us to terminate this study a year before the lizards reached maturity, we conducted a separate laboratory study to determine whether locomotor performance relates to social dominance in adult males (Garland *et al.* 1990*b*). Speed, but not endurance, was positively associated with dominance: males that were dominant in paired encounters in the laboratory were typically the faster of the (size-matched) pair.

5.3 Locomotor performance in canyon lizards

Beginning the summer of 1984, Dunham, Huey, Overall and Newman (unpublished data) have been measuring locomotor performance (speed, endurance, or both) of adult individuals (known age) from two marked populations of canyon lizards (*S. merriami*). We can score survivorship for two intervals, 1 month and 10 months after measurement; and we can partition data by age, sex and population. Moreover, because these lizards are also the subject of intensive focal-animal observations, we can search for correlations between locomotor performance and social dominance, territory size, movement rates, etc. This study is still continuing, and preliminary analyses have been conducted only for survivorship for the years 1984–8. So far, neither speed nor endurance correlate with survivorship; and the consistency of this pattern (among years, or between sexes and populations) suggests that the lack of detectable selection is not just an artefact of relatively small sample sizes.

5.4 Summary of selection studies

On the basis of these three studies, particularly with such different results, no general conclusions can be drawn concerning patterns of selection on locomotor performance. On the one hand, selection can occur (e.g. speed of snakes, egg size of lizards) and can be reasonably easy to detect; on the other, selection is not universally strong enough to be detected, and perhaps is not even common. Several different factors may account for these disparate observations.

First, the importance of locomotor performance may be highly population- or other taxon-specific, depending on local conditions of both biotic and abiotic factors. The fence lizard study was carried out near the northern limit of the species, where predators are few and rare. Indirect data (mortality rates, relative clutch mass of females, greater locomotor capacities) suggest that selection may well be stronger in more southern populations (Sinervo 1988). Perhaps a study on southern populations would detect selection on locomotor characteristics. However, we have not been able to detect selection on canyon lizards, in which mortality rates are very high (Dunham 1981). Secondly, perhaps selection on performance characteristics is usually too weak or variable to detect in a short-term study, but will nevertheless produce evolutionary change in performance over time (see Section 6). This is certainly not an unreasonable proposal. Selection on morphological traits is sometimes detectable only intermittently (e.g. Boag and Grant 1981; Grant and Grant 1989), and very weak selection – far too weak to be detectable – can of course lead to evolutionary change (Lewontin 1974). Thirdly, the different results for the snake versus lizard studies probably do not reflect a basic difference between these taxa, for an ongoing study with lizards of the genus *Urosaurus* has detected significant directional selection on size-corrected speed (D. B. Miles 1989, personal communication). Finally, in some groups, other kinds of behavior may compensate for limited performance capacity, such as increased wariness or aggressiveness (Rand 1964; Bauwens and Thoen 1981; Hertz *et al.* 1982; Crowley and Petruszka 1983). Such behavioral shifts might ameliorate any survival disadvantages of limited performance. These alternatives can only be discriminated by further studies.

5.5 A commentary on our general approach

Maximal locomotor performance is a complex trait that can be interpreted directly (e.g. as probability of escape from predation) or indirectly (e.g. as indices of physical as opposed to Darwinian fitness). Only a broadly based study is likely to derive insights into the evolution of such complex traits. Accordingly, we have developed and advocated a multi-step study that meets rigorous criteria:

1. The temporal and thermal repeatability of performance must be determined.
2. The heritability and genetic correlations involving performance must be known.
3. Correlations between performance and fitness must be measured in demographically established populations and over the life-cycle of the organism.
4. Supplementary manipulative experiments (e.g. Ferguson and Fox 1984; Mitchell-Olds and Shaw 1987; Marden 1989; Garland *et al.* 1990*b*; Sinervo and Huey 1990) should then validate or extend the results of descriptive analyses.

Although our field studies test explicit hypotheses (e.g. speed correlates positively with survivorship, growth rate or feeding success), we fully recognize that such analyses are necessarily descriptive and exploratory. Nevertheless, such exploratory approaches should be a key first step in fitness studies; and only subsequently (or perhaps concurrently) should manipulative experiments (sensu Mitchell-Olds and Shaw 1987) be used.

A complementary approach to cohort analysis involves the combination of performance data with focal animal observations (Pough 1989). As mentioned above, we are accumulating such data on *S. merriami* in Big Bend, but several focal animal studies on activity performance have already been completed on anuran amphibians (Wells and Taigen 1984; Sullivan and Walsberg 1985; Walton 1988). None of these studies has demonstrated significant correlations between performance and characters thought to be related to fitness. Whether this reflects biological reality or instead is an artefact of a problem inherent in most focal animal studies – small sample sizes and hence low statistical power – is unknown. In any case, focal animal studies that demonstrate significant correlations will still require both supplementary demonstrations of the genetic bases of the investigated characters and a search for trade-offs.

6. ANALYZING EVOLUTIONARY PATTERNS: COMPARATIVE STUDIES

Given the diversity of organismal form and function, there is no doubt that organismal features, including locomotor ability, do evolve. Comparisons of functional structures and mechanisms found among organisms living in different environments have been the traditional method of examining patterns of evolutionary diversification in physiology (Prosser 1986; Bartholomew 1987). To be maximally informative, such studies should be undertaken on closely related groups of organisms, so that factors extraneous to the comparison can be minimized (Gould and Lewontin 1979; Huey and Bennett 1986; Huey 1987). One recent extension of traditional comparative approaches involves making comparisons with explicit reference to phylogeny, so, for example, that evolutionary directionality (e.g. primitive vs. derived condition) may be inferred from the pattern (Gittleman 1981; Lauder 1981; Ridley 1983; Felsenstein 1985; Huey and Bennett 1986, 1987; Huey 1987). Here we review several types of comparative studies that evaluate evolutionary patterns of locomotor capacity.

Locomotor capacity is strongly influenced by body temperature in ectotherms (Bennett 1980, 1990), increasing directly with body temperature up to a maximum level (at the optimal body temperature for locomotion, *sensu* Huey and Stevenson 1979), and then declining sharply at still greater temperatures. Lizards unable to achieve their optimal temperatures can be subject to high rates of predation (Christian and Tracy 1981) and have reduced foraging success (Avery *et al.* 1982). Consequently, if maximal locomotor performance is regularly important to fitness, then ectotherms should thermoregulate at temperatures that are near optimal for locomotor performance (van Berkum *et al.* 1986). The thermal dependence of speed has now been studied in more than 50 species of reptiles, mainly lizards (Huey, van Berkum, Bennett and Hertz, unpublished review). With a few interesting exceptions (e.g. van Berkum 1986; Huey and Bennett 1987; Huey *et al.* 1989*a*), most species thermoregulate at body temperatures that are very close to their optimal body temperatures for sprinting; this pattern suggests that the ability to run quickly is important over evolutionary time.

The above comparative pattern suggests that thermal preferences of reptiles are 'co-adapted' evolutionarily with optimal temperatures. Huey and Bennett (1987) tested this hypothesis by conducting a phylogenetic analysis of interspecific data on the thermal dependence of speed in Australian lygosomine skinks. These skinks show remarkable variation in laboratory thermal preferences and in the upper temperatures at which their righting response is lost (critical thermal maxima: Greer 1980; Bennett and John-Alder 1986). Using an independently derived phylogeny, Huey and Bennett (1987) found that evolutionary shifts in thermal preferences were positively correlated with shifts in optimal temperatures for sprinting, consistent with the hypothesis that the ability to run quickly is important, at least on an evolutionary time-scale. In one genus (*Eremiascincus*), however, shifts in thermal preference and in optimal temperature appear to have been in opposite directions. Obviously, evolution does not always favor high sprint performance.

An interspecific study on the thermal sensitivity of maximal jump distance in tree frogs (John-Alder *et al.* 1988) found a correspondence between biogeographic distribution and ability to jump at low body temperatures. Northern temperate species that breed early in the season can hop at body temperatures that incapacitate later breeding (sympatric) species; both are less affected by low temperatures than are southern temperate or tropical species.

Another type of interspecific comparison involves searching for correlations between activity levels in nature and maximal performance capacities in the laboratory. For example, the foraging mode of species of frogs correlates with their aerobic capacity for exercise (Taigen *et al.* 1982; Pough 1983; Taigen and Pough 1985). For example, lacertid lizards in the Kalahari Desert of Africa differ strikingly in forging mode (Huey and Pianka 1981): some ('widely foraging species') move about 50–60 per cent of the time they are out of their burrows, whereas others ('sit-and-wait species') move only about 15 per cent of the time. Locomotor capacity reflects these differences in foraging mode: widely foraging species have relatively high endurance, but relatively low acceleration and speed (Huey *et al.* 1984). These capacity differences are also reflected in organismal metabolic potentials (Bennett *et al.* 1984) and field metabolic rates (Nagy *et al.* 1984). The sit-and-wait mode appears evolutionarily derived in these lacertids (Huey and Bennett 1986).

In an important set of recent studies, Losos (1990*a,b*) has examined the relations between morphology (e.g. relative hindlimb length), locomotor capacity (e.g. speed, jumping ability) and field behavior (movement patterns) in Caribbean lizards of the genus *Anolis*. Because phylogenies are available for these lizards, Losos was able to conduct phylogenetic as well as multivariate comparisons. He found that evolutionary changes in morphology, performance and field behavior were correlated. For example, the evolution of short legs in a lineage is associated in the laboratory with decreased abilities to sprint quickly (Fig. 4) and to jump far. In nature, such lineages crawl relatively frequently, but jump rarely and for short distances.

Finally, several studies have compared locomotor capacities of populations believed subject to different rates of predation. Selection may have favored faster lizards when exposed to a higher risk of predation (Bakker 1983; Snell *et al.* 1988), although alternative ways of reducing risk (Endler 1986*b*) could of course be favored. Several studies are consistent with this expectation (Crowley 1985*a*; Snell *et al.* 1988; Sinervo 1988), but others are not (Huey and Dunham 1987; J. Herron and B. Wilson, personal communication).

Intersexual differences in locomotor capacities may reflect different selection pressures on the sexes. Male lava lizards (*Tropidurus albemarlensis*) in the Galapagos run faster than do the females. Snell *et al.* (1988) suggest that sexual selection for territorial defense by males increases their vulnerability to predation, leading to selection for longer limbs and greater speed. Alternatively, but not exclusively, the shorter limbs and slower speeds of the females may relate to biomechanical requirements associated with carrying heavy egg burdens.

This eclectic set of comparative studies suggests that locomotor capacity often reflects the behavioral ecology of a species, population or even sex. They are thus consistent with the hypothesis that maximal locomotor

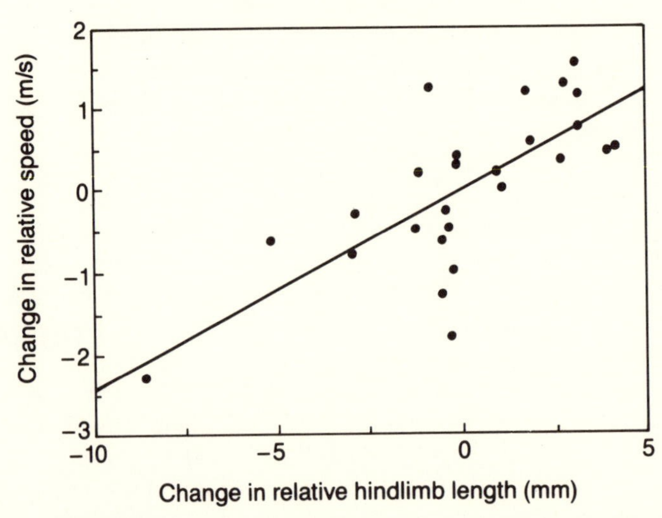

Fig. 4. A phylogenetic analysis of different species of *Anolis* lizards from the Caribbean showing that evolutionary changes in relative hindlimb length are correlated with evolutionary changes in relative speed. The effects of body size have been removed by regressing the variable against snout–vent length. Reprinted from Losos (1990*a*) with permission.

performance is evolutionarily important. Nevertheless, comparative studies show that reduced locomotor capacity can sometimes evolve (Huey and Bennett 1987; Huey *et al.* 1989).

7. CONCLUSIONS AND FUTURE DIRECTIONS

We have examined here a physiological character, specifically maximal locomotor capacity, from a variety of distinct but complementary evolutionary perspectives. We chose locomotor capacity because of its presumed ecological significance and the relative availability of information on the topic, but at least theoretically nearly any functional character could be subjected to a similar analysis. Because we are still developing and applying our approach, our chapter should be considered a progress report. To date we have found that locomotor capacity is highly variable among individuals, that this variability is stable through time, and that it has a genetic component. Therefore, the potential exists for studying the dynamics of selection on the character in real time and for detecting reponses to selection over evolutionary time. Intrapopulation studies of phenotypic selection, which examine the process of evolution, yielded conflicting results: one study found directional selection on locomotor performance in certain age classes, yet two others failed to detect such selection. Interspecific, interpopulation and intersexual comparative patterns, which document the results or the patterns of evolutionary change, generally suggest that maximal performance correlates with the behavior and ecology of reptiles in nature, thereby suggesting that maximal performance is important over evolutionary time-scales.

We believe that the integrative, multi-approach type of study that we have described here has potential not only to enrich physiological ecology and evolutionary biology, but also to promote interaction between them. For example, this approach can yield insights into the evolution of complex organismal functions, such as energy exchange, activity and growth. These sorts of characters are generally speculated to be of biological and evolutionary importance but are often too complex and environmentally dependent to be analyzed profitably in experimental laboratory systems or by traditional physiological or genetic techniques. Moreover, such studies have the promise of freeing physiological ecology from an implicit adaptationist program (cf. Gould and Lewontin 1979; Feder 1987; Pough 1989) by turning attention to the process of adaptation, rather than its simple assumption. These protocols also open hosts of new and interesting questions and firmly embed studies within the natural environments, demography and ecology of the organisms investigated.

The general approach outlined here is not the only one likely to generate novel insights into the evolution of physiology. Several 'gene to performance' studies (Arnold 1987) have been remarkably successful (Watt 1985; Koehn 1987; Powers 1987). Moreover, considerable potential exists for using selection experiments (Service 1987; Rose and Graves 1989; Hoffman and Parsons 1989; Huey and Kingsolver 1989; Bennett *et al.* 1990; Huey *et al.*, in press; Garland, personal communication), 'allometric' engineering (Sinervo 1990; Sinervo and Huey 1990) and genetic engineering (Powers 1987) as manipulative tools to probe the evolution of physiological performance.

Because new directions in evolutionary physiology are rapidly being developed, we expect a much broader dialogue among organismal biologists and evolutionary biologists than has occurred in the recent past. We expect to see many more explicit interactions between evolutionary geneticists and physiological ecologists and much more sophisticated analyses of the genetic contribution to observed character variation. We encourage a greater concentration on the examination of performance during the entire life-cycle of organisms, rather than a concentration just on a single stage (e.g. neonates, or adults only during breeding). Studies – descriptive and manipulative – that investigate the selection of complex organismal functions in natural populations have only just begun. Similarly, physiological ecologists are only beginning to take advantage of phylogenetic techniques to study comparative patterns of trait evolution. Wide open are opportunities to use selection experiments and related techniques to study the microevolution of physiological traits in the laboratory. If these new approaches are integrated with the powerful ones traditionally used by physiological ecologists to dissect the mechanistic bases of trait variation, the field of evolutionary physiological ecology will have an exciting future.

ACKNOWLEDGMENTS

Support for the authors was provided by NSF grants BSR8600066 and DCB8812028 to AFB, and BSR8415855 and BSR8718063 to RBH. We thank our collaborators for their many contributions to the studies summarized here: S. J. Arnold, A. E. Dunham, T. Garland, Jr, P. E. Hertz, B. C. Jayne, H. John-Alder, R. A. Newman, K. L. Overall, B. Sinervo, J. S. Tsuji and F. van Berkum. In particular, S. J. Arnold has long influenced our ideas. We also thank J. Herron, J. Losos, D. B. Miles, B. R. Sinervo and B. Wilson for sharing unpublished manuscripts or data.

REFERENCES

Arnold, S. J. (1981). Behavioral variation in natural populations. I. Phenotypic, genetic and environmental correlations between chemoreceptive responses to prey in the garter snake, *Thamnophis elegans*. *Evolution* **35,** 489–509.

—— (1983). Morphology, performance and fitness. *Amer. Zool.* **23,** 347–61.

—— (1986). Laboratory and field approaches of the study of adaptation. In *Predator–prey relationships: Perspectives and approaches from the study of lower vertebrates* (ed. M. E. Feder and G. V. Lauder), pp. 156–79. University of Chicago Press, Chicago.

—— (1987). Genetic correlation and the evolution of physiology. In *New directions in ecological physiology* (ed. M. E. Feder, A. F. Bennett, W. Burggren and R. B. Huey), pp. 189–215. Cambridge University Press, Cambridge.

—— (1988). Quantitative genetics and selection in natural populations: Microevolution of vertebral number in the garter snake *Thamnophis elegans*. In *Proceedings of the Second International Conference on Quantitative Genetics* (ed. B. S. Weir, E. J. Eisen, M. M. Goodman and G. Namkoong), pp. 619–36. Sinauer Associates, Sunderland, Mass.

—— and Bennett, A. F. (1984). Behavioural variation in natural populations. III. Antipredator displays in the garter snake *Thamnophis radix*. *Anim. Behav.* **32,** 1108–1118.

—— and Bennett, A. F. (1988). Behavioral variation in natural populations. V. Morphological correlates of locomotion in the garter snake, *Thamnophis radix*. *Biol. J. Linn. Soc.* **34,** 175–90.

Avery, R. A., Bedford, J. D. and Newcombe, C. P. (1982). The role of thermoregulation in lizard biology: Predatory efficiency in a temperate diurnal basker. *Behav. Ecol. Sociobiol.* **11,** 261–7.

Bakker, R. T. (1983). The deer flees, the wolf pursues: Incongruencies in predator–prey coevolution. In *Coevolution* (ed. D. J. Futuyma and M. Slatkin), pp. 350–82. Sinauer Associates, Sunderland, Mass.

Bartholomew, G. A. (1958). The role of physiology in the distribution of terrestrial vertebrates. In *Zoogeography* (ed. C. L. Hubbs), pp. 81–95. Publication 51. AAAS, Washington, D.C.

—— (1987). Interspecific comparison as a tool for ecological physiologists. In *New directions in ecological physiology* (ed. M. E. Feder, A. F. Bennett, W. Burggren and R. B. Huey), pp. 11–37. Cambridge University Press, Cambridge.

Bauwens, D. and Thoen, C. (1981). Escape tactics and vulnerability to predation associated with reproduction in the lizard *Lacerta vivpara*. *J. Anim. Ecol.* **50,** 733–43.

Bennett, A. F. (1980). The thermal dependence of lizard behaviour. *Anim. Behav.* **28,** 752–62.

—— (1987). Inter-individual variability: An underutilized resource. In *New directions in ecological physiology* (ed. M. E. Feder, A. F. Bennett, W. W. Burggren and R. B. Huey), pp. 147–69. Cambridge University Press, Cambridge.

—— (1989). Integrated studies of locomotor performance. In *Complex organismal functions: Integration and evolution in vertebrates* (ed. D. B. Wake and G. Roth), pp. 191–202. John Wiley, Chichester.

—— (1990). The thermal dependence of locomotor capacity. *Amer. J. Physiol.* **259**, R253–8.

—— and John-Alder, H. B. (1986). Thermal relations of some Australian skinks (Sauria: Scincidae). *Copeia* **1986,** 57–64.

——, Huey, R. B. and John-Alder, H. B. (1984). Physiological correlates of natural activity and locomotor capacity in two species of lacertid lizards. *J. Comp. Physiol.* **B154,** 113–18.

——, Garland, T., Jr and Else, P. L. (1989). Individual correlation of morphology, muscle mechanics, and locomotion in a salamander. *Amer. J. Physiol.* **256,** R1200–R1208.

——, Dao, K. M. and Lenski, R. E. (1990). Rapid evolution in response to high-temperature selection. *Nature* **346**, 79–81.

Boag, P. T. and Grant, P. R. (1981). Intense natural selection in a population of Darwin's finches (Geospizinae) in the Galapagos. *Science* **214,** 82–5.

Bouchard, C. and Malina, R. M. (1983*a*). Genetics for the sport scientist: Selected methodological considerations. *Exercise Sport Sci. Rev.* **11,** 274–305.

—— and Malina, R. M. (1983*b*). Genetics of physiological fitness and motor performance. *Exercise Sport Sci. Rev.* **11,** 306–339.

Bumpus, H. C. (1899). The elimination of the unfit as illustrated by the introduced sparrow, *Passer domesticus. Biol. Lect. Woods Hole Mar. Biol. Sta.* **6,** 209–216.

Caldwell, R. L. and Hegmann, J. P. (1969). Heritability of flight duration in the milkweed bug *Lygaeus kalmii. Nature* **223,** 91–2.

Charlesworth, B. (1980). *Evolution in age-structured populations.* Cambridge University Press, Cambridge.

Christian, K. and Tracy, C. (1981). The effect of the thermal environment on the ability of hatchling Galapagos land iguanas to avoid predation during dispersal. *Oecologia* **49,** 218–23.

Crespi, B. J. and Bookstein, F. L. (1989). A path-analytic model for the measurement of selection on morphology. *Evolution* **43,** 18–28.

Crowley, S. R. (1985*a*). Thermal sensitivity of sprint-running in the lizard *Sceloporus undulatus*: Support for a conservative view of thermal physiology. *Oecologia* **66,** 219–25.

—— (1985*b*). Insensitivity to desiccation of sprint running performance in the lizard, *Sceloporus undulatus. J. Herpetol.* **19,** 171–4.

—— and Pietruszka, R. D. (1983). Aggressiveness and vocalization in the leopard lizard (*Gambelia wislizennii*): The influence of temperature. *Anim. Behav.* **31,** 1055–60.

Curtsinger, J. W. and Laurie-Ahlberg, C. C. (1981). Genetic variability of flight metabolism in *Drosophila melanogaster*. I. Characterization of power output during tethered flight. *Genetics* **98,** 549–64.

Djawdan, M. and Garland, T., Jr (1988). Maximal running speeds of bipedal and quadrupedal rodents. *J. Mammal.* **69,** 765–72.

Dunham, A. E. (1981). Populations in a fluctuating environment: The comparative population ecology of the iguanid lizards *Sceloporus merriami* and *Urosaurus ornatus. Misc. Publ. Mich. Mus. Zool.* **158,** 1–62.

——, Grant, B. W. and Overall, K. L. (1989). Interfaces between biophysical and physiological ecology and the population ecology of terrestrial vertebrate ectotherms. *Physiol. Zool.* **62,** 335–55.

Elliot, J. P., Cowan, I. McT. and Holling, C. S. (1977). Prey capture by the African lion. *Can. J. Zool.* **55,** 1811–28.
Endler, J. A. (1986*a*). *Natural selection in the wild.* Princeton University Press, Princeton, N.J.
—— (1986*b*). Defense against predators. In *Predator–prey relationships: Perspectives and approaches from the study of lower vertebrates* (ed. M. E. Feder and G. V. Lauder), pp. 109–134. University of Chicago Press, Chicago.
Falconer, D. S. (1989). *Introduction to quantitative genetics*, 3rd edn. Longman, London.
Feder, M. E. (1987). The analysis of physiological diversity: The prospects for pattern documentation and general questions in ecological physiology. In *New directions in ecological physiology* (ed. M. E. Feder, A. F. Bennett, W. W. Burggren and R. V. Huey) pp. 38–75. Cambridge University Press, Cambridge.
——, Bennett, A. F., Burggren, W. W. and Huey, R. B. (ed.) (1987). *New directions in ecological physiology.* Cambridge University Press, Cambridge.
Felsenstein, J. (1985). Phylogenies and the comparative method. *Amer. Nat.* **125,** 1–15.
Ferguson, G. W. and Fox, S. F. (1984). Annual variation of survival advantage of large juvenile side-blotched lizards, *Uta stansburiana*: Its causes and evolutionary significance. *Evolution* **38,** 342–9.
Futuyma, D. J. (1986). *Evolutionary biology*, 2nd edn. Sinauer Associates, Sunderland, Mass.
Gaffney, B. and Cunningham, E. P. (1988). Estimation of genetic trend in racing performance of thoroughbred horses. *Nature* **332,** 722–4.
Garland, T., Jr (1984). Physiological correlates of locomotory performance in a lizard: An allometric approach. *Amer. J. Physiol.* **247,** R806–R815.
—— (1985). Ontogenetic and individual variation in size, shape, and speed in the Australian agamid lizard *Amphibolurus nuchalis. J. Zool., Lond.* **A207,** 425–39.
—— (1988). Genetic basis of activity metabolism. I. Inheritance of speed, stamina, and antipredator displays in the garter snake *Thamnophis sirtalis. Evolution* **42,** 335–50.
—— and Arnold, S. J. (1983). Effects of a full stomach on locomotory performance of juvenile garter snakes (*Thamnophis elegans*). *Copeia* **1983,** 1092–6.
—— and Bennett, A. F. (in press). Genetic basis of activity metabolism. II. Heritability of maximal oxygen consumption in a garter snake. *Amer. J. Physiol.*
—— and Else, P. L. (1987). Seasonal, sexual, and individual variation in endurance and activity metabolism in lizards. *Amer. J. Physiol.* **252,** R439–R449.
——, Else, P. L., Hulbert, A. J. and Tap, P. (1987). Effects of endurance training and captivity on activity metabolism of lizards. *Amer. J. Physiol.* **252,** R450–R456.
——, Bennett, A. F. and Daniels, C. B. (1990*a*). Heritability of locomotor performance and its functional bases in a natural population of vertebrates. *Experientia.*
——, Hankins, E. and Huey, R. B. (1990*b*). Locomotor capacity and social dominance in male lizards. *Funct. Ecol.* **4**, 243–50.
Gittleman, J. L. (1981). The phylogeny of parental care in fishes. *Anim. Behav.* **29,** 936–41.
Gleeson, T. T. (1979). The effects of training and captivity on the metabolic

capacity of the lizard *Sceloporus occidentalis*. *J. Comp. Physiol.* **129,** 123–8.

—— and Harrison, J. M. (1988). Muscle composition and its relation to sprint running in the lizard *Dipsosaurus dorsalis*. *Amer. J. Physiol.* **255,** R470–R477.

Goldspink, G. (1981). The use of muscles during flying, swimming, and running from the point of view of energy savings. *Symp. Zool. Soc. Lond.* **48,** 219–38.

Gould, S. J. and Lewontin, R. C. (1979). The spandrels of San Marco and the Panglossian paradigm: A critique of the adaptationist programme. *Proc. Roy. Soc. Lond.* **B205,** 581–98.

Grant, B. R. and Grant, P. R. (1989). Natural selection in a population of Darwin's finches. *Amer. Nat.* **133,** 377–93.

Greer, A. E. (1980). Critical thermal maxima temperatures in Australian scincid lizards: Their ecological and evolutionary significance. *Aust. J. Zool.* **28,** 91–102.

Hertz, P. E., Huey, R. B. and Nevo, E. (1982). Fight versus flight: Body temperature influences defensive responses of lizards. *Anim. Behav.* **30,** 676–9.

——, Huey, R. B. and Garland, T., Jr (1988). Time budgets, thermoregulation, and maximal locomotor performance: Are reptiles olympians or boy scouts? *Amer. Zool.* **28,** 927–38.

Hoffman, A. A. and Parsons, P. A. (1989). An integrated approach to environmental stress tolerance and life-history variation: Desiccation tolerance in *Drosophila*. *Biol. J. Linn. Soc.* **37,** 117–36.

Huey, R. B. (1987). Phylogeny, history, and the comparative method. In *New directions in ecological physiology* (ed. M. E. Feder, A. F. Bennett, W. W. Burggren and R. B. Huey), pp. 76–98. Cambridge University Press, Cambridge.

—— and Bennett, A. F. (1986). A comparative approach to field and laboratory studies in evolutionary biology. In *Predator–prey relationships: Perspectives and approaches from the study of lower vertebrates* (ed. M. E. Feder and G. V. Lauder), pp. 82–98. University of Chicago Press, Chicago.

—— and Bennett, A. F. (1987). Phylogenetic studies of coadaptation: Preferred temperatures versus optimal performance temperatures of lizards. *Evolution* **41,** 1098–1115.

—— and Dunham, A. E. (1987). Repeatability of locomotor performance in natural populations of the lizard *Sceloporus merriami*. *Evolution* **41,** 1116–20.

—— and Hertz, P. E. (1984*a*). Effects of body size and slope on acceleration of a lizard (*Stellio stellio*). *J. Exp. Biol.* **110,** 113–23.

—— and Hertz, P. E. (1984*b*). Is a jack-of-all-temperatures a master of none? *Evolution* **38,** 441–4.

—— and Kingsolver, J. G. (1989). Evolution of thermal sensitivity of ectotherm performance. *Trends Ecol. Evol.* **4,** 131–5.

—— and Pianka, E. R. (1981). Ecological consequences of foraging mode. *Ecology* **62,** 991–9.

—— and Stevenson, R. D. (1979). Integrating thermal physiology and ecology of ectotherms: A discussion of approaches. *Amer. Zool.* **19,** 357–66.

——, Schneider, W., Erie, G. L. and Stevenson, R. D. (1981). A field-portable racetrack and timer for measuring acceleration and speed of small cursorial animals. *Experientia* **37,** 1356–7.

——, Pianka, E. R. and Schoener, T. W. (ed.) (1983). *Lizard ecology: Studies of a model organism.* Harvard University Press, Cambridge, Mass.

——, Bennett, A. F., John-Alder, H. B. and Nagy, K. A. (1984). Locomotor

capacity and foraging behavior of Kalahari lacertid lizards. *Anim. Behav.* **32,** 41–50.
——, Niewiarowski, P. H., Kaufmann, J. and Herron, J. C. (1989). Thermal biology of nocturnal ectotherms: Is sprint performance of geckos maximal at low body temperatures? *Physiol. Zool.* **62,** 488–504.
——, Dunham, A. E., Overall, K. L. and Newman, R. A. (1990). Variation in locomotor performance in demographically known populations of the lizard *Sceloporus merriami. Physiol. Zool.* **63**, 845–72.
——, Partridge, L. and Fowler, K. (in press). Thermal sensitivity of *Drosophila melanogaster* responds rapidly to laboratory natural selection. *Evolution.*
Jayne, B. C. and Bennett, A. F. (1989). The effect of tail morphology on locomotor performance of snakes: A comparison of experimental and correlative investigations. *J. Exp. Zool.* **252,** 126–33.
—— and Bennett, A. F. (1990*a*). Scaling of speed and endurance in garter snakes: A comparison of cross-sectional and longitudinal allometries. *J. Zool., Lond.* **220**, 257–77.
—— and Bennett, A. F. (1990*b*). Selection on locomotor performance capacity in a natural population of garter snakes. *Evolution* **44**, 1204–29.
John-Alder, H. B. (1984). Seasonal variations in activity, aerobic energetic capacities, and plasma thyroid hormones (T_3 and T_4) in an iguanid lizard. *J. Comp. Physiol.* **154,** 409–419.
—— and Joos, B. (in press). Interactive effects of thyroxine and captivity on running stamina, organ and muscle masses, and intermediary metabolic enzymes in lizards. *Gen. Comp. Endocrinol.*
——, Morin, P. J. and Lawler, S. (1988). Thermal physiology, phenology, and distribution of tree frogs. *Amer. Nat.* **132,** 506–520.
Kaufmann, J. S. and Bennett, A. F. (1989). The effect of temperature and thermal acclimation on locomotor performance in *Xantusia vigilis*, the desert night lizard. *Physiol. Zool.* **62,** 1047–58.
Kephart, D. G. (1982). Microgeographic variation in the diets of garter snakes. *Oecologia* **52,** 287–91.
—— and Arnold, S. J. (1982). Garter snake diets in a fluctuating environment: A seven-year study. *Ecology* **63,** 1232–6.
Koehn, R. K. (1987). The importance of genetics to physiological ecology. In *New directions in ecological physiology* (ed. M. E. Feder, A. F. Bennett, W. W. Burggren and R. B. Huey), pp. 170–88. Cambridge University Press, Cambridge.
Lack, D. (1954). *The natural regulation of animal numbers.* Clarendon Press, Oxford.
Lande, R. and Arnold, S. J. (1983). The measurement of selection on correlated characters. *Evolution* **37,** 1210–26.
Langlois, B. (1980). Heritability of racing ability in thoroughbreds – a review. *Livestock Prod. Sci.* **7,** 591–605.
Lauder, G. V. (1981). Form and function: Structural analysis in evolutionary morphology. *Paleobiology* **7,** 430–42.
Lewontin, R. C. (1974). *The genetic basis of evolutionary change.* Columbia University Press, New York.
Losos, J. B. (1990*a*). Ecomorphology, performance capacities, and scaling of West

Indian *Anolis* lizards: An evolutionary analysis. *Ecol. Monogr.* **60**, 369–88.
—— (1990*b*). The evolution of form and function: Morphology and locomotor performance ability in West Indian *Anolis* lizards. *Evolution* **44**, 1189–1203.
Losos, J. and Sinervo, B. (1989). The effect of morphology and perch size on sprint performance in *Anolis* lizards. *J. Exp. Biol.* **145,** 23–30.
Marden, J. H. (1989). Bodybuilding dragonflies: Costs and benefits of maximizing flight muscle. *Physiol. Zool.* **62,** 505–521.
Mayr, E. (1982). *The growth of biological thought.* Harvard University Press, Cambridge, Mass.
Miles, D. B. (1987). Habitat related differences in locomotion and morphology in two populations of *Urosaurus ornatus. Amer. Zool.* **27,** 44A.
—— (1989). Selective significance of locomotory performance in an iguanid lizard. *Am. Zool.* **29**, 146A.
—— and Smith, R. G. (1987). A microcomputer-based timer and data acquisition device for measuring sprint speed and acceleration in cursorial animals. *Funct. Ecol.* **1,** 281–6.
Milstead, W. W. (1967). *Lizard ecology: A symposium.* University of Missouri Press, Columbia.
Mitchell-Olds, T. and Shaw, R. G. (1987). Regression analysis of natural selection: Statistical and biological interpretation. *Evolution* **41,** 1149–61.
Nagy, K. A., Huey, R. B. and Bennett, A. F. (1984). Field energetics and foraging mode of Kalahari lacertid lizards. *Ecology* **65,** 588–96.
Porter, W. P. (1989). New animal models and experiments for calculating growth potential at different elevations. *Physiol. Zool.* **62,** 286–313.
Pough, F. H. (1983). Amphibians and reptiles as low energy systems. In *Behavioral energetics* (ed. W. P. Aspey and S. I. Lustick), pp. 141–88. Ohio State University Press, Columbus.
—— (1989). Organismal performance and Darwinian fitness: Approaches and interpretations. *Physiol. Zool.* **62,** 199–236.
—— and Andrews, R. M. (1984). Individual and sibling group variation in metabolism of lizards: The aerobic capacity model for the origin of endothermy. *Comp. Biochem. Physiol.* **79A**, 415–19.
Powers, D. A. (1987). A multidisciplinary approach to the study of genetic variation within species. In *New directions in ecological physiology* (ed. M. E. Feder, A. F. Bennett, W. W. Burggren and R. B. Huey), pp. 106–134. Cambridge University Press, Cambridge.
Prosser, C. L. (1986). *Adaptational biology: Molecules to organisms.* John Wiley, New York.
Putnam, R. W. and Bennett, A. F. (1981). Thermal dependence of behavioural performance of anuran amphibians. *Anim. Behav.* **29,** 502–509.
Rand, A. S. (1964). Inverse relationship between temperature and shyness in the lizard *Anolis lineatopus. Proc. U.S. Nat. Mus.* **122,** no. 3595.
Ridley, M. (1983). *The explanation of organic diversity: The comparative method and adaptations for mating.* Oxford University Press, Oxford.
Rose, M. R. and Graves, J. L., Jr (1989). Minireview: What evolutionary biology can do for gerontology. *J. Gerontol.* **44**, B27–B29.
Ryan, J. E. (1975). The inheritance of track performance in greyhounds. Unpublished Master's Thesis, Trinity College, Dublin.

Schluter, D. (1989). Estimating the form of natural selection on a quantitative trait. *Evolution* **42,** 849–61.
—— and Smith, J. N. M. (1986). Natural selection on beak and body size in the song sparrow. *Evolution* **40,** 221–31.
Schmidt-Nielsen, K. (1972). *How animals work*. Cambridge University Press, Cambridge.
Seigel, R. A., Collins, J. T. and Novak, S. S. (ed.) (1987). *Snakes: Ecology and evolutionary biology*. Macmillan, New York.
Service, P. M. (1987). Physiological mechanisms of increased stress resistance in *Drosophila melanogaster* selected for postponed senescence. *Physiol. Zool.* **58,** 380–89.
Shaffer, H. B., Austin, C. C. and Huey, R. B. (in press). The physiological consequences of metamorphosis on salamander (*Ambystoma*) locomotor performance. *Physiol. Zool.*
Shine, R. (1980). 'Costs' of reproduction in reptiles. *Oecologia* **46,** 92–100.
Sinervo, B. (1988). The evolution of growth rate in *Sceloporus* lizards: Environmental, behavioral, maternal, and genetic aspects. Unpublished Ph.D. Thesis, University of Washington.
—— (1990). The evolution of maternal investment in lizards: An experimental and comparative analysis of egg size and its effects on offspring performance. *Evolution* **44**, 279–94.
—— and Huey, R. B. (1990). Allometric engineering: An experimental test of the causes of interpopulational differences in performance. *Science* **248**, 1106–9.
Smith, J. N., Arcese, P. and Schluter, D. (1986). Song sparrows grow and shrink with age. *Auk* **103,** 210–12.
Snell, H., Jennings, R., Snell, H. and Harcourt, S. (1988). Intrapopulation variation in predator-avoidance performance of Galapagos lava lizards: The interaction of sexual and natural selection. *Evol. Ecol.* **2,** 353–69.
Sullivan, B. K. and Walsberg, G. E. (1985). Call rate and aerobic capacity in Woodhouse's toad (*Bufo woodhousei*). *Herpetologica* **41,** 404–407.
Taigen, T. L. (1983). Activity metabolism of anuran amphibians: Implications for the origin of endothermy. *Amer. Nat.* **121,** 94–109.
—— and Pough, F. H. (1985). Metabolic correlates of anuran behavior. *Amer. Zool.* **25,** 987–97.
—— and Wells, K. D. (1985). Energetics of vocalization by an anuran amphibian (*Hyla versicolor*). *J. Comp. Physiol.* **144,** 247–52.
——, Emerson, S. B. and Pough, F. H. (1982). Ecological correlates of anuran exercise physiology. *Oecologia* **52,** 49–56.
——, Wells, K. D. and Marsh, R. L. (1985). The enzymatic basis of high metabolic rates in calling frogs. *Physiol. Zool.* **58,** 719–26.
Tsuji, J. S., Huey, R. B., van Berkum, F. H., Garland, T., Jr and Shaw, R. G. (1989). Locomotor performance of hatchling fence lizards (*Sceloporus occidentalis*): Quantitative genetics and morphometric correlates. *Evol. Ecol.* **3,** 240–52.
van Berkum, F. H. (1986). Evolutionary patterns in the thermal sensitivity of sprint speed in *Anolis* lizards. *Evolution* **40,** 594–604.
—— and Tsuji, J. S. (1987). Inter-familial differences in sprint speed of hatchling lizards (*Sceloporus occidentalis*). *J. Zool., Lond.* **212,** 511–19.

——, Huey, R. B. and Adams, B. A. (1986). Physiological consequences of thermoregulation in a tropical lizard (*Ameiva festiva*). *Physiol. Zool.* **59,** 464–72.

——, Huey, R. B., Tsuji, J. S. and Garland, T., Jr (1989). Repeatability of individual differences in locomotor performance and body size during early ontogeny of the lizard *Sceloporus occidentalis. Funct. Ecol.* **3,** 97–107.

Walton, B. M. (1988). Relationships among metabolic, locomotor, and field measures of organismal performance in the Fowler's toad (*Bufo woodhousei fowleri*). *Physiol. Zool.* **61,** 107–118.

Watt, W. B. (1985). Bioenergetics and evolutionary genetics: Opportunities for new synthesis. *Amer. Nat.* **125,** 118–43.

Webb, P. W. (1986). Locomotion and predatory–prey relationships. In *Predator–prey relationships: Perspectives and approaches from the study of lower vertebrates* (ed. M. E. Feder and G. V. Lauder), pp. 24–41. University of Chicago Press, Chicago.

Wells, K. D. and Taigen, T. L. (1984). Reproductive behavior and aerobic capacities of male American toads (*Bufo americanus*): Is behavior constrained by physiology? *Herpetologica* **40,** 292–8.

Wilson, B. S. and Havel, P. J. (1989). Dehydration reduces the endurance running capacity of the lizard *Uta stansburiana. Copeia* **1989,** 1052–6.

Plant consumers and plant secondary chemistry: past, present and future

MAY R. BERENBAUM

1. 'SECONDARY PRODUCT' DEFINED

The chemical diversity of angiosperm plants did not become apparent until the nineteenth century, when developments in chemical methods allowed for the first time the isolation, characterization and identification of plant constituents. From these extensive chemical studies, patterns began to emerge – primarily, the constancy of production of particular chemicals within a species regardless of locality (deCandolle 1816), and irregularities in the distribution of chemical classes among plant families (Rochleder 1854); Sachs (1882) remarked on the idiosyncratic distribution of laticifers and secretory structures in plants. The erratic distribution implied that these compounds (and the structures in which they are produced and stored) cannot be essential to the physiological function of the plant. Pfeffer (1897) confirmed this suspicion experimentally by chemically removing tannins from *Spirogyra* without affecting its growth. The secondary nature of these constituents was succinctly described by Kossel (1891):

> Just as microscope research has succeeded in stripping the cell of its non-essential accessories and in separating its casing and the reserves stored in it from the actual life-carriers, so now chemistry must attempt to separate those compounds which are present, without exception, in a protoplasma capable of developing, and to recognise the substances which are either incidental or not absolutely necessary for life. . . . I propose calling these essential components of the cell primary components, and those which are not found in every cell capable of developing, secondary. To decide whether a substance belongs to the primary or secondary components is extremely difficult in some cases (as translated by Mothes 1980).

Despite the fact that Kossel declared that 'finding and describing those atom complexes to which life is bound comprises the most important basis for the investigation of life processes', interest in the taxonomically inconsistent and physiologically enigmatic secondary compounds increased. Within 30 years after Kossel had explicitly distinguished between physiologically essential primary products and the 'incidental' secondary products of plants, the third edition of Czapek's *Plant biochemistry* (1921) devoted 547 of its 730 pages to such secondary plant products.

Today, known secondary compounds number in the tens, if not hundreds, of thousands. Despite awesome structural complexity and diversity, these compounds all share certain general characteristics. These general features include:

1. Taxonomic restriction in distribution – individual compounds, or even classes of compounds, may be restricted to a handful of families, genera or species.
2. Formation by specific enzymes, or from precursors formed by specific enzymes, with no other known substrates.
3. Regulation, either environmental or developmental, of the amount and activity of enzymes involved in the biosynthesis.
4. Compartmentalization of enzymes, precursors, intermediates and products, and formation of specialized cells or subcellular structures as synthesis sites, and/or storage sites during developmental specialization (Luckner 1980). These include the multitudinous glands, trichomes, vacuoles, laticifers, canals, and other such structures that are so typical of angiosperm plants.

Thus, in theory, biosynthetic criteria exist for the identification of 'secondary products'; in practice, however, recognition of the secondary nature of plant products is not as straightforward as it might appear to be.

2. BIOSYNTHETIC RELATIONSHIPS AMONG PHYTOCHEMICALS: WHEN ARE SECONDARY COMPOUNDS SECONDARY?

Secondary metabolites almost invariably are derived via pathways involved in the production of primary metabolites. Calvin cycle products initiate pathways involved in the synthesis of several major groups of secondary metabolites. Shikimic acid, derived from products of glycolysis and the pentose phosphate shunt, is a major intermediate in the biosynthesis of aromatic compounds, including aromatic amino acids (phenylalanine, tyrosine and tryptophan). The aromatic amino acids, in addition to serving as primary metabolites in plant physiological processes, are themselves precursors for a number of secondary biochemical pathways, including those leading to the formation of cyanogenic glycosides, glucosinolates and many alkaloids (Table 1). In addition, deamination via phenylalanine ammonia lyase of aromatic amino acids leads to the formation of cinnamic acid derivatives, including phenylpropanoids and coumarins (Fig. 1).

Acetyl CoA is another crucial intermediate in both primary and secondary pathways. Acetyl CoA derives from the glycolysis product pyruvate via oxidative decarboxylation, catalyzed by the pyruvate dehydrogenase

phenylalanine → 1 → cinnamic acid → 2 → *p*-coumaric acid → 3 →

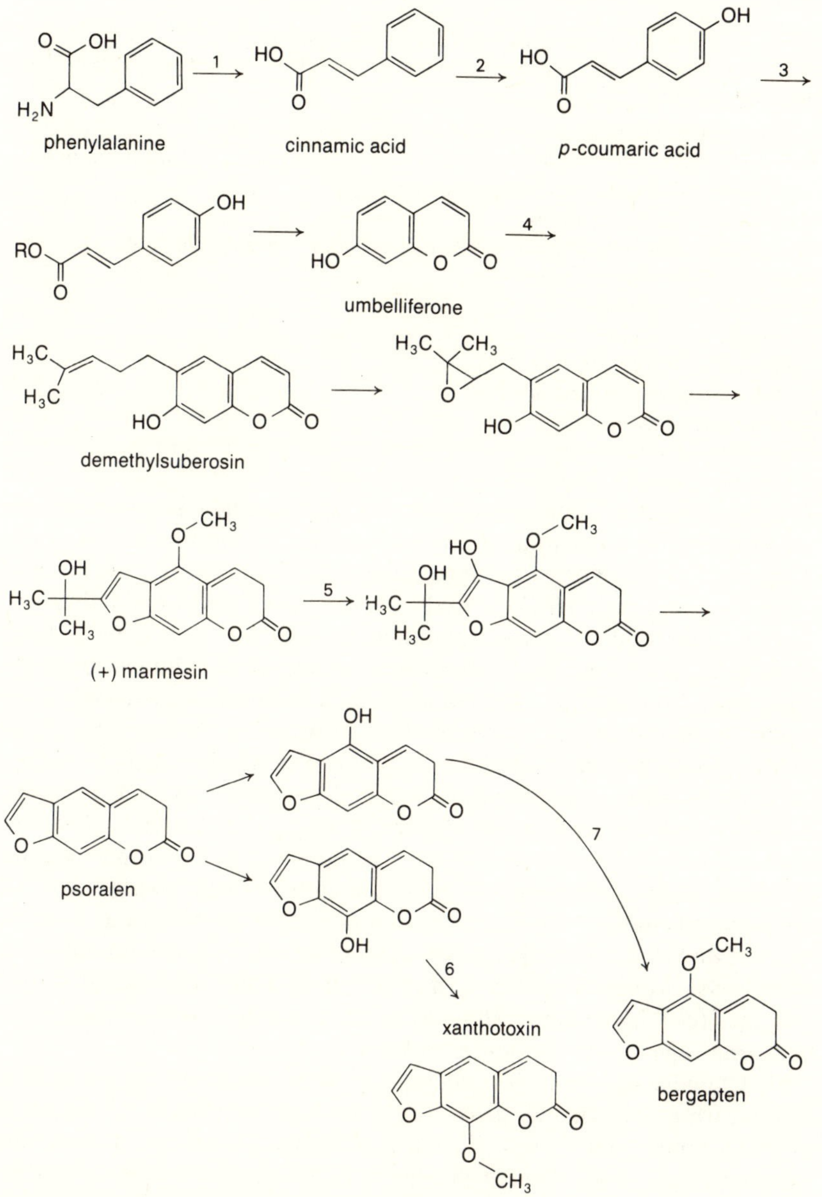

→ umbelliferone → 4 →

demethylsuberosin →

(+) marmesin → 5 →

psoralen

7

6

xanthotoxin

bergapten

Fig. 1. Biosynthesis of coumarins and furanocoumarins (Brown 1983; Ebel 1986; Hauffe *et al.* 1986. 1, phenylalanine ammonia-lyase; 2, cinnamate-4-hydroxylase; 3, 4-coumarate: CoA ligase; 4, dimethylallyl diphosphate: umbelliferone dimethyltransferase; 5, psoralen synthase; 6, *S*-adenosyl-L-methionine: xanthotoxol *O*-methyltransferase; 7, *S*-adenosyl-L-methionine: bergaptol *O*-methyltransferase.

Table 1
Alkaloids derived from amino acids

Phenylalanine: ephedra alkaloids, cytochalasins, gliotoxin, taxus alkaloids, Lythraceae alkaloids, lunaria alkaloids
Tyrosine/DOPA: phenylethylamines, isoquinoline alkaloids, benzylisoquinoline alkaloids, colchicine, emetine, betalains
Tryptophan: indole alkaloids
Anthranilic acid: quinoline alkaloids, quinazolines, acridine alkaloids
Histidine: dolicotheline, pilocarpine
Isoleucine: pyrrolizidine alkaloids
Ornithine: nicotine, tropane alkaloids, pyrrolizidine alkaloids, phenanthroindolizidine alkaloids
Lysine: mimosine, securinine
Aspartic acid: nicotine

system. Acetyl CoA contributes not only to catabolic energy production but also to the formation of all major cellular constituents in plants (Lehninger 1972). Fatty acid synthesis results from the condensation of acetyl CoA and malonyl CoA, itself derived form acetyl CoA and carbon dioxide. Subsequent 2-carbon additions lead to the formation of long-chain fatty acids and their derivatives, universal constituents of living organisms. Esterification of these long-chain fatty acids to alcohols leads to the formation of wax esters, surface constituents of many plants. In addition to primary metabolites, other acetate derivatives include unsaturated fatty acids, cyclopropanes and cyclopropenes, as well as metabolically modified low molecular weight aldehydes, alcohols, ketones, acids and hydrocarbons; many of these compounds are major components of essential oils. Acetylenes are also ultimately derived from acetyl CoA via dehydrogenation of polyunsaturated fatty acid double bonds to triple bonds. Polyketides form via condensation of acetyl CoA without reduction of carbonyl groups. These in turn play a role in the synthesis of certain quinones (e.g. hypericin in *Hypericum perforatum*) as well as some alkaloids (e.g. coniine in *Conium maculatum*).

Mevalonic acid forms via condensation of three acetyl CoA molecules; this 6-carbon compound gives rise via decarboxylation to 5-carbon isoprene units, which form the basis of terpene synthesis. Isopentenyl pyrophosphate and dimethylallylpyrophosphate are interconvertible derivatives of mevalonic acid. Terpenes, the most ubiquitous class of secondary compounds, are found in units of C5 (hemiterpenes), C10 (monoterpenes), C15 (sesquiterpenes), C20 (diterpenes), C25 (sesterpenes), C30 (triterpenes) and C40 (tetraterpenes). Terpenoid metabolism is involved in both primary and secondary metabolism (Table 2), contributing not only to the production of plant hormones and pigments but also to the production of a vast array of deterrents, repellents and toxins.

3. ECOLOGICAL FUNCTIONS OF SECONDARY COMPOUNDS: WHEN ARE SECONDARY COMPOUNDS ALLELOCHEMICALS?

Distinguishing between primary and secondary metabolites is far from an easy task, particularly when no clear biosynthetic demarcations exist between these two functional categories. In describing the distribution of pyridine and quinoline bases, Czapek (1921) suggested that:

> the sporadic distribution of these bases, the irregularity of their occurrence in closely related species, perhaps indicate that the production of such substances involves processes not inherent to every cell plasma but more of a secondary character. Caution is, however, necessary in drawing such conclusions (as translated by Mothes 1980).

This lack of a clear distinction was probably a significant contributing factor in the general reluctance to ascribe an ecological function to those metabolites for which no primary physiological function could be found. One early explanation for the existence of these compounds (Sachs 1882) was that they function as metabolic wastes, 'excretory products' stored in the plant only by virtue of the fact that plants lack means of actively excreting them (Blasdale 1947; Muller 1969; Lutz 1928). Inherent difficulties with this hypothesis (notably the structural complexity of so-called 'waste products' and the physiological regulation of biosynthesis) led to a revision of the concept, according to which secondary metabolites function as repositories for 'unnecessary metabolites' (Robinson 1974). In support of this hypothesis is cited work demonstrating phenological and ontogenetic variations in distribution and abundance of secondary substances (Seigler 1977; Seigler and Price 1976), ostensibly reflecting phenological or ontogenetic differences in metabolic demands. However, in many, if

Table 2
Representative terpenoids with primary and secondary functions

Class	Primary metabolite	Secondary metabolite
Sesquiterpene	abscisic acid	germacranolides, ligustulides
Diterpene	phytol, gibberellic acid	kaurene
Triterpene	*b*-sitosterol, sigmasterol	cucurbitacins, cardenolides

not most, cases, the amount of reserve material contained in such repositories is negligible in terms of plant physiological demands (Mothes 1980).

An alternate hypothesis for the existence of plant secondary compounds is that these compounds serve 'as means of protection, acquired in the struggle with the animal world. . . . Thus, the animal world which surrounds the plants deeply influenced not only their morphology but also their chemistry' (Stahl 1888, as translated by Fraenkel 1959). Whittaker and Feeny (1971) proposed the term 'allelochemic' to describe secondary metabolites to emphasize that these compounds function primarily in mediating interactions between the plants producing them and other organisms (Gr. *allēlon*, 'of one another, reciprocally'). Fraenkel (1959), in particular, emphasized the importance of insects as selective agents on plant chemistry.

Ehrlich and Raven (1964), based on patterns of host utilization among butterflies, proposed a scenario by which the phytochemical diversity of plants could be attributed in large part to selection pressure by insects. By random genetic events (such as mutation or genetic recombination), plants may synthesize a new compound. This compound may alter the suitability of the plant as food for its enemies; released from constraints of herbivory, plants possessing the 'biochemical novelty' enter a new adaptive zone and can speciate, until such time as, due to random genetic events, an enemy develops the capacity to tolerate the novel compound. This enemy, able to exploit new host plants, can then undergo its own adaptive radiation. The notion of reciprocal adaptive radiations, generated by reciprocal selection pressures, contributed substantially to subsequent efforts to document patterns of congruent phylogenesis between host and herbivore (or host and parasite) taxa (Mitter and Brooks 1983).

As for documenting the process of, as opposed to the patterns generated by, co-evolution, much of the subsequent work stimulated by the publication of Ehrlich and Raven's paper rests on three assumptions:

1. That phenotypic variation in secondary metabolism is commonplace.
2. That variation in secondary metabolism is under genetic control.
3. That variation in secondary metabolism differentially affects plant fitness in the presence of herbivores or pathogens.

This Darwinian triad of assumptions leads to the logical conclusion that natural selection, as imposed by plant herbivores or pathogens, is responsible at least in part for plant chemical diversity. These three assumptions differ in the ease with which they may be tested; accordingly, the amount of evidence in support of each assumption differs as well. Phenotypic variation in secondary metabolism was thoroughly documented even before Ehrlich and Raven put pen to paper and continues to be documented. At least 2000 distinct alkaloids had been characterized by 1972;

by 1982, that number had grown to 5500 (Goodwin and Mercer 1972; Harborne 1982) and continues to increase.

Evidence for the genetic control of variation in secondary metabolism was also available at the time of the publication of Ehrlich and Raven's 1964 paper, although neither Fraenkel (1959) nor Ehrlich and Raven (1964) cited extensively the plethora of studies in plant breeding journals meticulously documenting the genetic basis for differences among cultivars in phytochemical content. Ehrlich and Raven's suggestion that the evolution of chemically novel compounds results from 'occasional mutations and recombination' implies a genetic basis for qualitative, if not quantitative, variation between taxa.

Perhaps the most difficult of the three assumptions to prove is the notion that variation in plant secondary chemistry affects fitness in the presence of herbivores (and, in turn, that selection pressure by herbivores is responsible for maintaining that variation). Evidence for a selective impact of plant chemistry on an insect herbivore (i.e. reduced survival or fecundity) is not equivalent to evidence for a selective advantage of plant chemistry to the plant. Indeed, Janzen (1980) strongly objected to the misuse of the term 'co-evolution', in contexts in which reciprocity was not explicitly demonstrated. The notion that intraspecific variation in plant chemistry may result in differential plant fitness was perhaps suggested earliest by plant breeders (e.g. Collins and Kempthorne 1917); it remains today a difficult notion to nail down. In few studies documenting the genetic basis for resistance variation are chemical resistance factors identified (e.g. Simms and Rausher 1989); by the same token, in few studies documenting toxicological effects of plant chemicals on insects is the genetic basis for chemical variation established (Bowers 1988).

4. A CASE STUDY: FURANOCOUMARINS IN APIACEAE (= UMBELLIFERAE)

The furanocoumarins are a group of chemicals for which at least some shreds of evidence have accumulated in support of the assumptions underlying the co-evolutionary scenario; namely, that phenotypic variation is widespread, that it is genetically based, and that it is associated with differential reproductive success in the presence of insect herbivores. Furanocoumarins are benz-2-pyrone compounds with a furan ring attached at the 6,7 positions (in the case of linear furanocoumarins) or at the 7,8 positions (in the case of angular furanocoumarins) (Fig. 1). They are in many ways typical secondary compounds:

1. They are, as are most secondary compounds, taxonomically restricted in distribution, occurring in fewer than a dozen plant families (Murray *et al.* 1982). They are most widespread and structurally diverse in two

families: the Rutaceae, or citrus family, and the Apiaceae (= Umbelliferae), the carrot family.

2. Biosynthetically, they are formed by the action of specific enzymes with no other known substrates. Furanocoumarins derive ultimately from phenylalanine via substituted cinnamic acid esters (Fig. 1); thus, as products of general phenylpropanoid metabolism, early stages of their biosynthesis involve phenylalanine ammonia lyase (PAL), cinnamate-4-hydroxylase, and *p*-coumarate:CoA ligase. Subsequent biosynthetic transformations unique to furanocoumarin production involve dimethylallylpyrophosphate; umbelliferone dimethylallyltransferase, which attaches a dimethylallyl substituent to umbelliferone; 'psoralen synthase', a probable cytochrome P450 that catalyzes via oxidation the loss of a hydroxypropyl group from marmesin to produce psoralen; and two *O*-methyltransferases, which attach methyl groups to hydroxylated derivatives of psoralen to produce bergapten and xanthotoxin (Ebel 1986).

3. Furanocoumarin biosynthesis is regulated, both developmentally and environmentally. Developmental studies of parsley cotyledons (*Petroselinum sativum*) (Knogge *et al.* 1987) revealed that both furanocoumarins and furanocoumarin biosynthetic enzymes reach a peak level after 20 days of growth, approximately at the onset of cotyledon senescence. Foliar furanocoumarin content in *Pastinaca sativa* is relatively high as overwintering plants flower and declines with senescence (Berenbaum 1981). Seeds vary developmentally as well in this species. Seeds produced early in development (e.g. on primary and secondary umbels) contain significantly greater amounts of furanocoumarins than do seeds produced late in development (e.g. on tertiary umbels: Berenbaum and Zangerl 1986). As the reproductive parts mature, furanocoumarin content and composition change; as reproductive units progress from bud to male to female stages, furanocoumarin content tends to increase and a substantial increase occurs immediately after fertilization and upon the expansion of developing fruit (Nitao and Zangerl 1987). That these increases with developmental state cannot be attributable solely to age is evidenced by the fact that, within a single umbel, floral units that are developmentally more advanced contain higher concentrations of furanocoumarins (Nitao and Zangerl 1987).

   Environmental influences on furanocoumarin biosynthesis are numerous; biosynthesis and accumulation are induced by both abiotic factors (ultraviolet light: Zangerl and Berenbaum 1987; cold temperatures, exposure to copper sulfate: Beier and Oertli 1983; mechanical damage: Zangerl, in press) as well as biotic stress factors (fungal infection: Johnson *et al.* 1973; Ceska *et al.* 1986; Wu *et al.* 1972; insect feeding damage: Zangerl 1990). Regulation of the furanocoumarin-induced response has been studied most extensively in parsley (*Petroselinum sativum*). Continuous illumination of cell cultures resulted in an increase in phenylalanine ammonia lyase activity that was inhibited by transcription, translation, protein and RNA synthesis inhibitors (Hahlbrock and Ragg 1975). A similar increase in activity of both phenylalanine ammonia lyase and 4-coumarate:CoA ligase was effected by treatment of cell cultures with an elicitor preparation from *Phytophthora megasperma* var. *glycinea* (Pmg), a non-pathogen of parsley. The specificity of this induction was suggested by the fact that phenylalanine ammonia lyase mRNA activity increased at a greater rate in elicitor-treated cells than in irradiated cells and the fact that elicitor treatment did not (as did irradiation) induce activities of enzymes involved in flavonoid glycoside synthesis (Hahlbrock *et al.* 1981). Tietjen *et al.* (1983) identified the linear furanocoumarins psoralen, xanthotoxin and bergapten (as well as the benzodipyran-dione graveolone) as the Pmg elicitor-induced products in parsley, and in addition identified isopimpinellin and bergapten as products induced by elicitor prepared from *Alternaria carthami* in concentrations sufficient to inhibit growth of that fungus. Elicitor treatment again failed to induce flavonoid production.

The messenger RNAs coding for PAL and 4-coumarate:CoA lygase induction following irradiation and Pmg elicitor treatment appear to be similar, if not identical (Kuhn *et al.* 1984); treatment with fungal elicitor appears to induce more rapid changes in mRNA translational activities. Thus, treatment by fungal elicitors initiates increased transcription of genes controlling synthesis of furanocoumarins. The *cis*-acting promoter elements regulating responses of one PAL gene to elicitor or UV light have been identified and sequenced (Lois *et al.* 1989). Additional evidence of the specificity of the induced response is provided by the induction of additional enzymes specifically involved in furanocoumarin synthesis, including dimethylallylpyrophosphate: umbelliferone dimethylallyltransferase (Tietjen and Matern 1983), 'psoralen synthase' and two methyl transferases (Hauffe *et al.* 1986). Consistent with their position in the biosynthetic pathway, these enzymes are induced several hours after induction of PAL and 4 coumarate:CoA ligase.

4. Furanocoumarin production is also compartmentalized; generally, furanocoumarins are localized in storage organs throughout the plant. In parsnip, for example, furanocoumarins in roots are sequestered in oil channels, in foliage they are restricted to vessels closely aligning vascular tissue and in seeds they are stored in vittae or oil tubes (Zangerl *et al.* 1989; Zangerl 1990; Ladygina *et al.* 1979).

Furanocoumarins, then, by all rights can be considered 'secondary compounds'. That they are allelochemicals as well – natural products involved in mediating interactions among organisms – is suggested by their distinctive toxicological properties. Furanocoumarins are relatively unusual among secondary compounds in that they are photosensitizers; they can absorb photons of light energy to form an excited state that can proceed to react with biomolecules of various descriptions. Among the more toxicologically important reactions include covalent crosslinking of DNA, cycloadduct formation with unsaturated lipids and denaturation of proteins (Murray *et al.* 1982). Thus, not surprisingly, toxicity of furanocoumarins at naturally occurring concentrations has been reported against viruses, bacteria, fungi, snails, nematodes, insects, birds and mammals (Murray *et al.* 1982).

The question remains, however, as to whether the diversity and abundance of furanocoumarins in any given plant species reflects adaptive variation resulting from environmental selection pressures or simply reflects 'neutral variation' of no particular consequence to survival of the species. Documenting qualitative and quantitative variation in furanocoumarin chemistry is relatively straightforward; ascertaining the significance of that variation is another matter altogether. There is no easy procedure for documenting the adaptive significance of variation in any trait. The existence of significant additive genetic variation in chemical resistance traits may indeed result from contemporary selection pressures; if so, stabilizing, directional and balancing selection all generate different patterns of variation, complicating interpretation of existing variation in the absence of direct evidence of the selective impact of the putative selection agent. However, such variation may also be the result of a transient polymorphism, rather than the result of ongoing selection. On the other hand, attributing between-taxa variation in the distribution and abundance of genetically controlled chemical traits to past selection regimes is inferential at best.

To determine the extent to which phytochemical variation represents adaptive variation, it is helpful to focus on a single species, and its environmental context, so that such variation can be better understood. Phenotypic variation in furanocoumarin production has been exhaustively documented in *Pastinaca sativa*, the wild parsnip. Populations vary both on a geographic scale (Table 3) and on a relatively narrow local scale (Table 4) in both constitutive and induced foliar furanocoumarin content. Within populations, individual variation in both foliage and seeds is high (Table 5). Within a plant, different plant parts differ significantly in furanocoumarin content and composition (Table 6) – in general, the furanocoumarin content of reproductive tissues is highest, foliar content intermediate and root content lowest. Even within a single plant part, furanocoumarin production is not homogeneous. In seeds, furanocoumar-

Table 3
Geographic and varietal differences in Pastinaca sativa[a]

Furanocoumarin	Range (% dry weight)
Xanthotoxin	0.02–1.55
Bergapten	0.03–0.40
Imperatorin	0.04–0.90
Isopimpinellin	0.01–0.25
Sphondin	0.00–0.11
Total	0.12–2.63

[a]From Berenbaum (1981).

Table 4
Comparison of furanocoumarin characteristics of two populations of Pastinaca sativa *in east-central Illinois*[a]

	Race Street	Phillips Tract	Population effect (p=)	h^2 Race	h^2 Phillips
Constitutive (ug/cm²)					
Imperatorin	0.82	0.75	0.36	0.75	0.85[b]
Bergapten	0.25	0.18	0.01	0.64	0.85[b]
Xanthotoxin	1.95	1.67	0.02	0.78[b]	0.43
Sphondin	0.12	0.13	0.82	0.87[b]	0.86[b]
Induced (ug/cm²					
Imperatorin	1.03	1.03	0.83	0.62	0.75
Bergapten	0.37	0.30	0.06	0.67[b]	0.64
Xanthotoxin	2.51	2.31	0.20	0.66[b]	0.78[b]
Sphondin	0.20	0.21	0.86	0.94	0.87[b]

[a]From Zangerl and Berenbaum (1990).
[b]Significantly different from zero (see Zangerl and Berenbaum 1990).

ins are sequestered in vittae, or oil tubes. Fruits of *P. sativa* are schizocarps, consisting of two seed-like mericarps suspended from a carpophore; the two seeds separate from each other at maturity. The two large vittae on the inner face of the seeds contain over nine times the furanocoumarins as do the more numerous vittae on the outer face of the seeds (Berenbaum and Zangerl 1986).

Table 5
Between-plant variation in furanocoumarins within a single population (Perkins Rd., Champaign Co., Illinois)

	μg/seed	% of total	h^2
Imperatorin	2.2–45.1	29.3–50	0.70[a]
Bergapten	1.1–19.1	8.7–22.6	0.54
Isopimpinellin	0.6–8.0	5.3–16.9	0.98[a]
Xanthotoxin	2.1–25.3	22.7–40.7	0.61
Sphondin	0.0–0.4	0.3–9.6	0.62[a]

[a]Significantly different from zero (see Berenbaum *et al.* 1986).

Table 6
Average furanocoumarin content of various organs of Pastinaca sativa *during development*[a]

Organ	% dry weight
Leaves	0.076–0.646
Stems	0.035–0.143
Roots	0.001–0.014
Terminal umbel (buds, flowers, seeds)	0.650–3.529

[a]From Berenbaum (1981).

Phenotypic differences in furanocoumarin content would be largely irrelevant in the context of natural selection if they are entirely determined by environmental variation. However, ample evidence exists as to the genetic control of furanocoumarin variation, particularly in *Pastinaca sativa*. Heritabilities of the seed content of bergapten, sphondin and imperatorin are significantly different from zero; moreover, in the case of sphondin, concentration is genetically correlated with vittae area, suggesting a genetic constraint on localization of this compound (Zangerl *et al.* 1989).

Within a population, individuals vary both qualitatively and quantitatively in furanocoumarin content. The total furanocoumarin content can vary by more than an order of magnitude between conspecific individuals (Table 5). Half-sib heritabilities of the absolute and relative amounts of furanocoumarins in both seeds and foliage are significantly different from zero, indicating at least the partial genetic regulation of these chemical traits (Berenbaum *et al.* 1986).

Between populations of *P. sativa*, genetically based variation in furanocoumarin content and composition has been documented. In one study in east-central Illinois, the seed content of furanocoumarins in one population growing in a shady wooded area was three-fold higher than the seed content of furanocoumarins in an oldfield population approximately 6 km northeast (Zangerl and Berenbaum, in press). Significant differences were found in constitutive concentrations of bergapten, xanthotoxin and sphondin in foliage. Heritability estimates from common gardens for these traits indicated that populational differences in furanocoumarin levels are at least in part attributable to genetic factors.

The importance of plant chemicals as defenses against insect herbivores has been established primarily through indirect means. Such indirect evidence includes the existence of chemical polymorphisms with respect to environmental mosaics of herbivore intensity (e.g. Dolinger *et al.* 1973), as well as correlations of phenotypic chemical characters with survival and fecundity (Berenbaum 1981). Such patterns are suggestive of the action of natural selection (Primack and Kang 1989). An extension of such evidence was the development of optimal defense theory (McKey 1979), the idea that, given resource limitation on plant secondary metabolism, the allocation of secondary compounds among tissues within an individual plant should reflect relative risk of attack or relative value of those tissues to the plant. Although demonstrations of proportional allocation of secondary compounds relative to tissue value or risk are suggestive, they are far less satisfying than a direct demonstration of differential plant fitness based on genetically based differences in secondary chemistry. The adverse effects of furanocoumarins on insect herbivores have been abundantly demonstrated; they are, for example, toxic as well as antifeedant to a variety of not only generalized feeders (Yajima *et al.* 1977; Berenbaum 1978; Muckensturm *et al.* 1981; Klocke *et al.* 1989), but also to specialists that feed exclusively on umbelliferous plants (Berenbaum and Feeny 1981; Berenbaum *et al.* 1989).

One very likely selective agent on parsnip furanocoumarin chemistry is the parsnip webworm, *Depressaria pastinacella* (Lepidoptera: Oecophoridae). It is the chief (and at times the sole) insect herbivore associated with the plant throughout much eastern North American, into which both the plant and insect were introduced from Europe (Berenbaum *et al.* 1986). The parsnip webworm feeds primarily on flowers and fruits of *P. sativa* and closely related species in the genus *Heracleum*. Overwintering females oviposit on plants shortly before they send up a flowering stalk; hatching larvae feed briefely on foliage before moving to developing buds, flowers and fruits, which are webbed together and consumed as the caterpillar develops. That the parsnip webworm may be an important selective agent on parsnip chemistry is suggested by the fact that, as a seed-feeding herbivore, it is capable of reducing seed production in the primary umbel by over 90 per cent (Thompson 1978; Berenbaum *et al.* 1986). Both wild parsnip and the parsnip webworm were introduced into North America from Europe; relatively few native herbivores have successfully colonized wild parsnip and the parsnip webworm has virtually no other host plants in North America (Berenbaum *et al.* 1986).

Physiologically, the parsnip webworm is extraordinarily tolerant of certain furanocoumarins. It is capable of metabolizing xanthotoxin 10 to 300 times more efficiently than species that rarely or never eat plants containing furanocoumarins (Nitao 1989). However, the cytochrome P450 monooxygenases that metabolize xanthotoxin are less effective against bergapten and sphondin. As a result, bergapten significantly reduces growth rates and approximate digestibility in *D. pastinacella* (Berenbaum *et al.* 1989). In addition, bergapten and sphondin act as competitive inhibitors of xanthotoxin metabolism, which is reduced by 60 per cent in the presence of bergapten (Zangerl and Berenbaum, personal observations).

Even within an individual plant, furanocoumarin allocation patterns are consistent with a defensive function. Although *P. sativa* can to some extent compensate for the removal of floral units, this compensatory reproduction is limited in extent and is restricted to plants in certain size classes (Hendrix 1979). The removal of successively more mature floral units has increasingly greater effects on fitness; the removal of buds, for example, from the primary umbel, does not have as pronounced an effect on reproduction as does the removal of fertilized female flowers. The pattern of furanocoumarin contribution parallels the fitness contributions of the floral unit (Nitao and Zangerl 1987). Even in non-reproductive tissues, furanocoumarin production is highly correlated with the nitrogen content of foliage, an association consistent with a defensive function (Berenbaum 1981).

One of the most compelling indirect arguments for the adaptive significance of furanocoumarin variation in wild parsnip in the context of insect herbivory is the existence of parthenocarpic fruits in parsnip umbels (Zangerl *et al.* 1991). These fruits are entirely lacking in embryo and endosperm yet are of similar shape and size as normal fruits. The furanocoumarin content of such fruits is significantly lower than is the furanocoumarin content of developed fruits (Table 7). Despite the significantly lower nutritional content of these fruits, they are preferentially consumed by parsnip webworm larva. Accordingly, growth rates and efficiencies of conversion of food are significantly reduced on this overwhelmingly preferred food item. One possible explanation for such quixotic behavior is that lower furanocoumarin concentration, encountered by the larva as they first contact the seed coat, entice caterpillars to consume nutritionally

Table 7
Nutritional and chemical characters of filled and empty fruits of wild parsnip[a]

	Filled fruits	Empty fruits
Total nitrogen (μg)	404	245
Calories	56	33
Imperatorin (μg)	130	56
Bergapten (μg)	21	10
Isopimpinellin (μg)	11	7
Xanthotoxin (μg)	40	24
Sphondin (μg)	1.4	0.4
Relative growth rate (parsnip webworm) mg/mg/day	0.201	0.072

[a]From Zangerl *et al.* (1991).
Note: For all comparisons by one-way analysis of variance, fruit effects were significant.

substandard food. The loss of parthenocarpic (and thus inviable) fruit has no negative fitness impact on the plant; rather, diverting herbivory away from viable fuits may actually increase fitness in the presence of seed-feeding herbivores by reducing the loss of viable fruits.

Possibly, the best direct evidence for an adaptive function of furanocoumarin variation with respect to herbivore selection pressure comes from a study examining the seed production of several parsnip families in the presence of natural populations of *Depressaria pastinacella*. Among these families, amounts of two furanocoumarins, bergapten and sphondin, in both seeds and leaves, accounted for almost 75 per cent of the variation in resistance (as measured by viable seed production) of *P. sativa* to *Depressaria pastinacella* (Berenbaum *et al.* 1986). Thus, individuals with high concentrations of bergapten and sphondin in seeds – compounds known to have detrimental effects on the growth and development of the parsnip webworm – have higher fitness in the presence of this seed-feeding herbivore. The concentrations of bergapten and sphondin in seeds in this environment have heritabilities significantly different from zero, i.e. they are traits that possess additive genetic variance and thus are potentially able to respond to selection by this herbivore. That such a selection response may have indeed taken place is evidenced by a comparison between two populations of wild parsnip with different histories of webworm population densities; the population with a history of predictable heavy attack averages significantly higher constitutive and induced levels of furanocoumarins (Zangerl and Berenbaum 1990).

Thus, evidence in support of adaptive variation in furanocoumarin production is both indirect and direct. Indirect evidence of past selective impacts include:

1. Phenotypic variation consistent with environmental stress factors, i.e. induction following damage by herbivores.
2. Non-random association of particular furanocoumarin characters with particular herbivory regimes (high levels of bergapten and sphondin, furanocoumarins that decrease the growth rate and survival of parsnip webworms, in poplations historically subjected to high levels of webworm attack).
3. Within-plant allocation patterns consistent with tissue value (as measured by nitrogen content or by removal effects on fitness).

More direct evidence of the ongoing selective impact of insect herbivory on plant chemistry includes:

4. Substantial additive genetic variation in populations in the content and composition of furanocoumarins in foliage and seeds, including those furanocoumarins correlated with resistance (fitness in the presence of herbivores).
5. Phenotypic correlations between furanocoumarin characters and plant fitness in the presence of an insect herbivore (which is known to be adversely affected by those furanocoumarins).

All this is not to say that every genetically based variation in furanocoumarin chemistry is adaptive, although the temptation to construct *post-hoc* adaptive scenarios is powerful. Furanocoumarin variation may be attributable to pleiotropic effects of selection on entirely unrelated traits or may result from transitory random variation. General metabolic considerations may influence the expression of genetically influenced chemical traits; indeed, high concentrations of furanocoumarins that may appear adaptive in the context of herbivory may in reality be maladaptive if such secondary metabolism effects a metabolic cost (Berenbaum *et al.* 1986). It is important to emphasize, however, that variation in plant chemistry in and of itself is worthy of study. Just as the presence of 'secondary products' was once regarded as non-essential to the 'survival of species', quantitative variation in plant chemistry is even today overlooked in favor of analyses that consider only the presence or absence of particular chemical classes. The additive genetic variation underlying exceedingly fine-grained differences in the temporal and spatial variation in plant chemistry suggests, if not the action of natural selection, then at least the potential for natural selection to act. Not only, as Janzen (1979) states, is it 'apparent that Latin binomials do not contain secondary compounds, but rather plant parts do', but it is also apparent that the chemical content of those plant parts is likely to be affected by genotype; phytochemical protocols purporting to document ecological phenomena must take genetic variation into account and establish the degree to which genes determine the chemical phenotype.

5. PATTERNS OF PLANT CHEMICAL DEFENSE IN THE CONTEXT OF PEST CONTROL

Far from being the esoteric obsession of a tiny handful of ivory-tower biologists, establishing the patterns of chemical variation in plants and the effects of that variation on insect herbivores is of major importance in developing rational approaches to insect pest control. Understanding the nature of the selective regime imposed by insects on plants and, reciprocally, the selective regime imposed by plant chemicals on insects, may provide insights into the design and implementation of more viable strategies of chemical control of insect pests.

Plant secondary chemicals, by their nature, are highly variable. Indeed, this variability is why they were attributed secondary status in the first place; their presence is not required in all plant tissues at all times, in all plants in a population, in all populations of a species, or in all species of a genus. It is difficult to imagine an angiosperm plant bereft of its secondary chemistry surviving for any length of time against inevitable environmental unpleasantries. In this sense, secondary compounds can hardly be considered secondary; chemical variation translates into fitness variation. Genetically based chemical variation may enable plants to minimize losses to herbivory – by reducing herbivore damage directly or by diverting herbivory away from vital tissues.

In contrast with natural populations, crop plants often experience devastating levels of damage, particularly by insect enemies. The changes wrought by natural selection in plant chemical defense stand in contrast with the genetic manipulations of plant chemistry brought about by artificial selection. Genetic changes effected by plant domestication, either deliberate or accidental, may in part be responsible for the susceptibility of crop plants to insect attack relative to their wild or feral relatives. Artifical selection, with the objective of increasing palatability, generally results in the loss of toxic defensive compounds (with the notable exception of domesticated drug plants). In addition to selecting for lower concentrations of toxins, even in non-edible parts, humans tend to select for uniformity and phenological synchrony as well. Such uniformity and synchrony facilitates planting, cultivation and harvesting. Selection for synchronous fruit production in all probability results in increased synchrony in developmentally dependent phytochemical changes. Finally, inbreeding (not uncommon in plant breeding protocols) tends to reduce genetic diversity overall and in all probability underlying chemical diversity as well. For example, even though no documented artificial selection program specifically called for reducing concentrations of seed furanocoumarins (plant parts that are not consumed by humans and which therefore present no risks of toxicity), seeds of cultivated varieties of parsnips have substantially lower concentrations of furanocoumarins as well as a smaller range of variation in the furanocoumarin content. Whereas seeds of wild plants contain an average 37.9 μg of furanocoumarin (ranging from 17.0 to 60.1), seeds of cultivated parsnips contain on average only 13.7 μg (ranging from 8.5 to 22.2) (Berenbaum *et al.* 1984). Thus, artificial selection renders the timing and distribution of chemical defenses predictable and uniform and renders the chemicals themselves less abundant and diverse (Table 8). Evolution of resistance to such defenses is no doubt greatly accelerated.

As for chemical control programs designed and implemented by humans, comparisons with natural systems of chemical defense provide a number of striking contrasts as well. Chemical control programs (i.e. application of exogenous synthetic organic insecticides) have historically been stimulus-independent, in the sense that pesticides are applied by the calendar and not according to timing and extent of damage (Metcalf and Flint 1962). Pesticides are also traditionally applied in a broadcast fashion, whereas plant allelochemicals are highly variable spatially. The plant parts at greatest risk of attack can contain the highest concentrations of allelochemicals. No sprayer has yet been devised, for example, that can differentially protect buds, male flowers, and pre- and post-fertilization female flowers; furanocoumarin allocation in the umbels of wild parsnip reflects these developmental differences. Finally, pesticides generally consist of simple mixtures of very few compounds; in contrast, allelochemical diversity is very high (Table 9). No fewer than five furanocoumarins can be found in variable ratios within the vittae of a single parsnip seed (Zangerl *et al.* 1989).

It appears likely, then, that agronomic practices combine to increase the vulnerability of crop plants to insect pests. The study of natural systems

Table 8

Comparison of effects of artificial vs. natural selection on plant chemistry

	Artificial selection	Natural selection
Objective	palatability	non-palatability
Timing	predictable	variable
Distribution	predictable	variable
Composition	low diversity	high diversity

Table 9
Comparison of plant and human chemical defense strategies

	Plant	Human
Objective	impact minimization	insect death
Timing	damage-dependent	prophylactic
Distribution	proportional allocation according to risk	indiscriminate
Composition	variable mixtures	single toxin

can provide at least some novel approaches for testing, in comparison with standard technology. Constructive methodological changes, derived from a study of naturally occurring allelochemical variation in plant populations, might include:

1. Selectively breeding plants for higher levels of defensive chemicals in non-economic parts that nonetheless harbor insect populations.
2. Selectively breeding for inducible resistance.
3. Applying pesticides only where and when damage warrants their use.
4. Increasing the diversity of chemical control agents in use in any given crop system.

Current efforts to manipulate plants genetically should also incorporate patterns generated by natural selection. The insertion of the delta-endotoxin gene of *Bacillus thuringiensis* in crop plants, for example, is likely to lead rapidly to the evolution of resistance in herbivorous insects, because, as currently engineered, the toxin acts in the same manner as a synthetic organic insecticide – present at all times, and lethal to a large percentage of the population (Gould 1988). Inserting genes that regulate biosynthesis could greatly enhance the environmental longevity of such control strategies (Gould 1988). Such genes might include those involved in:

1. Producing a variety of analogues that can interact synergistically (e.g. the transferase genes involved in methylation of hydroxypsoralen derivatives to produce xanthotoxin and bergapten, structural analogues which are competitive inhibitors of webworm detoxification enzymes).
2. Regulating environmental responsiveness (e.g. promoters that respond to UV or elicitors).
3. Synchronizing biosynthesis with metabolic status such that nutrient limitations or reduction in tissue value (as due to senescence or

fatal injury) would result in the reduced production of defensive compounds.

The existence of such genes has been demonstrated; what remains is the development of technology to locate and transfer them from one plant genome to another while maintaining their regulatory function. The development of genetically engineered plants that can produce defensive compounds commensurate with environmental risk or demand would not only decrease reliance on synthetic organic insecticides but may also prove to be a long-lived efficacious system for plant protection. The feasibility of such an approach is beyond question; the evidence of its success is everywhere apparent in the angiosperm plants that dominate our planet's surface. Humans stand to benefit greatly from learning principles developed over 150 million years or so of genetic engineering research by these chlorophyll-laden "chemists".

ACKNOWLEDGMENTS

I thank, as always, my colleague and research associate Arthur Zangerl for information, insights and inexhaustible enthusiasm. I also thank Drs Janis Antonovics and Doug Futuyma for their interest in asking for this manuscript, their patience in waiting for it, and for their insightful suggestions for improving it. This work was supported by NSF BSR 8806015.

REFERENCES

Beier, R. C. and Oertli, E. H. (1983). Psoralen and other linear furanocoumarins as phytoalexins in celery. *Phytochem.* **22,** 2595–7.

Berenbaum, M. (1978). Toxicity of a furanocoumarin to armyworms: A case of biosynthetic escape from insect herbivores. *Science* **201,** 532–4.

—— (1981). Patterns of furanocoumarin production and insect herbivory in a population of wild parsnip (*Pastinaca sativa* L.). *Oecol.* **49,** 236–44.

—— and Feeny, P. (1981). Toxicity of angular furanocoumarins to swallowtails: Escalation in the coevolutionary arms race. *Science* **212,** 927–9.

—— and Zangerl, A. R. (1986). Variation in seed furanocoumarin content within the wild parsnip (*Pastinaca sativa*). *Phytochem.* **25,** 659–61.

——, Zangerl, A. R. and Nitao, J. K. (1984). Furanocoumarins in seeds of wild and cultivated parsnip (*Pastinaca sativa*). *Phytochem.* **23,** 1809–1810.

——, Zangerl, A. R. and Nitao, J. K. (1986). Constraints on chemical coevolution: Wild parsnips and the parsnip webworm. *Evol.* **40,** 1215–28.

——, Zangerl, A. R. and Lee, K. (1989). Chemical barriers to adaptation by a specialist herbivore. *Oecol.* **80,** 501–506.

Blasdale, W. C. (1947). The secretion of farina by species of *Primula*. *J. Roy. Hort. Soc.* **72,** 240–45.

Bowers, M. D. (1988). Chemistry and coevolution: Iridoid glycosides, plants, and herbivorous insects. In *Chemical mediation of coevolution* (ed. K. Spencer), pp. 133–65. Academic Press, New York.

Ceska, O., Chaudhary, S., Warrington, P., Poulton, G. and Ashwood-Smith, M. (1986). Naturally occurring crystals of photocarcinogenic furocoumarins on the surface of parsnip roots sold as food. *Experientia* **42**, 1302–4.

Collins, G. N. and Kempthorne, J. H. (1917). Breeding sweet corn resistant to the corn earworm. *J. Agr. Res.* **11,** 549–72.

Czapek, F. (1921). *Biochemie der Pflanzen,* 2. Aufl. 3 Bd. G. Fischer, Jena.

DeCandolle, A. P. (1916). *Essai sur les Proprietes Medicales des Plantes, Comparees avec leur Formes Exterieures et leur Classification Naturelle*, 2nd edn, Paris.

Dolinger, P. M., Ehrlich, P. R., Fitch, W. L. and Breedlove, D. E. (1973). Alkaloid and predation patterns in Colorado lupine populations. *Oecol.* **13,** 191–204.

Ebel, J. (1986). Phytoalexin synthesis: The biochemical analysis of the induction process. *Ann. Rev. Phytopathol.* **24,** 235–64.

Ehrlich, P. R. and Raven, P. H. (1964). Butterflies and plants: A study in coevolution. *Evolution* **18,** 586–608.

Fraenkel, G. S. (1959). The *raison d'etre* of secondary plant substances. *Science* **129,** 1466–70.

Goodwin, T. W. and Mercer, E. (1972). *Introduction to plant biochemistry.* Pergamon Press, New York.

Gould, F. (1988). Evolutionary biology and genetically engineered crops. *BioScience* **38,** 26–33.

Hahlbrock, K. and Ragg, H. (1975). Light-induced changes of enzyme activities in parsley cell suspension cultures. *Arch. Biochem. Biophys.* **166,** 41–6.

——, Lamb, C. J., Purwin, C., Ebel, J., Fautz, E. and Schaefer, E. (1981). Rapid response of suspension-cultured parsley cells to the elicitor form *Phytophthora megasperma* var. *sojae. Plant Physiol.* **67,** 768–73.

Harborne, J. B. (1982). *Introduction to ecological biochemistry.* Academic Press, New York.

Hauffe, K. D., Hahlbrock, K. and Scheel, D. (1986). Elicitor-stimulated furanocoumarin biosynthesis in cultured parsley cells: *S*-adenosyl-L-methionine:bergaptol and *S*-adenosyl-L-methionine: xanthotoxol *O*-methyltransferases. *Z. Naturforsch.* **41c,** 228–39.

Hendrix, S. (1979). Compensatory reproduction in a bienniel herb following insect defloration. *Oecol.* **42,** 107–118.

Janzen, D. H. (1979). Horizons in the biology of plant defenses. In *Herbivores: Their interaction with secondary plant metabolites* (ed. G. Rosenthal and D. H. Janzen), pp. 331–50. Academic Press, New York.

—— (1980). When is it coevolution? *Evolution* **34,** 611–12.

Johnson, C., Brannon, D. R. and Kuc, J. (1973). Xanthotoxin: A phytoalexin of *Pastinaca sativa* root. *Phytochem.* **12,** 2961–2.

Klocke, J. A., Balandrin, M. F., Barnby, M. A. and Yamasaki, R. B. (1989). Limonoids, phenolics, and furanocoumarins as insect antifeedants, repellents, and growth inhibitory compounds. In *Insecticides of plant origin* (ed. J. T. Arnason, B. J. R. Philogene and P. Morand), pp. 136–49. ACS Symposium Series 387.

Knogge, W., Kombrink, E., Schmelzer, E. and Hahlbrock, K. (1987). Occurrence of phytoalexins and other putative defense-related substances in uninfected parsley plants. *Planta* **171,** 279–87.

Kuhn, D. N., Chappell, J., Boudet, A. and Hahlbrock, K. (1984). Induction of phenylalanine ammonia-lyase and 4-coumarate:CoA ligase mRNAs in cultured plant cells by UV light or fungal elicitor. *Proc. Natl. Acad. Sci.* **81,** 1102–1106.

Ladygina, E. Y., Makarova, V. A. and Ignat'eva, N. S. (1979). Morphological and anatomical description of *Pastinaca sativa* fruit and localization of the furocoumarins in them. *Farmatsyia (Moscow)* **19,** 39–46 (in Russian, CA 1970: 74: 61588x.)

Lehninger, A. L. (1970). *Biochemistry*. Worth Publishing, New York.

Lois, R., Dietrich, A., Hahlbrock, K. and Schulz, W. (1989). A phenylalanine ammonia-lyase gene from parsley: Structure, regulation, and identification of elicitor and light responsive *cis*-acting elements. *EMBO J.* **8,** 1641–8.

Luckner, M. (1980). Expression and control of secondary metabolism. In *Secondary plant products* (ed. E. A. Bell and B. V. Charlwood), pp. 23–63. Springer-Verlag, New York.

Lutz, L. (1928). Sur le role biologique du tannin dans la cellule vegetale. *Bull. Soc. Bot. Fr.* **75,** 9–18.

McKey, D. (1979). The distribution of secondary compounds within plants. In *Herbivores: Their interaction with secondary plant metabolites* (ed. G. A. Rosenthal and D. H. Janzen), pp. 56–133. Academic Press, New York.

Metcalf, C. L. and Flint, W. P. (1962). *Destructive and useful insects*, 4th edn. McGraw-Hill, New York.

Mothes, K. (1980). Historical introduction. In *Secondary plant products* (ed. E. A. Bell and B. V. Charlwood), pp. 1–10. Springer-Verlag, New York.

Muckensturm, B., Duplay, D., Robert, P. C., Simonis, M. T. and Kienlen, J.-C. (1981). Substances antiappetante pour insectes phytophages presentes dans *Angelica silvestris* et *Heracleum sphondylium. Biochem. Syst. Ecol.* **9,** 289–92.

Muller, C. H. (1969). The 'co-' in coevolution. *Science* **164,** 197–8.

Murray, R. D. H., Mendez, J. and Brown, S. A. (1982). *The natural coumarins*. John Wiley, Chichester.

Nitao, J. K. (1989). Enzymatic adaptation in a specialist herbivore for feeding on furanocoumarin-containing plants. *Ecology* **70,** 629–35.

—— and Zangerl, A. R. (1987). Floral development and chemical defense allocation in wild parsnip (*Pastinaca sativa*). *Ecology* **68,** 521–9.

Pfeffer, W. (1897). *Pflanzenphysiologie*, 2. Aufl. I. Bd., S. 991. W. Engelmann, Leipzig.

Primack, R. B. and Kang, H. (1989). Measuring fitness and natural selection in wild plant populations. *Ann. Rev. Evol. Sys.* **20,** 367–96.

Robinson, T. (1974). Metabolism and function of alkaloids in plants. *Science* **184,** 430–35.

Rochleder, F. (1854). *Phytochemie.* W. Engelmann, Leipzig.

Sachs, J. (1882). *Vorlesungen uber Pflanzenphysiologie*, S. 203–222. W. Engelmann, Leipzig.

Seigler, D. S. (1977). Primary roles for secondary compounds. *Biochem. Syst. Ecology* **5,** 195–9.

—— and Price, P. (1976). Secondary compounds in plants: Primary functions. *Amer. Nat.* **110,** 101–105.

Simms, E. L. and Rausher, M. D. (1989). The evolution of resistance to herbivory in *Ipomoea purpurea*. II. Natural selection by insects and costs of resistance. *Evolution* **43,** 573–85.

Stahl, E. (1888). Pflanzen und Schnecken. *Jenz. Z. Med. u. Naturw.* **22,** 559–684.

Thompson, J. N. (1978). Within-patch structure and dynamics in *Pastinaca sativa* and resource availability to a specialized herbivore. *Ecology* **59,** 1112–19.

Tietjen, K. G. and Matern, U. (1983). Differential response of cultered parsley cells to elicitors from two non-pathogenic strains of fungi 2. Effects on enzyme activities. *Eur. J. Biochem.* **131,** 409–413.

——, and Matern, U. (1983) Differential response of cultured parsley cells to elicitors frm two non-pathogenic strains of fungi. II. Effects on enzyme activities. *Ecol. J. Biochem.* **131,** 409–13.

——, Hunkler, D. and Matern, U. (1983). Differential response of cultured parsley cells to elicitors from two non-pathogenic strains of fungi. I. Identification of induced products as coumarin derivatives. *Eur. J. Biochem.* **131,** 401–407.

Whittaker, R. H. and Feeny, P. P. (1971). Allelochemics: Chemical interactions between species. *Science* **171,** 757–70.

Wu, C. M., Koehler, P. E. and Ayers, J. C. (1972). Isolation and identification of xanthotoxin (8-methoxypsoralen) and bergapten (5-methoxypsoralen) from celery infected with *Sclerotinia scerlotiorum*. *Appl. Microbiol.* **23,** 852–6.

Yajima, T., Kato, N. and Munakata, K. (1977). Isolation of insect antifeeding principles in *Orixa japonica* Thunb. *Agr. Biol. Chem.* **41,** 1263–8.

Zangerl, A. R. (1990). Furanocoumarin induction in wild parsnip: Evidence for an adaptive induced defense. *Ecology* **71**, 1926–32.

—— and Berenbaum, M. R. (1987). Furanocoumarins in wild parsnip: Effects of photosynthetically active radiation, ultraviolet light, and nutrients. *Ecology* **68,** 516–20.

—— and Berenbaum, M. R. (1990). Furanocoumarin induction in wild parsnip: Genetic and populational variation. *Ecology* **71**, 1933–40.

——, Berenbaum, M. R. and Levine, E. (1989). Genetic control of seed chemistry and morphology in wild parsnip (*Pastinaca sativa*). *J. Hered.* **80,** 404–407.

——, Nitao, J. K. and Berenbaum, M. R. (1991). Parthenocarpic fruits in wild parsnip: decay defense against a specialist herbivore. *Evolutionary Ecol.* (in press).

Index

OXFORD SURVEYS IN EVOLUTIONARY BIOLOGY

Volume 1: 1984

Edited by R. Dawkins and M. Ridley

Contents

OXFORD SURVEYS IN EVOLUTIONARY BIOLOGY

Volume 2: 1985

Edited by R. Dawkins and M. Ridley

Contents

OXFORD SURVEYS IN EVOLUTIONARY BIOLOGY

Volume 3: 1986

Edited by R. Dawkins and M. Ridley

Contents

OXFORD SURVEYS IN EVOLUTIONARY BIOLOGY

Volume 4: 1987

Edited by P. H. Harvey and L. Partridge

Contents

OXFORD SURVEYS IN EVOLUTIONARY BIOLOGY

Volume 5: 1988

Edited by P. H. Harvey and L. Partridge

Contents

OXFORD SURVEYS IN EVOLUTIONARY BIOLOGY

Volume 6: 1989

Edited by P. H. Harvey and L. Partridge

Contents